Pyrolysis of Biomass for Fuels and Chemicals

Pyrolysis of Biomass for Fuels and Chemicals

Akwasi A. Boateng

ELSEVIER

ACADEMIC PRESS
An imprint of Elsevier

Academic Press is an imprint of Elsevier
125 London Wall, London EC2Y 5AS, United Kingdom
525 B Street, Suite 1650, San Diego, CA 92101, United States
50 Hampshire Street, 5th Floor, Cambridge, MA 02139, United States
The Boulevard, Langford Lane, Kidlington, Oxford OX5 1GB, United Kingdom

Notices
Knowledge and best practice in this field are constantly changing. As new research and experience broaden our understanding, changes in research methods, professional practices, or medical treatment may become necessary.

Practitioners and researchers must always rely on their own experience and knowledge in evaluating and using any information, methods, compounds, or experiments described herein. In using such information or methods they should be mindful of their own safety and the safety of others, including parties for whom they have a professional responsibility.

To the fullest extent of the law, neither the Publisher nor the authors, contributors, or editors, assume any liability for any injury and/or damage to persons or property as a matter of products liability, negligence or otherwise, or from any use or operation of any methods, products, instructions, or ideas contained in the material herein.

British Library Cataloguing-in-Publication Data
A catalogue record for this book is available from the British Library

Library of Congress Cataloging-in-Publication Data
A catalog record for this book is available from the Library of Congress

ISBN: 978-0-12-818213-0

For Information on all Academic Press publications
visit our website at https://www.elsevier.com/books-and-journals

Publisher: Brian Romer
Acquisitions Editor: Peter Adamson
Editorial Project Manager: Ali Afzal-Khan
Production Project Manager: Kamesh Ramajogi
Cover Designer: Christian J. Bilbow

Typeset by MPS Limited, Chennai, India

Dedication

To Maame

Contents

About the author

Dr. Boateng is a product of Opoku Ware and St. Augustine Schools in Ghana. He received a combined BS/MS degree in mechanical engineering in the former Soviet Union, an MS in thermochemical engineering, and a PhD in metals and materials engineering from the universities of New Brunswick and British Columbia of Canada, respectively. He taught thermo-fluids engineering at the University of Guyana in South America and at Swarthmore College in Pennsylvania before transitioning to industry and eventually to the U.S. Department of Agriculture (USDA) research laboratories where, for over 16 years, he led Congress-appropriated projects as lead scientist, including "Distributed-Scale Pyrolysis of Agricultural Biomass for Production of Refinable Crude Bio-Oil and Valuable Coproducts" (2009–14); "Farm-Scale Pyrolysis Biorefining" (2014–19); and "Thermo-Catalytic Biorefining" (2019–) all aimed to fulfill US energy independence and mitigate climate change through the replacement of fossil resources with biorenewable, fungible fuels under the 2007 US Energy Independence & Security Act. He was the 2012 recipient of the USDA-US Department of Energy (DOE) biomass R&D Initiative (BRDi) competitive grant and served as the Principal Investigator (PI) for a 5-year project "FarmBio^3" (Distributed On-Farm Bioenergy, Biofuels & Biochemicals). Aside from this, he was co-PI on the first BRDi grant awarded to a Honeywell-UOP led consortium, "Stabilization of Fast Pyrolysis Oils" (2009–12) that was followed up by the DOE-led National Advanced Biofuels Consortium funded under Green Stimulus of the American Recovery Plan (2010–13); he was also co-PI on the 2011 BRDi grant to Metabolix, "Renewable Enhanced Feedstocks for Advanced Biofuels and Bioproducts (REFABB)," (2011–14); co-PI, with other Agricultural Research Service scientists, on CenUSA, a USDA National Institute of Food and Agriculture (NIFA) sponsored, Coordinated Agricultural Project (CAP) investigating the creation of a Midwestern United States sustainable biofuels and bioproducts systems (2010–17); co-PI on the Penn State university CAP, NEWBio Bioenergy Consortium, a 5-year USDA-NIFA-funded effort to develop the sustainable biofuels industry in the Northeast United States (2011–18). He garnered over $11M in research grants as PI or co-PI in competitive grants and over $90 million in nationwide group funding during his tenure at the USDA. He has published over 165 scientific papers, 135 of which are in peer reviewed journals; 133 scientific conference abstracts; and an inventor on six issued patents with two patents pending. He has served on several editorial boards as associate editor most fittingly the American Institute of Chemical Engineers (AIChE) journal of Environmental Progress & Sustainable Energy and Springer's BioEnergy Research.

Prior to USDA, he spent 10 years in the minerals and materials industry as process design engineer, manager of research and production engineering, and chief engineer overseeing the design, installation, and operation of large, energy-intensive industrial furnaces for minerals beneficiation. He is the author of—*Rotary Kilns, Transport Phenomena & Transport Processes*—Butterworth-Heinemann Publishers and founder and principal of Alpha Thermal Process, LLC, a process engineering consulting that advises the industry on thermal process design, applications and optimization.

His scholarly awards include a Fulbright Fellow and visiting Scholar (Kansas State University, 1988); fellow of the American Society of Mechanical Engineers (2008); senior member of the AIChE; founding member of the North America Pyrolysis Network Group, including representing USDA on the observer status at the IEA Bioenergy Task force 34—pyrolysis (2007–15); visiting expert at EMPRAPA Agroenergy's science without borders program, (Brasilia, 2012–16); USDA-Foreign Agricultural Service consultant on Energy and Climate Change Partnership of the Americas, ECPA (Cenicafe, Colombia, 2013–15). He was the first recipient of the US symposium Series' Thermo-catalytic Conversion Science career achievement award for his contributions to thermal and catalytic sciences for biofuels and bio-based products (2018).

Abbreviations

Acronyms

AM	arbuscular mycorrhizal (fungus)
ARRA	American Recovery and Reinvestment Act
ARS	Agricultural Research Service
ASTM	American Society for Testing and Materials
ATCC	American Type Culture Collection
ATP	adenosine triphosphate, an energy molecule
BET	Brunauer, Emmett, and Teller
BF	breeding factor (exergy efficiency)
BRDi	Biomass Research and Development Initiative
BTG	Biomass Technology Group
BTEX	benzene, toluene, ethyl benzene, and xylene
BTX	benzene, toluene, xylene
CANMET	Canada Center for Mineral and Energy Technology
CAP	Coordinated Agricultural Project
CExD	cumulative exergy demand
CFD	computational fluid dynamics
CFP	catalytic fast pyrolysis
Cond	condenser
CoMo	cobalt–molybdenum
CRADA	Cooperative Research and Development Agreement
CRIPS	combustion reduction integrated pyrolysis system
DCFROR	discounted cash flow rate of return
DCR	WR Grace-developed FCC pilot plant
DDGS	distiller's dried grains with solubles
DEPT	distortionless enhancement by polarization transfer
DoD	Department of Defense
DOSY	diffusion-ordered NMR spectroscopy
EIA	Energy Information Administration
EISA	Energy Independence and Security Act
EPA	Environmental Protection Agency
EPAct	Energy Policy Act
ER	equivalence ratio
ERRC	Eastern Regional Research Center
EU	European Union
Eucal.	*Eucalyptus*
ESP	electrostatic precipitator
EtOH	ethanol (ethyl alcohol)
FCC	fluid catalytic cracking

FP	fast pyrolysis
FTS	Fischer–Tropsch synthesis
GB	guayule bagasse
GC	gas chromatography
GC–MS	gas chromatography–mass spectroscopy
GHG	greenhouse gas
GIS	geographic information system
GPC	gel permeation chromatography
GWP	global warming potential
HC	hydrocarbon
HCO	heavy cycle oil
HDO	hydrodeoxygenation
HDPE	high-density polyethylene
HDT	hydrotreating (catalyst)
HHV	higher heating value
HLB	hydrophilic–lipophilic balance
HPLC	high-performance liquid chromatography
HT	hydrotreating
HTL	hydrothermal liquefaction
HYP	hydropyrolysis
IC	internal combustion
IRR	internal rate of return
KF	Karl Fischer
KIT	Karlsruhe Institute of Technology
KOH	potassium hydroxide
LC	liquid chromatography
LCA	life cycle analysis
LCI	life cycle inventory
LCO	light cycle oil
LDPE	low-density polyethylene
LHSV	liquid hourly space velocity
LHV	lower heating value
LG-HF	lignin gasification for hog fuel
LPG	liquid petroleum gas
Manu.	manure
MFI	micro-flow imaging
MFSP	minimum fuel selling price
MIT	Massachusetts Institute of Technology
MLO	mid-level oxygen content
MTBE	methyl tertiary butyl ether
MTPD	metric tons per day
MW	molecular weight
MWCNT	multiwalled carbon nanotube
NABC	National Advanced Biofuels Consortium
NASA	National Aeronautics and Space Administration
NCAUR	National Center for Agricultural Utilization Research
NCG	noncondensable gas
NIFA	National Institute of Food and Agriculture
NMR	nuclear magnetic resonance spectroscopy

NOx oxides of nitrogen
NPK nitrogen, phosphorous, and potassium
ORNL Oak Ridge National Laboratory
PA Pennsylvania
PC principal component
PCA principal component analysis
PDU process development unit
PET polyethylene terephthalate
Pet-Coke petroleum coke
PKM passenger-kilometer
PNNL Pacific Northwest National Laboratory
PP polypropylene
PS polystyrene
PVC polyvinyl chloride
PyGas pyrolysis gas
Py-GC/MS pyrolysis coupled with gas chromatography/mass spectroscopy
Py-Oil pyrolysis oil
R&D research and development
RCFP reactive catalytic fast pyrolysis
RFS renewable fuel standard
RTI Research Triangle Institute
RTP rapid thermal process
SEM scanning electron microscopy
Si/Al silica–alumina ratio
SMD Sauter mean diameter
SP slow pyrolysis
SG, SwG switchgrass
TAN total acid number
TCI total capital investment
TEA technoeconomic analysis
TEM transmission electron microscopy
TGA thermogravimetric analysis
TGRP tail gas reactive pyrolysis
TPD tons per day
TPEC total purchase equipment cost
TPR temperature-programmed reduction
TRL technology readiness level
UNCCD United Nations Convention to Combat Desertification
UOP Universal Oil Products
USDA United States Department of Agriculture
US-DOE United States Department of Energy
VGO vacuum gas oil
VM volatile matter
WTW well-to-wheel
XRD X-ray diffraction

Nomenclature

A	preexponential coefficient; or an empirical coefficient accounting for the interaction of energy between the two components
d_o	diameter of nozzle orifice
DH	degree of hydrogenation
E	activation energy (J/mol)
H	enthalpy (J/kg)
S	entropy (J/K)
k	first-order reaction rate constant (1/s)
m_b	mass of biomass (kg)
m_c	mass of char
m_o	initial mass of biomass
m_{oil}; $m_{oil,i}$	final and initial mass of the oil phase
P	protein fragment
R	gas constant (J/mol K)
t	time (s)
T	temperature (K)
TEU	total extent of condensed phase upgrading
U_R	velocity of the αιρ relative to the fuel
V	molar volume of liquid
W_a	air mass flow rate (kg/s)
W_l	liquid mass flow rate (kg/s)
X	mass fraction of biomass at time (t)
x_i	mole or mixture fraction of component, i
x_o; $x_{o,i}$	final oxygen and initial oxygen content
Z_i	elemental compositional species mass fraction
$\%DO_{rel}$	relative deoxygenation extent
$\%DO_{abs}$	absolute deoxygenation extent

Symbols

β	expression relation exergy and LHV
μ	viscosity (pa s)
ρ	density (kg/m^3)
σ	surface tension (N/m)

Foreword

When I met the author, Dr. Akwasi Boateng (Kwesi, as we call him), we both had ants in our pants. It was in 2006, before the modern implementation of hydrofracking and subsequent windfall of shale oil and gas, and the country's federal agencies were very serious about alternate energy. The occasion of our meeting was an interagency summit of the Biomass Research and Development Initiative (BRDi), attended by the likes of the U.S. Department of Agriculture (USDA), the Departments of Energy, Commerce, Defense, and the Interior, the Environmental Protection Agency, and the National Science Foundation. The purpose of the summit was to update the National Action Plan on Biofuels. At that time the plan for alternate transportation fuel was essentially "ethanol-only," and the latest discussions were on improving the theoretical yield of ethanol by using not only the corn grain but also the nonfood parts of the plant; hence, "cellulosic ethanol."

While the refinement of the National Action Plan to improve fuel yields by recycling agricultural and forest waste was and still is a very worthwhile goal, Kwesi and I were convinced that the plan was missing a main target. He, a mechanical engineer by training and an expert in kiln operation, had been working at USDA on the pyrolysis of corn stover and other agricultural waste. I was a new director of the Catalysis program in the Engineering Directorate at National Science Foundation (NSF) and had tapped into the catalysis community's expertise on biomass conversion. Pyrolysis was discussed at the summit but was viewed only as an option for dealing with a particularly intractable fraction of the plant fiber—lignin—which could not be turned into ethanol. What we both realized was that nobody was talking about hydrocarbon biofuels. Kwesi knew that his pyrolysis process could convert not just lignin but whole biomass into hydrocarbon intermediates, and I realized that not only pyrolysis but also other catalytic processes such as gasification and liquid phase processing could be used to convert whole nonfood waste biomass into hydrocarbon products. These would be drop-in replacements to gasoline, diesel, and jet fuel with significantly higher energy density and would have complete infrastructure compatibility compared to cellulosic ethanol.

In fact, over the next few years a sea change in federal funding for biofuels research did occur, toward "green gasoline." This shift was abetted by the NSF-sponsored, George Huber–chaired workshop and roadmap report "Breaking the Chemical and Engineering Barriers to Lignocellulosic Biofuels: Next Generation Hydrocarbon Biorefineries," in which Kwesi was an invited and mainstay participant. Over the next several years, I had the pleasure of promulgating this roadmap at conferences around the world. I was Kwesi's greatest cheerleader.

As I returned to academia and have kept an interested eye on alternate energy developments, I have witnessed many start-up companies come and most go. With any new development in pyrolysis, be it at a university, national lab, startup, or even an attempted commercialization, my reflex has always been "What does Kwesi think about this one?" The heart of perhaps the greatest collection of pyrolysis expertise in the world is found within USDA's Eastern Regional Research Center in Wyndmoor PA, in Kwesi's pyrolysis process development unit, otherwise known as "The Kwesinator," and in his truck-mounted mobile pyrolysis unit, named the combustion reduction integrated pyrolysis system. You will find both of these units described in Chapter 7, Pyrolysis Conversion Technology Systems and Integration, of this book. Studying pyrolysis in various modes in both units, along with working on pyrolysis oil upgrading, has given the author ample scientific depth with which to compose the earlier chapters of the book. Having directed research teams that include experts in techno-economic analysis has given Dr. Boateng further background to compose the latter chapters of this book.

It has always been the figurative case that Akwasi Boateng has written the book on biomass pyrolysis. Now he has done it literally.

JR Regalbuto
University of South Carolina, Columbia, SC, United States
January 30, 2020

Preface

This book brings you the state of research and the art of pyrolysis biorefinery for the production of fuels and chemicals led by the U.S. Department of Agriculture (USDA) during the period leading to the renewable fuels drive to energy independence and to replace fossil fuel with advanced lignocellulosic biofuels in the United States as charted by the renewable-fuels standards Act on US Biofuels Policy. Agriculture is by all means the largest source of food, fiber, and lignocellulosic biomass, including agricultural residues and energy crops. Thus, biomass is by far the largest source of renewable carbon for biofuels production when it comes to the pyrolysis conversion platform. Because of its synergy with petrochemical technologies and its compact design footprint, the pyrolysis technology fits well into the farming framework as well as the USDA vision, whereby integrated on-farm bioenergy systems can produce biocrude at site leaving the biochar coproduct behind for soil amendment, sequestering carbon, and building soil quality to increase agricultural productivity so food and bioenergy crop can be sustainably harvested. Other conversion technologies, including biochemical conversion, fall short of this promise. The narrative outlined herein is the story of the research efforts spearheaded by the author and the knowledge base he helped built during his 16-year tenure as the lead scientist of the USDA Pyrolysis Biorefining Program.

Cellulosic biofuels (i.e., biofuels produced from the structural fibers of plants) are in their infancy, with a limited number of commercial facilities operating in the United States. Amidst 20 + years of strong R&D funding attempts to commercialize, pyrolysis biorefining facilities have not been fully realized having faced challenges leading to some false starts and outright closures. The work presented here is part of the research and development efforts purported to address these challenges so as to reduce technology uncertainty and the risk of scaling up to long-term commercial operations.

The outline begins with Chapter 1, Introduction, an introduction to the biofuels agenda and where general bioconversion including biochemical and thermochemical technology platforms are defined making a case for the pyrolysis biorefinery appeal for on-farm and/or in-forest setting. Chapter 2, Thermal Pyrolysis, discuses thermal-only fast pyrolysis, which looks at pyrolysis as feedstock agnostic and bio-oil yields in excess of 60 wt.% but along with this high yielding is a slew of stability issues that creates barriers to its utility as fuel. In Chapter 3, Catalytic Fast Pyrolysis, catalytic pyrolysis research that addresses some of the barriers is discussed drawing synergies with petroleum refining where the use of heterogeneous cracking catalysts such as zeolites have received a widespread application. Chapter 4, Reactive Pyrolysis, provides some background and research results that demonstrate that use of reactive atmospheres, with or without catalyst, which is equally capable of forming reduced or mild-level oxygen bio-oils resulting in better stability even though, like in catalytic pyrolysis, yields are compromised for better quality. However, these products can be a better starting liquid feed for a biorefinery. Chapter 5, Condensed-Phase Pyrolysis Oil Upgrading, picks up from how the formed bio-oils can be upgraded to gasoline equivalent fungible fuels using some of the existing technologies in the petrochemical infrastructure such as hydrodeoxygenation and hydrodesulfurization and challenges faced due purely to the liquid being biomass derived. Here, it becomes evident why the synergy between the petroleum and biorefinery has not lived up to the full potential leading to commercializing challenges. Chapter 6, Combustion Applications of Pyrolysis Liquids, discusses how the bio-oil stability issues propagate to the use of bio-oil for combustion in compression-ignition engines, residential heaters, and industrial furnaces. One would think that here is an area particularly in the northeast United States where there is a need for residential heat run on fossil fuel and an abundance of biomass that could provide a renewable biofuel replacement. Combustion of "neat" pyrolysis oil is problematic, but nozzle designs are evolving toward those that can generate turbulent diffusion flames. Some companies are pursuing neat bio-oil as bridge fuel for industrial furnace combustion applications. Our research results on flames and emissions patterns are presented herein. In Chapter 7, Pyrolysis Conversion Technology Systems and Integration, pyrolysis oil production systems and their integration are viewed in terms of technology readiness level that guide good engineering practice for new technological developments such as pyrolysis biorefineries. Various scales are discussed and so are distributed/satellite and mobile systems that can go to the biomass source as against large centralized facilities based on economies of scale. Chapter 8, Biorefinery Performance Measurements, discusses bioenergy performance metrices that include mass, energy, and exergy balances along with techno-economic analysis and life cycle assessments for various processes in case studies that include stand-alone, colocated at the biomass source, and integration into existing industries, including petroleum infrastructure. Their end products minimum selling price as well as product life cycle are assessed in terms of global warming potential. The techno-economics of new trends where pyrolysis oils are partially cofed with vacuum gas oil are discussed. In Chapter 9, Energy Crops—Biomass Resources and Traits, biomass resources such as energy crops and how their compositional traits and harvest time affect pyrolysis oil yield and product distribution are discussed. These offer tips to farmers growing energy crops such as switchgrass on consequences of cropping and harvesting for biofuels production. This book completes with Chapter 10, Pyrolysis Solid Coproducts and Usage, by addressing fast pyrolysis solid coproduct applications. While biochar has been touted for soil amendment, agronomic applications have shown mixed results because not all biochars are created equal. While fast pyrolysis that maximizes bio-oil comes with biochar coproduct, their characteristics are different from that produced by slow pyrolysis in terms of functional groups that have the unique potential for use in biocidal applications.

Many thanks go to the many collaborators, including fellow research scientists at the agency, coprincipal investigators from various university and research institutions on his grants, engineers and technicians, postdoctoral fellows, and predoctoral students, including high-school students in the STEM program who graced his laboratory in Wyndmoor, PA and worked under his direction through appropriated and competitive grant projects.

Akwasi A. Boateng
Lead Scientist (Retired), USDA-ARS,
Royersford, PA, United States
2020

Introduction

1

1.1 Biomass conversion technologies

Lignocellulosic biofuels and biochemicals are fuels and chemicals such as ethanol or hydrocarbons produced from structured fibers of plants comprising cellulose, hemicellulose, and lignin polymers. These polymers need to be deconstructed thermally (thermochemical) or biologically (biochemical) and treated to produce transportation fuels and chemicals. Both technological routes are often categorized as bioconversion because biomass is the primary precursor, a biodegradable, organic, or renewable feedstock. Before describing thermochemical conversion technologies, a brief introduction of biochemical conversion is necessary to decipher the differences between these technologies.

1.1.1 Biochemical conversion

To produce cellulosic ethanol as an energy carrier, the biomass is broken down to release the carbohydrate that is, in turn, subjected to enzymatic or bacterial degradation, the most common process being fermentation, the oxidation/reduction of organic compounds that takes place in the absence of external electronic acceptors (Drapcho et al., 2008). Fermentation of simple sugars such as glucose can be accomplished by yeast or bacteria to form alcohols such as ethanol along with the release of CO_2.

$$C_6H_{12}O_6(+ATP) = 2C_2H_5OH + 2CO_2 + 2ATP \tag{1.1}$$

where 1 mol of glucose gives 2 mol of ethanol and 2 mol of carbon dioxide + energy. Before the US Renewable Fuel Standards (RFS) rule that placed some restrictions on sourcing food for biofuels was ratified in 2005, ethanol was the only known renewable fuel in the market and it was derived primarily from grains, particularly, corn. Two technologies could be discerned to produce ethanol from corn depending on how the starch is released; these include the wet and dry milling technologies. The main difference between the two is that, in the dry milling process, the whole corn is ground and fed to the fermenter, whereas in the wet milling process, the corn is first fractionated to release the starch that is the only fraction of the grain used. Another major source of ethanol production besides grains is sugar where raw juices and molasses from sugarcane and sugar beets are fermented to ethanol, the so-called Melle–Boinot process (Drapcho et al., 2008). To move away from food resources, the new US Renewable Fuel Standards established in 2007

Pyrolysis of Biomass for Fuels and Chemicals. DOI: https://doi.org/10.1016/B978-0-12-818213-0.00001-1

(RFS2) required the use of biomass, a lignocellulose, the structural composition of which comprise cellulose, hemicellulose, and lignin polymers. The technology for the biochemical or enzymatic conversion of lignocellulosic feedstocks to ethanol is what has become known as the sugar platform. In this process the lignin may be separated in a pretreatment of the biomass as is done in the pulp and paper industry to isolate it from the carbohydrates, that is, the cellulose (C6 sugars) and the hemicellulose (C5 sugars), which, in turn, may be hydrolyzed to fermentable sugars to produce ethanol via the aforementioned fermentation process. The fermentable sugars include glucose (from cellulose), xylose, arabinose, galactose, and mannose. Suffice it to say that the bulk of the process challenges and cost lie in the pretreatment and the hydrolysis process steps. Hydrolysis of carbohydrates may be carried out using acids or enzymes and is made easier by the type of the pretreatment method, which may include steam explosion, acid pretreatment, or digestive methods, the goal of which is to break down the polymer structure to make the carbohydrate fractions more susceptible to acid or enzymatic attack. While making ethanol from corn and sugarcane is considered matured technology, producing ethanol from lignocellulosic conversion is not and is complicated by all the preprocess steps which are cumbersome, with large design footprints, that succumb to economies of scale and, hence, economically challenged. As said earlier, only the carbohydrate fraction of the biomass is utilized with the lignin being discarded as a waste product to the core process. In fact, lignin is considered a recalcitrant, being a liability to contend with, as by nature its job is to protect the carbohydrate, the nutrient source to the plant cell wall. Various recombinants are required for various feedstocks making preprocessing a formidable task.

1.1.2 Thermochemical conversion

Unlike biochemical conversion, thermochemical conversion is the heating of the organic matter (biomass) to produce energy such as heat or energy carriers such as burnable gas (syngas or producer gas), liquid fuel intermediates, or carbonaceous solid similar to coal. Thermochemical conversion technologies may utilize all the three polymer components of the biomass feedstock and may offer smaller engineering design footprints if they can be optimized for biofuels conversion. The most well-known thermochemical conversion of biomass is combustion. Direct combustion of biomass has been practiced for centuries and, at one time, was the only source of energy for humanity. It is still the source of heat that millions in the developing world rely upon today for their energy needs. Like its fossilized counterpart, coal, when biomass is subjected to heat, it decomposes at some threshold temperature, into volatile compound and a fixed carbon. If the volatile matter (seen as smoke) is not harnessed (as pyrolysis vapor/liquid product), it will interact with the surrounding oxygen (atmospheric or supplied) to undergo subsequent reactions such as gasification and combustion. The combustion reaction is an oxidation reaction in excess of air or oxygen to generate heat, while gasification is a partial oxidation coupled with intergas–solid reactions to generate burnable producer gas, at times called syngas. The primary sequential processes encompassing the thermochemical conversion of biomass therefore are: (1) pyrolysis, the heating in the absence of oxygen; (2) gasification, heating under partial oxygen conditions; and (3) combustion, the heating in excess oxygen environment. If the heating is carried out in a solvent environment, it is called (4) solvent liquefaction or hydrothermal liquefaction if said solvent is water. Fig. 1.1 shows the layout of the thermochemical process flow to energy and/or energy carriers.

As Fig. 1.1 indicates, the differences between pyrolysis, gasification, and combustion is the extent of air or oxidant in the environment, something that can be perceived in the framework of the fire triangle (Fig. 1.2). The fire we visually see is a result of several processes beginning from pyrolysis. To make fire, one requires a fuel source, in this case biomass, heat, and oxygen (O_2) to coexist. However, the visible flame associated with fire occurs only after pyrolysis has occurred to release the volatiles which, upon encountering the surrounding or supplied oxygen at the appropriate temperature, will ignite. The extent of the flaming depends on the available O_2; and it is seen because part of the electromagnetic wave spectrum falls within visible light frequency (0.4–0.7 μm) (Boateng, 2015). With little O_2 in the surrounding, for example, in substoichiometric quantities, CO, CO_2, and H_2O are produced that react with the fixed carbon in a series of reactions, the so-called

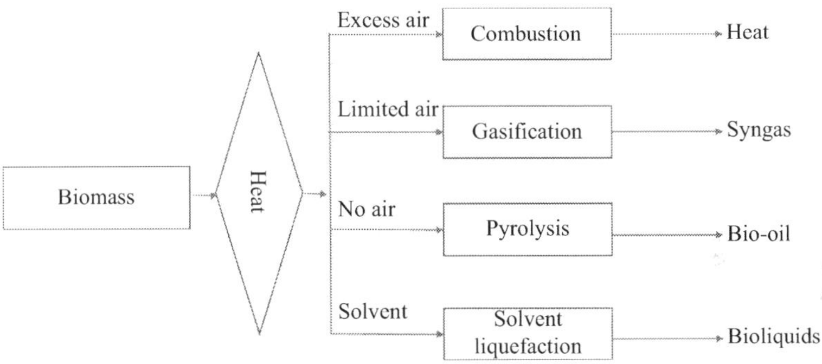

Figure 1.1 Thermochemical conversion and products.

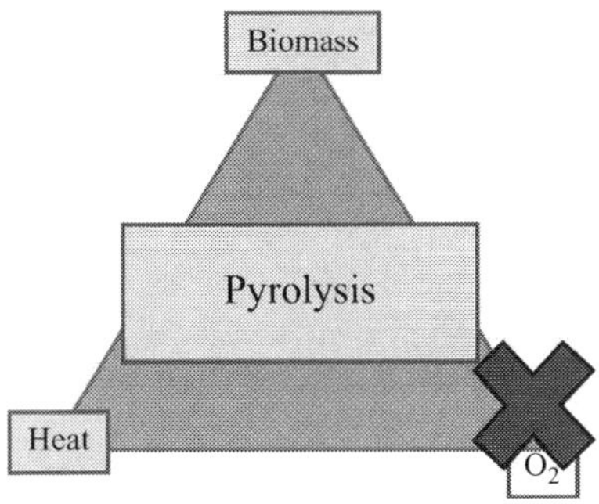

Figure 1.2 Pyrolysis and the fire triangle.

Boudouard and water–gas shift reactions to form synthesis or producer gas. However, when the environment is in excess of air/oxygen combustion products, along with a visible flame, result. A standalone pyrolysis-to-fuels technology is therefore a thermal conversion process in which the O_2 is completely devoid and the volatiles are not allowed to ignite nor cracked nor gasified but rather harnessed as liquefied smoke or pyrolysis oil, which is also known as bio-oil. Accompanying the bio-oil are solid and gaseous products in various fractions that depend on the biomass source and the process conditions, that is, whether the heat rate is high (fast pyrolysis) or low (slow pyrolysis).

Done properly, thermochemical liquefaction pathways, such as pyrolysis and solvent-assisted reactions, offer a unique opportunity to directly produce hydrocarbon fuel intermediates, that is, refinery fuel blend stocks that can potentially enter the existing petroleum refinery flow streams without added expense of creating a whole new refinery infrastructure, that is, drop-in fuels and chemicals.

During pyrolysis the large complex hydrocarbon molecules of the biomass polymer break down into relatively smaller and simpler molecules of gas, liquid, and char (biochar). A typical fractional distribution of fast pyrolysis products is as depicted in Fig. 1.3. Herein, the bio-oil is produced in large quantities that may serve as a feedstock for producing second-generation transportation fuels (Bridgwater and Peacocke, 2000; Granatstein et al., 2009; Huber, 2008; Mason et al., 2009; Woolf et al., 2010). Pyrolysis has similarities to and some overlaps with thermolysis processes such as cracking, devolatilization, carbonization, and distillation, but it is distinguished from gasification and combustion by lack of oxygen (Fig. 1.1). Another distinction is that pyrolysis is carried out at relatively low temperature range of 300°C–650°C compared to 800°C–1000°C for gasification and even higher for combustion.

As we saw earlier, biomass gasification as a standalone technology may be viewed as a two-stage process starting with pyrolysis followed by gasification reactions under partial oxidation or steam to produce syngas. While syngas can be a source as a standalone burnable producer gas, a purified syngas rich in CO and H_2 can also be fermented to produce ethanol or may be further synthesized to renewable gasoline and diesel via the Fischer–Tropsch synthesis (FTS). While this technology is matured for coal, biomass-to-gas-to-liquids is still at the research stages as extensive syngas cleanup needs to be carried out to rid of the tars, mostly

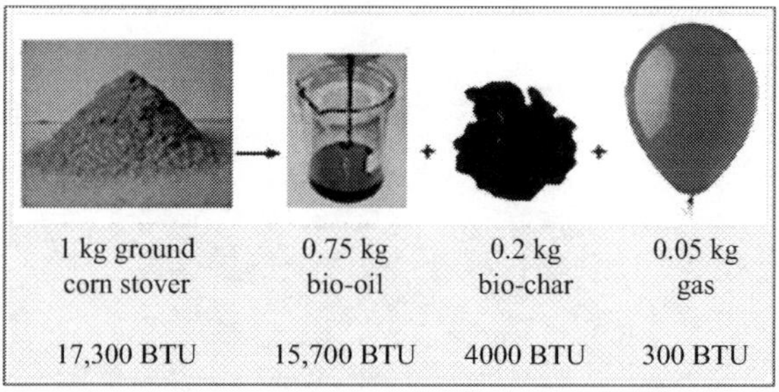

Figure 1.3 Three-phase product distribution.

polynuclear hydrocarbons, and particulate such as ash particles before the FTS process (Mahmoudi et al., 2017).

Hydrothermal liquefaction technology involves the process of thermally cracking the long carbon chains of the biomass polymer under wet conditions via the dehydration/hydroxylation to remove bound water and followed by decarboxylation to CO_2 resulting in liquid fuel intermediates with high H/C ratio. This pathway is not different from pyrolysis except that the latter is done in dry conditions curtailing the water-bound reactions found in the former.

This book is intended to address advances made in pyrolysis biorefining. Among the various thermochemical conversion routes, pyrolysis is about the only route that has the full potential to offer small design footprints easily adaptable to the farm setting. Pyrolysis also has the potential to handle a wide variety of agricultural residues, crop residues, energy crops, manure, and even agricultural plastics and can convert them into liquid fuels, chemicals, and valuable coproducts via condensed-phase upgrading.

The content in the ensuing chapters will describe the research and developments toward producing fuels and chemicals via pyrolysis biorefining spearheaded by the U.S. Department of Agriculture (USDA) during the period leading to the renewable fuels drive toward energy independence in the United States with a goal to address the barriers leading to uncertainties and the risk of scaling up the technology to long-term commercial operations.

1.2 Biomass decomposition and pyrolysis products

Biomass pyrolysis decomposition can be represented by the generalized reaction:

$$C_nH_mO_p(\text{Biomass}) \xrightarrow{\text{Heat}} \sum\nolimits_{\text{liquid}} C_xH_yO_z + \sum\nolimits_{\text{gas}} C_aH_bO_c + H_2O + C(\text{char}) \tag{1.2}$$

where the biomass, the reactant in this equation, is represented by the elemental CHO composite of the polymers of hemicellulose, cellulose, lignin, and minor amounts of other organics. However, each of the biomass polymers decomposes or pyrolyzes at different rates and by different mechanisms and pathways. Thermogravimetric studies (Fig. 1.4) teach us that lignin decomposes over a wider temperature range compared to the carbohydrate (cellulose and hemicellulose) that rather rapidly degrade over narrower temperature ranges, hence the apparent thermal stability of lignin during pyrolysis. The rate and extent of decomposition of each of these components depends on the process parameters set by the reactor (pyrolysis) temperature, biomass heating rate, and pressure. The degree of secondary reactions (and hence the product yields) of the gas/vapor products depends on the time–temperature history to which they are subjected before the collection point, which includes the influence of the reactor configuration (Yang et al., 2007).

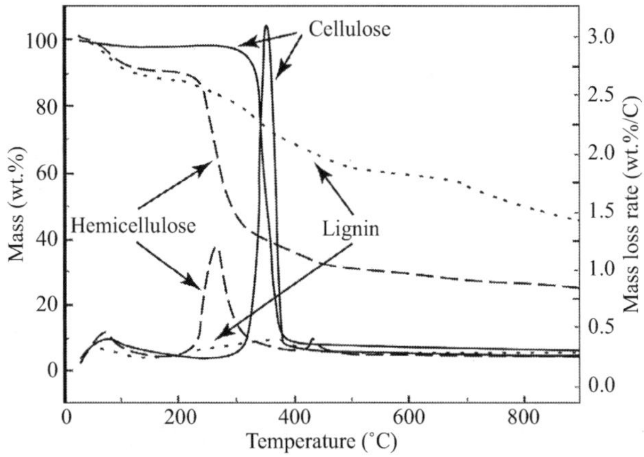

Figure 1.4 Thermogravimetric analysis of the pyrolysis of cellulose, hemicellulose (xylan), and lignin at constant heating rate (10°C/min) with *N* (99.9995%) sweep gas at 120 mL/min. *Source*: Adapted from Yang, H., Yan, R., Chen, H., Lee, D.H., Zheng, C., 2007. Characteristics of hemicellulose, cellulose and lignin pyrolysis. Fuel 86, 1781–1788.

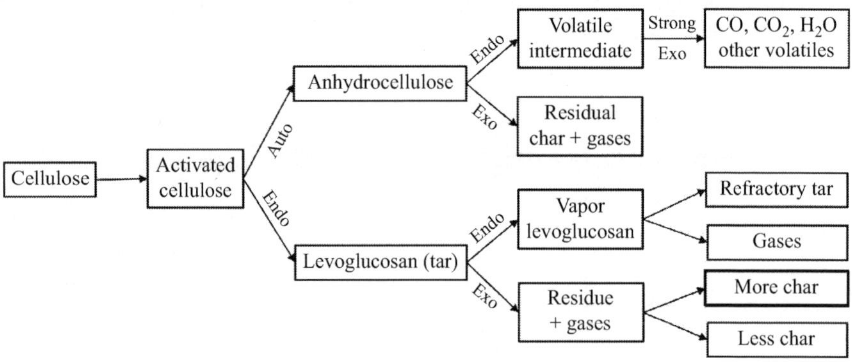

Figure 1.5 Reaction pathways for cellulose decomposition.

The reaction pathway during pyrolysis is complex as each individual component takes their own path depending upon the reaction conditions. The complexity of the reaction can be envisioned by Fig. 1.5, an illustration of the process steps when following cellulose alone, one component and the simplest of the three polymers. Herein, the primary and secondary reactions follow pathways that determine product yields and composition. The products of pyrolysis depend on the temperature and the heat rate. The heat rate is important as it determines the intermediate route and distinguishes what the product distribution will be.

As shown in Fig. 1.3, pyrolysis products are classified into three principal form types: solid, mostly char (carbon and ash), and liquid formed from condensable gases (tars comprising heavier hydrocarbons and oxygenates and water) and noncondensable gases (NCGs) (CO, CO_2, C_xH_y). The relative amount of each fraction depends on the heating rate and the final temperature reached by the biomass. Pyrolysis maximizes biochar production when the heat rate is low (slow pyrolysis) and maximizes condensable gas (bio-oil) production when the heat rate is high (fast pyrolysis). The generally accepted rate for slow pyrolysis is when the heat rate is $<100°C/s$ and fast pyrolysis when it is $>1000°C/s$; any rate in-between would be classified as intermediate pyrolysis.

While pyrolysis and devolatilization are similar, the pyrolysis product should not be confused with the *volatile matter* of a solid fuel, for example, coal as determined by its *proximate analysis*. Devolatilization as employed in proximate analysis is the release or a measure of volatile matter physically bound to the solid leaving a "fixed carbon," while pyrolysis creates a volatile matter via the breaking of some covalent bonds of the biomass polymer leaving fixed and formed carbon from the vapor-phase reactions. Since the relative fraction of both devolatilization and pyrolysis yields depends on many operating factors, the determination of the volatile matter of a solid fuel such as coal is different from pyrolysis and it requires the use of standard conditions as specified in test codes such as ASTM D-3172 and D-3175. The latter standard, for example, follows a protocol of heating the specified sample of the fuel in a furnace at 950°C for 7 minutes to measure its volatile matter (Basu, 2010).

1.2.1 Slow pyrolysis products

When the pyrolysis process is used with the aim to produce charcoal/biochar and not bio-oil, the process is typically called carbonization. Carbonization follows slow pyrolysis and is slightly exothermic. Pyrolysis can be an endothermic or exothermic reaction depending upon the temperature of the reactants, becoming increasingly exothermic as the reaction temperature decreases. The exothermicity of the slow pyrolysis reaction per unit of biochar yield is reported to range from 2.0 to 3.2 kJ/g biochar. Hence, because the fixed carbon content of biomass is high, biochar formation commences at low temperatures where autogenous pyrolysis begins. The self-sustaining nature of the low-temperature reaction explains why earlier traditional methods of making charcoal, that is, where the biomass was buried underground or in mound kilns (Fig. 1.6), could proceed for several days (Spokas et al., 2012).

For as long as human history has been recorded, heating or carbonization of wood for the purpose of manufacturing biochar or charcoal has been practiced (Lehmann and Joseph, 2015). Carbonization is as old as civilization itself; however, in the ancient times the production of charcoal was not the only intention. It appears that ancient peoples were equally acquainted with the methods of recovering the associated liquid coproduct or tars, as they are now known, in the same way as is done today for pyrolysis oil (bio-oil). The pyrolysis reactions, whether they are used in producing tar/liquid or biochar/solid products, occur simultaneously with one or the other always as a coproduct.

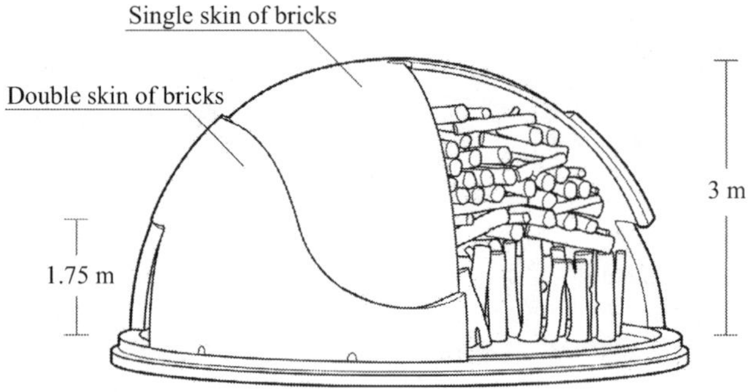

Figure 1.6 Mound kiln.
Source: Adapted from Veitch, F.P., 1907. Chemical Methods for Utilizing Wood Including Destructive Distillation, Recovery of Turpentine, Rosin and Pulp and the Preparation of Alcohols and Oxalic Acid. Washington Government Printing Office (Veitch, 1907).

1.2.2 Fast pyrolysis products

Fast pyrolysis provides a condition to maximize the production of the liquid fraction of the products, that is, bio-oil (pyrolysis oil, biocrude, etc.) with the primary goal of making a renewable fuel intermediate, a pathway to producing finished hydrocarbon fuels from biomass. Most of the material presented in this book therefore pertains to fast pyrolysis products and their postprocess upgrading in the context of a biorefinery, that is, pyrolysis biorefinery. The optimum fast pyrolysis condition is achieved by heating the biomass very rapidly (1000°C/s–10,000°C/s) at a peak temperature in the 450°C–550°C to maximize the liquid yield. Such high heat rates make the fast pyrolysis process heat transfer limited, that is, a high heat transfer resistance needs to be overcome to reach the biomass decomposition temperature. Heat transfer limitation is therefore dictated by the reactor type and operational conditions. The contact-type reactors such as fluidized bed vessels have become the equipment of choice to create fast pyrolysis conditions due to their high heat transfer coefficients associated with the bubbling nature of the medium although other reactors have been successfully used in pyrolysis applications.

Maximizing liquid production requires very short vapor residence time typically 1 second within the reactor vessel so as to minimize secondary reactions such as the cracking of the large hydrocarbon molecules to NCGs, although acceptable yields can be obtained at residence times of up to 5 seconds if the vapor temperature is kept lower than the optimum, for example, below 400°C. After cooling and condensation of the vapor products (smoke), a dark brown free-flowing liquid is formed, which looks like crude oil, although it is not. Rather, it is biomass in its liquid form comprising oxygenated tar compounds of the polymer breakdown that has a heating value of about half that of conventional fuel oil. Hence, fast pyrolysis is an advanced form of a carefully controlled pyrolysis process that yields high volumes of condensable liquids. A typical process flow arrangement for the

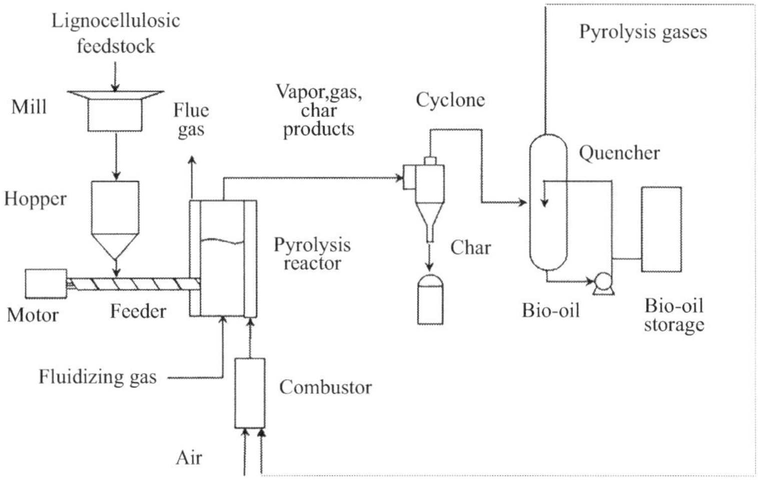

Figure 1.7 Conceptual fluid bed fast pyrolysis process (Bridgwater et al., 1999).

production of bio-oil using a fluidized bed technology and accompanying accessories is depicted in Fig. 1.7 showing feeding and collection points (Bridgwater et al., 1999).

1.3 Pyrolysis kinetics

Knowledge of the kinetics of the pyrolysis process provides pertinent information for the engineering and design of reactors (pyrolyzers). It also explains how different process conditions in a pyrolysis reactor affect product yields and distribution. Pyrolysis rates are influenced by chemical reaction kinetics, heat transfer, and mass transfer. Kinetic models of the pyrolysis of lignocelluloses may be broadly classified into three categories. The first is a one-stage global single reaction kinetics model where pyrolysis is modeled by assuming a one-step reaction; the second is a one-stage, multiple reactions where several parallel reactions are assumed; and third, a two-stage semiglobal reaction is assumed, which includes primary and secondary reactions following, for example, the pathways depicted in Fig. 1.5. For simplicity and in the interest of time the one-stage global model is described here (Basu, 2010).

$$\text{Biomass} \rightarrow \text{Volatile} + \text{Char} \tag{1.3}$$

The rate of pyrolysis depends on the unpyrolyzed mass of biomass, that is,

$$\frac{dm_b}{dt} = -k(m_b - m_c) \tag{1.4}$$

where m_c is the mass of char remaining after complete conversion in kg, k is the first-order reaction rate constant (1/s), and t is time in seconds.

The fractional change, X, the mass of biomass at any time may be expressed as:

$$X = \frac{(m_0 - m_b)}{(m_0 - m_c)} \tag{1.5}$$

where m_0 is the initial mass of biomass (kg). Substituting fractional conversion for the mass of biomass into Eq. (1.4) yields:

$$\frac{dX}{dt} = k(1 - X) \tag{1.6}$$

A solution of this equation gives the extent of biomass pyrolysis conversion as:

$$X = 1 - A\exp(-kt) \tag{1.7}$$

where

$$k = A\exp\left(-\frac{E}{RT}\right) \tag{1.8}$$

where A is the pre-exponential coefficient and E is the activation energy (J/mol). R is the gas constant (J/mol K) and T is temperature in Kelvin. Owning to the difficulties in extracting data from dynamic thermogravimetric analysis (TGA), reliable data on A and E are not easily available for fast pyrolysis but are needed to establish the rates of depolymerization as functions of the peak temperature and time. Looking at it from thermodynamics point of view, the biomass particle temperature is different from the reactor environment temperature by a factor that is dependent upon the heat penetration within the particle itself and, in turn, on the particle size as well as the heat transfer resistance within the reactor, that is, the heat transfer coefficient of the environment. The heat penetration through the particle, and hence the rate of decomposition, depends on the organic material remaining once the depolymerization begins following the shrinking core model (Fig. 1.8). Herein, heat and volatiles must diffuse through to and from the reaction front and the biochar layers, so the larger the surface area, that is, the smaller the biomass particles, the least the resistance, leading to isothermal temperature, and faster kinetics. While the true fast pyrolysis kinetic parameter such as activation energies are lacking, they may be approximated by modifying TGA data typically established at slower heat rates. Some of these are presented in Table 1.1 for select biomass.

1.4 Reactor technologies for fast pyrolysis

We mentioned that in order to maximize the liquid yield, high heat rate and short contact times are necessary. Therefore the reactors employed for the application of

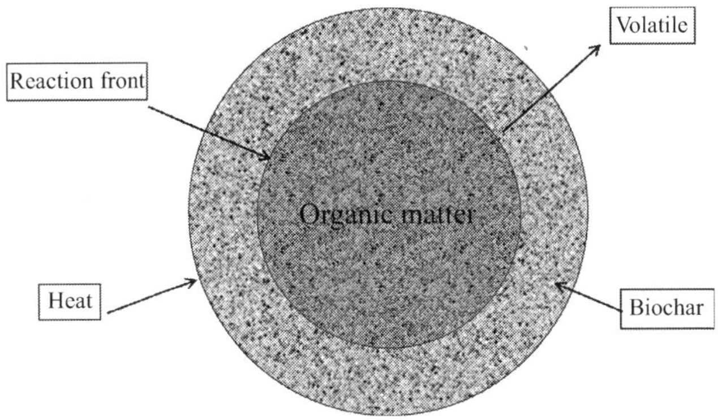

Figure 1.8 Depolymerization following shrinking model.

Table 1.1 Kinetic rate constants for select biomass.

Biomass	Temperature (K)	E (kJ/mol)	A (1/s)	Ref.
Cellulose	520–1270	166.4	3.9×10^{11}	Basu (2010)
Hemicellulose	520–1270	123.7	1.45×10^{9}	Basu (2010)
Lignin	520–1270	141.3	1.2×10^{8}	Basu (2010)
Wood	321–720	125.4	1.0×10^{8}	Basu (2010)
Almond shells	730–880	95–121	1.8×10^{6}	Basu (2010)
Beech sawdust	450–700	84 ($T > 600$K)	2.3×10^{4}	Basu (2010)
Cellulose	553–593	200–203	1.94×10^{15}	Capart et al. (2004)
Rice hull	460	181.75	1.67×10^{13}	Boateng et al. (1992)
Barley hull (thoroughbred)	873–1173	83	10.24×10^{6}	Boateng et al. (2007)
Frequency factor, A, is based on yield of noncondensable gas: CO, CO_2, CH_4, C_3H_8				
Switchgrass	873–1173	6.97–16.17	1.8336–4.1052	Boateng et al. (2006a,b)
Alfalfa	873–1173	2837.82	1.2911	Boateng et al. (2006a,b)
Reed canarygrass	873–1173	3427.8	0.9143	Boateng et al. (2006a,b)
Eastern gamagrass	873–1173	3418.82	1.0644	Boateng et al. (2006a,b)

fast pyrolysis must have such capabilities. There have been a considerable number of various reactor types that have been successfully employed to perform this task; some of these are similar in design as those used for gasification and even combustion.

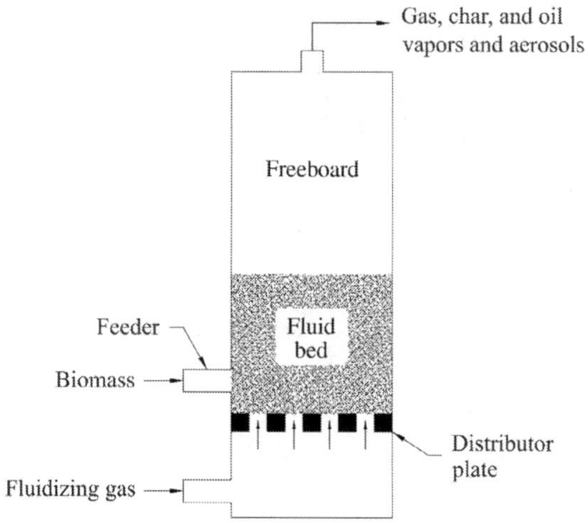

Figure 1.9 Bubbling fluidized bed pyrolysis reactor (Sadaka and Boateng, 2009).

Among the technologies that have found a widespread application are bubbling fluidized bed reactors (Fig. 1.9) because they are simpler to design and to construct; but importantly, they are capable of generating high heat transfer rates. Their rapid gas-to-solids contact times promote higher than typical heat transfer rates, good temperature control, temperature uniformity, and a large heat storage capacity. In a fluidized bed pyrolyzer a heated sand medium in an oxygen-deficient environment quickly heats the biomass (feedstock) to the desired temperature (450°C–500°C), where it is decomposed into solid char, gas, vapors, and aerosols that exit the reactor by the conveying fluidizing gas stream. After exiting the reactor zone the charcoal can be removed by a cyclone separator and stored as we saw in Fig. 1.7. The scrubbed gases, vapors, and aerosols enter a direct quenching system where they are rapidly cooled (<50°C) directly with a liquid immiscible in bio-oil or indirectly using chillers (heat exchanger). The condensed bio-oil is collected and stored, whereas the NCG may be recycled or used as a fuel source to heat the reactor. High liquid yields (about 60%–70% weight of biomass on a dry basis) can be typically recovered with bubbling fluidized bed pyrolyzers. Small feedstock particle sizes are needed (<2–3 mm) to ensure that the high heat rate requirement is fulfilled. The particle heating rate is the major factor limiting the rate of the pyrolysis reaction. Prior to recycling the NCG and residual bio-oil, aerosol droplets may be further scrubbed in an electrostatic precipitator to remove finer particulates and aerosols. The excess NCG that is typically a medium Btu gas may be exhausted, flared, or burned to provide necessary heat to the reactor.

Circulating fluidized bed pyrolyzers (Fig. 1.10) are similar to bubbling fluidized bed reactors but have shorter residence times for chars and vapors. The short residence times encountered in the reactor result in higher gas velocities, faster vapor and biochar escape, and higher biochar content in the bio-oil than bubbling fluidized beds. They have higher processing capacity, better gas–solid contact, and

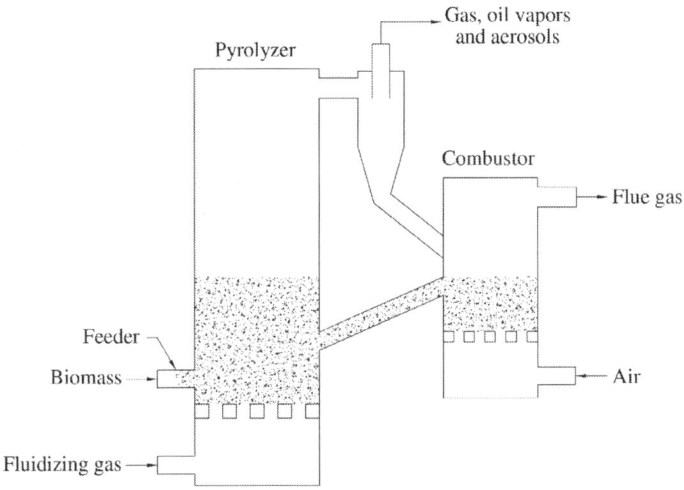

Figure 1.10 Circulating fluidized bed pyrolysis reactor (Sadaka and Boateng, 2009).

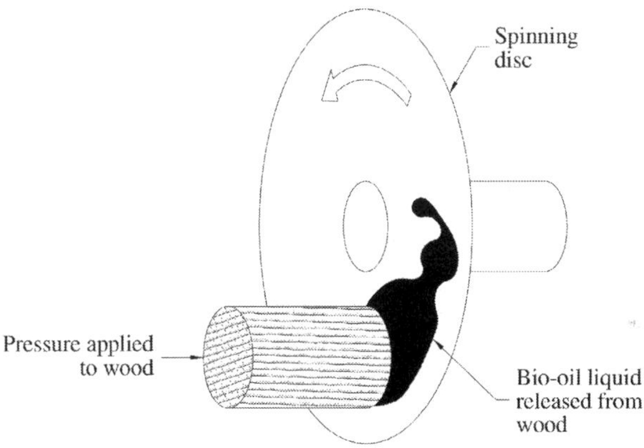

Figure 1.11 Rotary pyrolysis reactor.

improved ability to handle solids that are more difficult to fluidize. However, they are less commonly used because of the high fan power needed to suspend and circulate all the sand/catalyst medium. The heat for the endothermic pyrolysis reactions is typically generated in a secondary biochar combustor. This design is very beneficial in the case of catalytic pyrolysis where the sand medium is replaced with a heterogeneous catalyst because the used catalyst can be continuously regenerated in the secondary chamber without process interruption.

Rotating plates have been employed as a fast pyrolysis reactor in Germany and elsewhere (Fig. 1.11). They function on the premise that, while under pressure, heat transferred from a hot surface can soften and vaporize the feedstock in contact with it in a quick time allowing the pyrolysis reaction to move through the biomass in

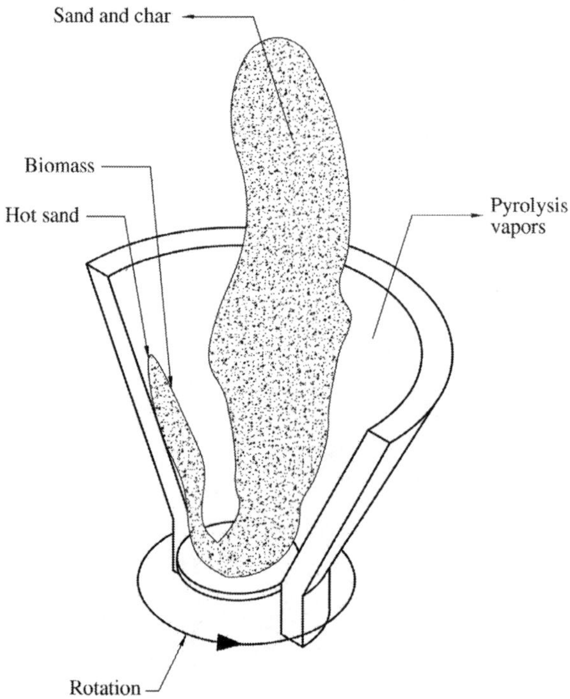

Figure 1.12 Rotating cone pyrolysis reactor.

one direction. With this arrangement, larger particles, including logs, can be pyrolyzed without pulverizing them. The most important feature is that there is no requirement for an inert gas medium thereby resulting in smaller processing equipment and more intense reactions. However, the process is dependent upon surface area, so scaling can be an issue for larger facilities.

Another technology is the rotating cone reactor. In a rotating cone pyrolysis reactor (Fig. 1.12), biomass particles at room temperature and hot sand are introduced near the bottom of a cone at the same time. They are mixed and transported upwards by the rotation of the cone. The pressures of outgoing materials are slightly above atmospheric levels. Rapid heating and short gas-phase residence times can be easily achieved in this reactor.

There are others such as auger systems, vacuum pyrolyzers, and rotary kilns (Boateng, 2015), but these do not have the short contact times that the abovementioned reactors have. They therefore result in medium, rather than fast, pyrolysis and are not discussed in great lengths in this book.

1.5 Liquid product distribution

As mentioned earlier, the primary product of interest in the pyrolysis process step is the bio-oil (biocrude) that serves as the feedstock to the upgrading step of the

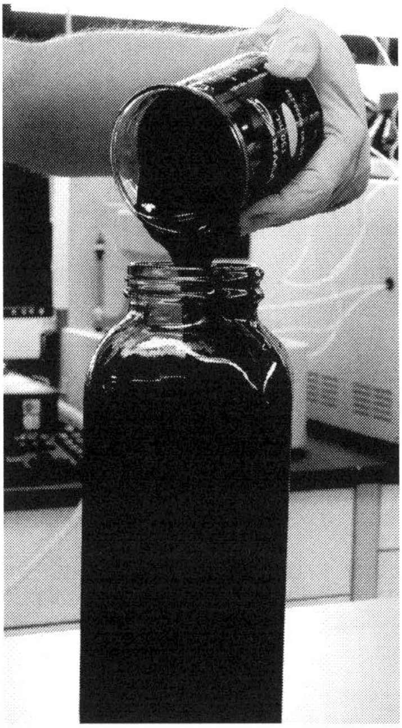

Figure 1.13 Bio-oil, pyrolysis oil, or biocrude.

biorefinery. Bio-oils are dark brown, free-flowing liquids with an acrid or smoky odor (Fig. 1.13). In terms of its elemental composition, bio-oil is not different from the parent biomass containing 35–45 wt.% oxygen except that it is the liquid form of the biomass that concentrates the energy of the bulky biomass (energy dense) that can be easily shipped or moved at longer distances. Chemically, bio-oils comprise quite a lot of water, more or less solid particles, and hundreds of organic compounds, including acids, alcohols, ketones, aldehydes, phenols, ethers, esters, sugars, furans, and nitrogen and multifunctional compounds. The molecular weights of these compounds vary significantly, from as low as 18 (water) to as high as 5000 or more (pyrolytic lignin).

The average molecular weight varies in the range of 370–1000 g/mol. Over 400 organic compounds have been identified in different bio-oils with most of the compounds found in low concentrations. The oil "as is" is therefore unstable because of its high oxygen content and the carboxylic acids present. Stabilization and/or upgrading is required if any blending into existing refinery fuel is expected.

1.6 Pyrolysis and the US national biofuels agenda

In the light of the foregoing attributes, a compelling argument could be made in support of pyrolysis biorefining as a viable pathway within the framework of the

US national biofuels agenda. But it was not always easy as the conversion of ethanol or the biochemical process had been the state of the art for renewable fuels production since the early 1980s. The background of this grounding is that in the 1990s, the US Federal Government had begun to look at ethanol as a means to combating air pollution from vehicle emissions. This is at the same time that The Clean Air Act Amendments of 1990 had created mandates for added oxygenates in gasoline to address CO and ozone problems in urban areas. At the time, the most well-known and cost-effective oxygenate was an ether, a petroleum derivative, methyl tertiary butyl ether popularly known as MTBE. But the Energy Policy Act (EPAct) of 1992 amended the motor fuels tax exemption and the blenders' credit to improve ethanol's ability to compete with MTBE. By 1999 US consumption of ethanol in gasoline-equivalent gallons had increased by some 36% leading to the phasing out of MTBE. The 2008 Farm Bill that followed, called The Food, Conservation and Energy Act included several provisions that encouraged biomass production of fuels under which the Biomass Crop Assistance Program and the Bioenergy Program for Advanced Biofuels were established (National Research Council, 2011).

The steep rise in oil prices, the growing concern over energy security and greenhouse gas (GHG) emissions, along with the desire to support domestic farm and rural economies in the mid-2000s reinvigorated the use of biofuels. This resulted in the EPAct of 2005 that established the first national Renewable Fuel Standard (RFS1) mandating an increased use of renewable fuels from 4 billion gallons per year in 2006 to 7.5 billion gallons per year in 2012. The 2007 Energy Independence and Security Act (EISA) established increased mandated volumes at 9 billion gallons in 2008 to reach 36 billion gallons in 2022 for conventional biofuels, then mainly ethanol sourced from corn, to include advanced or cellulosic biofuels, defined as fuels that reduce GHG emissions by greater than 50% or 60%, respectively (Fig. 1.14). These rules were the second set of Renewable Fuel Standards (RFS2) to be issued, following RFS1 that was established under the EPAct of 2005. In 2012 only about 20,000 gallons of cellulosic biofuels were produced in the United States, a mere fraction (0.004%) of the initial 500 million gallon target (U.S. Energy Information Administration (EIA) Report, 2013). With production lagging behind the anticipated amounts, the US-EPA issued proposed rules for 2014 that slashed the mandate for advanced and cellulosic biofuels by over 40%. This action demonstrated the urgency of developing technologies that would be capable of producing drop-in (hydrocarbon) advanced and cellulosic biofuels in lieu of ethanol if the United States was to meet its biofuels and GHG emissions reduction goals.

Complicating the situation was a move away from biofuels sourced from grains and, by 2004 the food-versus-fuel arguments had spurred interest in alternatives such as cellulosic ethanol. When President George W. Bush, the 44th US president, touted the potential use of switchgrass (lignocellulosic material) for ethanol in a state-of-the-union address, the calculus changed. By the time the RFS2 mandates (Fig. 1.14) were established as part of EISA in 2007, the national action plan shifted to cellulosic ethanol (Regalbuto, 2009, 2011), but it was soon realized that even with cellulosic ethanol, these mandates would not be forthcoming. Some of the

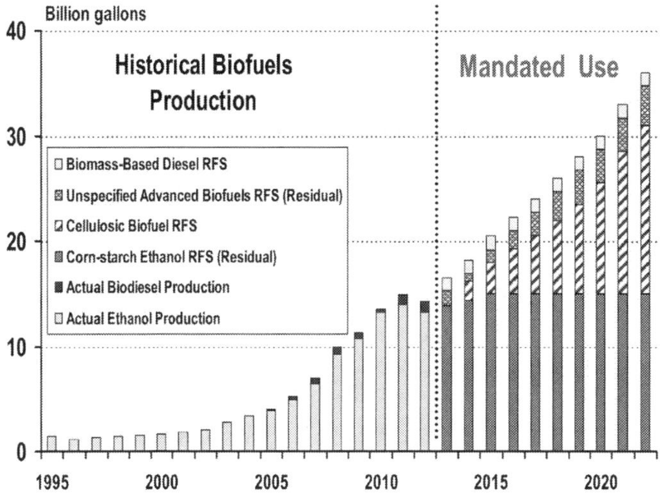

Figure 1.14 US-EPA Renewable Fuel Standard (2007) mandates.
Source: National Academy of Sciences.

challenges included, among others, various recombinants that are needed for the pretreatment of lignocellulosic biomass resources, the recalcitrance of lignin toward fermentation of lignocelluloses, and large economies of scale that go with biological processes mentioned earlier in this chapter. In addition to the technical hurdles, and with the US ethanol blend wall set at 10% of gasoline for the foreseeable future, there was limited room in the market for both grain and cellulosic ethanol, which necessitated the need for hydrocarbon biofuels (Regalbuto, 2009, 2011). At that time, thermal pyrolysis and gasification had shown promise to deliver transportation fuels, in part due to the South African experience at Sasol (Lamprecht, 2007) as well as proven catalytic upgrading processes that already existed in petrochemical refineries and could be easily adoptable to biomass processing. Four years later, the emphasis shifted from ethanol to high energy density, infrastructure-compatible biofuels—hydrocarbons. This placed pyrolysis of biomass to fungible fuels as a viable pathway to biorefinery. One of the R&D areas identified was to engineer pyrolysis systems to produce biofuels from cellulosic feedstocks at high efficiency rates while understanding the molecular species present and the cross-reactions of species during pyrolysis oil upgrading to fungible fuels.

Apart from the United States, the governments of Canada and the EU had at this time been supporting pyrolysis-based biomass conversion research and had funded demonstration projects. Unlike the United States, however, these countries had focused on the development of fast pyrolysis technologies for production of bio-oil in high yields but with a lesser emphasis on deoxygenation and stabilization during the pyrolysis step than found in the US research (Meier et al., 2013), which could result in transportation fuels. In Canada, Ensyn, ABRI-Tech (25 kg/h), and Agri-Therm had each developed fast pyrolysis systems with ABRI-Tech marketing several 1 ton/day units that were being used in various demonstrations of bio-oil production from wood waste. In Europe, major demonstration/precommercial operations included one by the company Bioliq (Germany), which focused on the production of dimethylether and gasoline from syngas produced by gasification of a bio-oil/biochar slurry. Wood based bio-oil production plants with heat and power applications in Europe had included a consortium made up of Metso, Fortum, UMP, and VTT (Finland) that went on to build a plant in Joensuu, Finland, followed by a joint venture of BTG Bioliquids and Tree Power BV in the Netherlands. These developments in thermochemical conversion technologies placed pyrolysis biorefining in the forefront of directly producing hydrocarbon molecules from biomass whereby the entirety of the biomass is utilized unlike just the extraction of sugars out of the carbohydrate portion of the biomass for cellulosic ethanol. The logistics of pyrolysis technologies with smaller design footprint reactors that can go to the biomass source and produce the liquid intermediate at the biomass site (distributed processing) and shipping the dense liquid to a centralized upgrading sites appeared to be economically feasible models and even attractive given the issues of shipping bulky biomass to large centralized processing plants. In addition, the upgrading technologies such as hydrodeoxygenation that rejects oxygen from oxygenated pyrolysis compounds to hydrocarbon molecules have worked well in existing petrochemical refineries affording some synergies and potential integration of biorefinery into already existing petrochemical infrastructure.

1.7 Feedstock challenges and U.S. Department of Agriculture approach

If lignocellulosic biorefineries are to thrive as envisioned under the RFS2, there should be a continuous supply of cellulosic feedstock. In a recent survey by researchers at the USDA's Agricultural Research Service (ARS), Orts and McMahan (2016) argued that the number one cost to a biorefinery is the biomass feedstock cost. It is important that research into biorefinery strategies be closely coupled to advances in crop science that account for crop yield and crop quality. To reach the RFS targets, stepwise progress in biorefinery technology was needed, if the industry was to move from corn ethanol toward utilizing a wider array of lignocellulose-based biomass feedstocks. In 2010 the USDA created five regional biomass research centers to optimize production, collection, and conversion of crops to bioenergy, thus building a network that fosters collaboration among researchers to improve the biorefinery industry. An important component of the five centers is the four USDA ARS regional utilization laboratories located across the country created by an act of Congress in 1938. These USDA ARS labs were originally set up by their commodities, whereby, in broad terms, the Northern Lab, now the National Center for Agricultural Utilization Research (NCAUR), focused on corn and soy; the Eastern Lab on oils, leather, dairy, and meats; the Southern Lab on cotton, sugars, and fibers; and the Western Lab on other grains, including wheat and specialty crops. Each lab's traditional expertise in these respective core commodity crops has been maintained as biofuel research came to the fore, but with the addition of new crops and biotechnological expertise, these labs collaborated with each other to develop germplasms for new energy crop lines, feedstock harvest, and logistics to support the bioeconomy. The USDA efforts on developing switchgrass as an energy crop and the compositional traits as the affect fast pyrolysis biorefining will be discussed later.

1.8 Pyrolysis technologies logic model

In the ensuing chapters, we will cover thermal pyrolysis and its product distribution. As shown earlier, the high oxygen content of thermal-only pyrolysis resulting from high concentrations of oxygenated compounds is not desirable. A major problem is that the liquid product has high levels of organic acids and other reactive oxygenates and is therefore unstable and corrosive. Over time, oligomerization and condensation reactions increase average molecular weight, viscosity, and water content causing storage and pumping problems. Such heterogeneities afford bio-oil end product very limited market as it is unsuitable as fuel for diesel engines and requires stabilization and/or upgrading before use in existing refineries. Research work needed to produce stable and upgradable pyrolysis liquids that can be seamless with existing hydrocarbon fuels, the so-called drop-in fuels, and challenges faced to establish pyrolysis as a viable technology for fuels and chemicals will be covered. Following thermal pyrolysis, we cover some of the technologies that mitigate the stability issues. These include catalytic pyrolysis, addressing stability through feedstock selection, and reactive and coreactant pyrolysis, and then cover current efforts at the pilot and field scales all following the logic model for the production of bioenergy and biochemicals as shown in Fig. 1.15 in the framework of

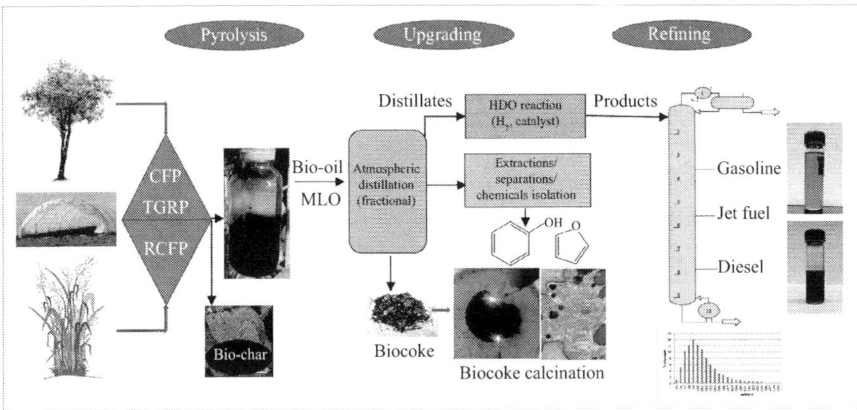

Figure 1.15 Pyrolysis biorefinery logic model. *CFP*, Catalytic fast pyrolysis; *RCFP*, reactive catalytic fast pyrolysis; *TGRP*, tail gas reactive pyrolysis.

pyrolysis biorefinery. The book will be completed with technoeconomic analysis and product life cycles for a set of pyrolysis processes described in case studies.

References

Basu, P., 2010. Biomass gasification and pyrolysis. Practical Design and Theory. Academic Press, Burlington, MA.

Boateng, A.A., 2015. Rotary kilns, Transport Phenomena & Transport Processes, second ed. Elsevier, Butterworth-Heinemann Publishers, Amsterdam, Oxford, Boston, ISBN: 978-0-1280-3780-5, 320 p.

Boateng, A.A., Fan, L.T., Walawender, W.P., Chee, C.S., Chern, S.M., 1992. Kinetics of rice hull char burnout in a fluidized-bed reactor. Chem. Eng. Commun. 113, 117–131.

Boateng, A.A., Jung, H.J.-G., Adler, P.R., 2006a. Pyrolysis of energy crops including alfalfa stems, reed canarygrass, and eastern gamagrass. Fuel 85, 2450–2457.

Boateng, A.A., Hicks, K.B., Vogel, K.P., 2006b. Pyrolysis of switchgrass (*Panicum virgatum*) harvested at several stages of maturity. Anal. Appl. Pyrolysis 75, 55–64.

Boateng, A.A., Hicks, K.B., Flores, R.A., Gutsol, A., 2007. Pyrolysis of hull-enriched byproducts from scarification of barley (*Hordeum vulgare* L.). J. Anal. Appl. Pyrolysis 78, 95–103.

Bridgwater, A.V., Peacocke, G.V.C., 2000. Fast pyrolysis of biomass. Renew. Sustain. Energy Rev. 4, 1–73.

Bridgwater, A.V., Meier, D., Radlein, D., 1999. An overview of fast pyrolysis of biomass. Org. Geochem. 30, 1479–1493.

Capart, R., Khezamia, L., Burnham, A.K., 2004. Assessment of various kinetic models for the pyrolysis of a microgranular cellulose. Thermochim. Acta 417, 79–89.

Drapcho, C.M., Nghiem, P.N., Walker, T.H., 2008. Biofuels Engineering Process Technology. McGraw Hill, New York.

Granatstein, D., Kruger, C., Collins, H., Garcia-Perez, M., Yoder, J., 2009. Use of biochar from the pyrolysis of waste organic material as a soil amendment. In: Final Project of Interagency Agreement C0800248.

Huber, G.W., 2008. Breaking the Chemical and Engineering Barriers to Lignocellulosic Bio-Fuels: Next Generation Hydrocarbon Biorefineries. A Research Roadmap for Making Lignocellulosic Bio-fuels a Practical Reality. <www/ecs.umass.edu/biofuels/Images/Roadmap2-08.pdf>.

Lamprecht, D., 2007. Fischer – Tropsch fuel for use by the U.S. Military as battlefield-use fuel of the future. Energy Fuels 21 (3), 1448–1453.

Lehmann, Joseph (Eds.), 2015. Biochar for Environmental Management: Science, Technology and Implementation. second ed. Routledge, Taylor & Francis, New York.

Mahmoudi, H., Mahmoudi, M., Doustdar, O., Jahangiri, H., Tsolakis, A., Gu, S., et al., 2017. A review of Fischer Tropsch synthesis process, mechanism, surface chemistry and catalyst formulation. Biofuels Eng. 2, 11–31.

Mason, L.C., Gustafson, R., Calhoun, J., Lippke, B.R., Raffaeli, N., 2009. Wood to energy in Washington. Report to the Washington State Legislature. The College of Forest Resources, University of Washington. <www/ruraltech.org/pub/reports/2009/wood_to_energy/index.asp>.

Meier, D., van Beld, B., Bridgwater, A.V., Elliott, D.C., Oasmaa, A., Preto, F., 2013. State-of-the-art of fast pyrolysis in IEA bioenergy member countries. Renew. Sustain. Energy Rev. 20, 619–641.

National Research Council. Renewable Fuel Standard, 2011. Potential Economic and Environmental Effects of U.S. Biofuel Policy. The National Academic Press, Washington, D.C.

Orts, W.J., McMahan, C.M., 2016. Biorefinery developments for advanced biofuels from a sustainable array of biomass feedstocks: survey of recent biomass conversion research from Agricultural Research Service. Bioenerg. Res. 9, 430–446.

Regalbuto, J.R., 2009. Cellulosic biofuels – got gasoline? Science 325, 822–824.

Regalbuto, J.R., 2011. The sea change in US biofuels' funding: from cellulosic ethanol to green gasoline. Biofuels, Bioprod. Biorefin. 5, 495–504.

Sadaka, S., Boateng, A.A., 2009. "Pyrolysis and Bio-Oil" Fact Sheet. Printed by University of Arkansas Cooperative Extension Service Printing Services. <http://www.uaex.edu/Other_Areas/publications/PDF/FSA-1052.pdf>.

Spokas, K.A., Cantrell, K.B., Novak, J.M., Archer, D.W., Ippolito, J.A., Collins, H.P., et al., 2012. Biochar: a synthesis of its agronomic impact beyond carbon sequestration. J. Environ. Qual. 41 (4), 973–989.

U.S. Energy Information Administration (EIA) Report, 2013. Annual Energy Outlook 2013 With Projections to 2040. Available from: <www.eia.gov/forecasts/aeo>.

Veitch, F.P., 1907. Chemical Methods for Utilizing Wood Including Destructive Distillation, Recovery of Turpentine, Rosin and Pulp and the Preparation of Alcohols and Oxalic Acid. Washington Government Printing Office.

Woolf, D., Amonette, J.E., Street-Perrott, A., Lehmann, J., Joseph, S., 2010. Sustainable biochar to mitigate global climate change. Nat. Commun. 1, 1–56.

Yang, H., Yan, R., Chen, H., Lee, D.H., Zheng, C., 2007. Characteristics of hemicellulose, cellulose and lignin pyrolysis. Fuel 86, 1781–1788.

Thermal pyrolysis

2

2.1 Introduction

Nonassisted or noncatalytic pyrolysis is what we might call "thermal-only" pyrolysis (as against catalytic pyrolysis), a pyrolysis process in which the medium is inert with no catalyst intentionally added. Suffice it to say that some mineral ashes that naturally occur in biomass can, under certain favorable conditions, catalyze pyrolysis reactions. Regardless, we will cover noncatalytic pyrolysis, its yields, product distribution, and appropriate uses particularly as a fuel intermediate or a precursor to producing biofuels and commodity chemicals.

Thermal-only pyrolysis is the primary thermolysis process in its pure form that yields bio-oil (pyrolysis oil or biocrude), biochar, and noncondensable (permanent) gases in various fractions. While this process has been in existence for centuries, recent resurgence toward renewable liquid fuels has placed emphasis on aspects of the pyrolysis process that maximizes the bio-oil fraction. Bio-oils can be a starting intermediate or used as a hydrocarbon biocrude that, and depending upon its quality, can enter, perhaps, seamlessly into existing petrochemical refineries. Therefore modern-day research in support of the RFS2 mandates has focused on fast pyrolysis in contrast with slow pyrolysis, a carbonization technology for biocharcoal production. It was mentioned in Chapter 1, Introduction, that a particle temperature of about 500°C is required to maximize the bio-oil yield. Also, the heat rate must be rapid, and the gas contact time must be short. The bio-oil resulting from this set of conditions may be high yielding but whether its product distribution meets the quality of fossil fuel is what the chapter attempts to discuss. To engineer a reactor with a capability to provide high heat rate conditions besides just reducing feedstock particle size is not an easy task. With heat transfer coefficients equivalent to boiling water, the bubbling fluidized bed technology provides an environment ideal for such conditions. The practical nature of such device, its ease of operation along with ease of scale-up opportunities compared with others, has made the fluidized bed reactor the equipment of choice for fast pyrolysis. It by far surpasses industrial furnaces such as the rotary- and tunnel-type kilns that tend to have longer contact times although these have found bio-oil applications as intermediate rather than fast pyrolysis reactors (Mahmood et al., 2013). Fluidization is when a granular material in a static, solid-like state at rest is transformed to a dynamic state. This occurs when a fluid gas is passed through the granular material at rest and its velocity increased to overcome the static pressure. The velocity at incipient fluidization is what is called a minimum fluidization velocity defined by the particle size. Upon increasing the fluidizing gas velocity beyond the minimum, the energy in the gas

Pyrolysis of Biomass for Fuels and Chemicals. DOI: https://doi.org/10.1016/B978-0-12-818213-0.00002-3

goes, not to increase the pressure drop, but to expand the bed volume causing the granules to bubble with a behavior similar to a boiling liquid. Fully fluidized media have gas–particle short contact times and are characteristically well mixed promoting high heat and mass transfer rates. Short contact times are preferred because long contact times such as that characterized by static beds, rotary kilns, etc., are likely to promote cracking of tars or condensable vapors that make up bio-oil and transform them into permanent gases the way gasifiers are intended. Most of the data and discussions presented in this book pertain to the fluidized bed pyrolysis technology as the reactor choice for developing and producing pyrolysis for fuels and chemicals at most developmental establishments, including the U.S. Department of Agriculture (USDA).

Like many others elsewhere before it, including the Bioeconomy Institute of Iowa State University and The Bioenergy Research Group at Aston University (Bridgwater, 2012), the bubbling fluidized bed process development unit (PDU) at the USDA Agricultural Research Service (ARS) Laboratory in Wyndmoor near Philadelphia, named "The Kwesinator" has been the workhorse for the Agency's research on the pyrolysis of agricultural residues and energy crops such as switchgrass and other herbaceous grasses (Boateng, et al., 2007). The process flow arrangement of the PDU (Fig. 2.1) may offer some insights into the technology. The system operation typically uses N_2 as a fluidizing medium but other inert gases such as CO_2 can be employed. The fluidized bed material is usually silica sand that serves as a heat flywheel by storing heat from the external heaters and transferring it quickly to the incoming biomass to initiate and sustain pyrolysis. Upon pyrolysis, char is dedusted in a cyclone downstream of the reactor while the condensable vapors are condensed further downstream in water-jacketed canisters. The condensate fractions are mainly water-soluble compounds that leave the gas stream while the aerosols are collected in an electrostatic precipitator (ESP). Although the bio-oil collection system shown here has been successfully used in collecting bio-oil fractions at various boiling points, there are alternative methods for bio-oil collection. For example, the US-DOE's National Renewable Energy Laboratory collects bio-oil over a stream of dodecane, which is later separated as dodecane is immiscible with pyrolysis liquids (Ringer et al., 2006).

Fast pyrolysis of biomass is an endothermic process requiring added heat to be supplied to the biomass to initiate depolymerization at the reaction temperature. Since the fluidized bed environment must be at or around the reaction temperature, the process is heat limited with the bulk of the energy supplied to heat the sand and maintained above the pyrolysis temperature for the incoming biomass. Biomass must be ground finely to optimize heat transfer from the sand bed under the high heat transfer rates made available by the bubbling bed of sand. Optimum operating conditions for high liquid yields from organic materials have been established by various researchers (Bridgwater, 2012). It has become a common practice that fast pyrolysis is carried out at optimum temperatures between 450°C and 550°C (Fig. 2.2) at about atmospheric pressure with biomass particle size of 2–3 mm to obtain liquid yields of 50–70 wt.%. This high liquid yield produced at near atmospheric pressure makes fluidized bed fast pyrolysis one of the most preferred

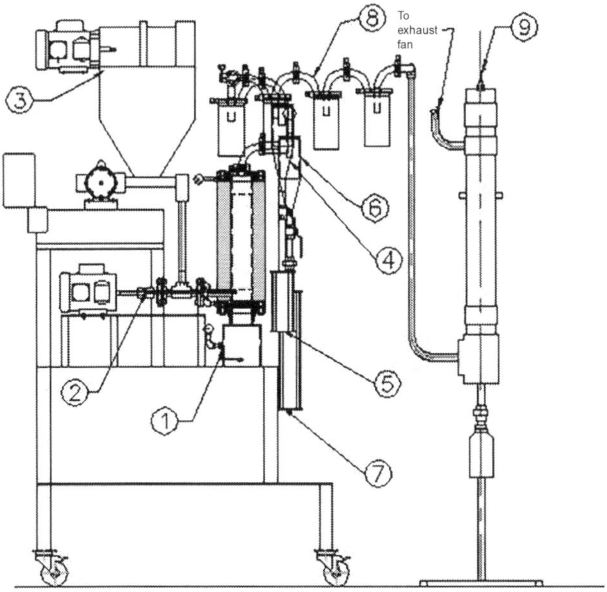

Pyrolysis reactor assembly

List of materials		
Item	Qty	Description
1	1	Reactor
2	1	Biomass feed injector
3	1	Gravimetric feeder
4	1	Primary cyclone
5	1	Primary char receptacle
6	1	Secondary cyclone
7	1	Secondary char receptacle
8	1	Condenser
9	1	Electrode assembly

Figure 2.1 Schematic diagram of the USDA-ARS thermal pyrolysis PDU, "The Kwesinator" (Boateng et al., 2008). *PDU*, Process development unit.

systems in the biorefinery although others listed in Chapter 1, Introduction, have been explored perhaps equally successfully. This condition gives optimum bio-oil yield but not necessarily the optimal product distribution.

2.2 Space-time evolution of pyrolysis products in a fluidized bed

The evolution of pyrolysis products in a fluidized bed was modeled using the design features of the PDU at the USDA (Boateng and Mtui, 2012) to predict the

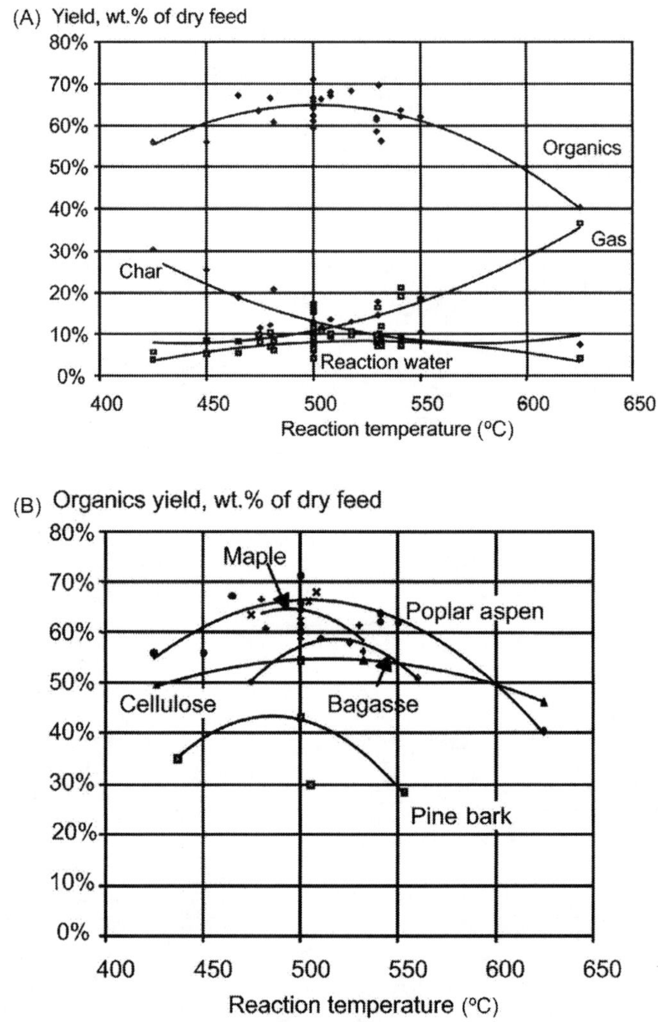

Figure 2.2 (A) Product variation with temperature from aspen poplar pyrolysis (From Bridgwater, 2012). (B) Organic yields from various feedstocks (From Bridgwater, 2012).

spontaneous emergence of the primary pyrolysis products, that is, condensable vapors, char, and noncondensable (permanent) gases. The model employed unsteady constitutive transport equations for inert gas flow and biomass fast pyrolysis decomposition kinetics using the commercial computational fluid dynamics (CFD) software FLUENT-12. The Eulerian–Eulerian multiphase continuum model system was built for a fluidized bed of sand externally heated to the optimum pyrolysis temperature prior to the introduction of biomass. Data from the PDU collected for corn cob, corn stover, and switchgrass pyrolysis were modeled to predict temporal evolution of solids and gaseous species concentrations as well as temperature distribution (Fig. 2.3) and compared with experimental observation. The predictions confirmed that the kinetics of the process are fast and that bio-oil vapor evolution is accomplished in a few seconds, and occupying two-thirds of the spatial volume of the reactor as widely reported in the open literature.

The model results show that the distribution of sand in the reactor (7.5 cm internal diameter and 50 cm in height) takes only about 14 seconds from incipient fluidization to fully expanded the bed and that the bed is fully mixed after this time frame validating theoretical trends reported elsewhere.

Predicted temperature contours during heating (i.e., the preconditioning) of the reactor ranged between 0 and 30 seconds time frame indicating that the reactor temperature increased rapidly from 15 seconds (when the sand starts to fluidize vigorously) to 30 seconds when sand is fully fluidized resulting in isothermal or near-isothermal temperature distribution within the sand bed. The model could simulate the introduction of biomass starting after the 30 seconds initiating pyrolysis reactions at around 32 seconds where biomass feedstock decomposes to pyrolysis products. Temporal predictions show bio-oil vapor concentration (for the case of switchgrass) increases steadily from the time biomass enters the reactor at 31.30 seconds (Fig. 2.4) and by 35.28 seconds and thereafter, nearly 75% of all the

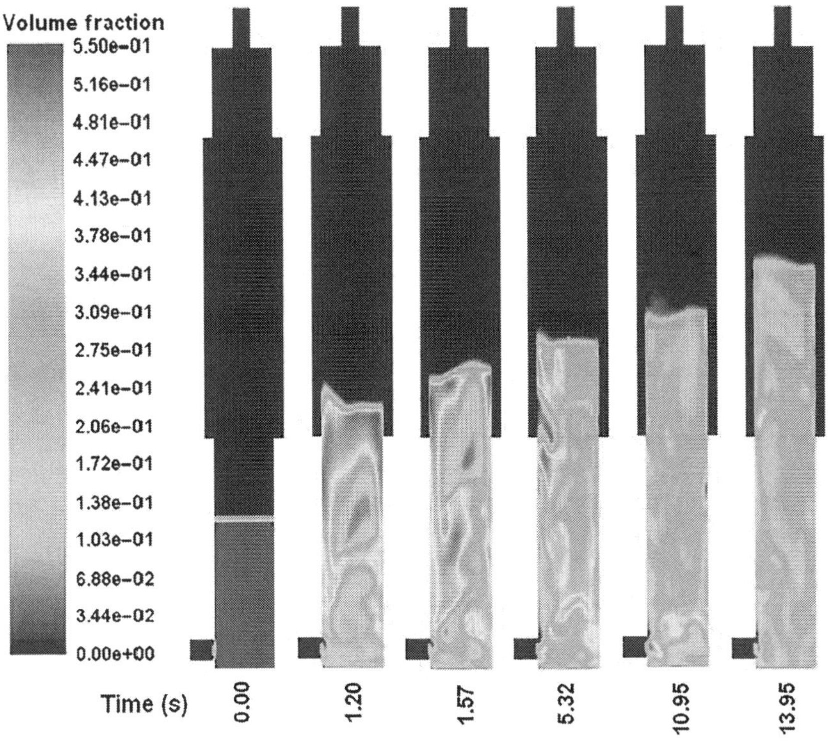

Figure 2.3 Distribution of sand in the reactor from static bed at 0 s to fully expanded bed at 13.95 s (Boateng and Mtui, 2012).

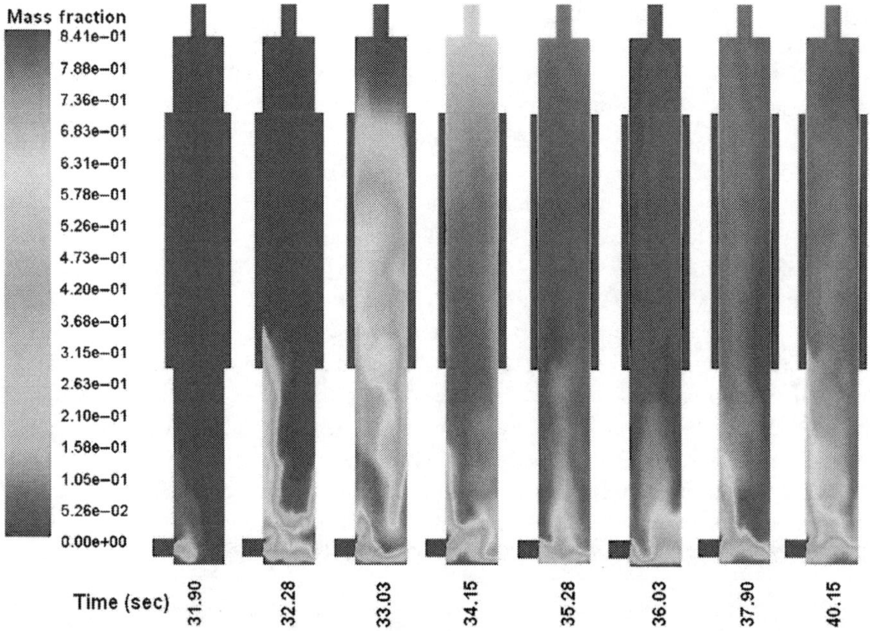

Figure 2.4 Bio-oil concentration (from switchgrass) increases steadily from the time biomass enters the reactor at 31.30 s. From 35.28 s onward the nearly 75% all the reactor volume is actively producing bio-oil (Boateng and Mtui, 2012).

reactor volume is actively occupied by bio-oil vapor. The concentration profiles for the gaseous species (noncondensable gases) indicate that the top section (freeboard) of the reactor is mostly filled with gases within ~14 seconds beyond the onset of devolatilization, while the bottom is partly occupied with the fluidizing sand medium. The predictions reveal that the kinetics of bio-oil production in fluidized bed reactor zones are uneven as per the expectation and that the reaction rate at the bottom section of the reactor is slower than other zones.

This exercise shows that a CFD model predicts product yields accurately within engineering errors of observed values showing that it can be a useful tool for scale-up design of fluidized bed fast pyrolysis reactors and their optimization to meet the economic scales required for biorefinery applications.

2.3 Pyrolysis product yield

The yield fractions that make up the products of pyrolysis vary in their amounts depending on the biomass for the same pyrolysis operating conditions. This is because of compositional differences in biomass such as its carbohydrate and lignin contents as well as ash content and their mineral constituents. There has been an extensive research on the role of inorganic minerals, including alkali and alkaline earth minerals on pyrolysis oil yield, and bio-oil quality, including its attributes such as stability and hydrocarbon content. It has been shown that pyrolysis of agricultural residues tends to result in less bio-oil yield fraction and more char coproduct than that of woody biomass (softwood or hardwood) due to their higher ash content (Patwardhan et al., 2010). In a USDA study on the pyrolysis of bioenergy crops such as switchgrass and *Miscanthus* grown in the Delmarva Peninsula, it was shown that the presence of potassium, K, in the biomass catalyzed fragmentation reactions as evidenced by the greater levels of CO_2 and CO production in the pyrolysis gaseous products (Boateng et al., 2015). The greater levels of CO_2 as K content increased indicated that more oxygen was removed from the biomass during pyrolysis under high levels of K, leading to more condensable gases that could formulate a bio-oil with higher C/O ratios. According to the "billion-ton" study conducted on the availability of biomass for biofuels production in the United States (Perlack, 2005, 2011), agricultural residues are considered a necessary part of the biomass feedstock of interest to drive the US bioeconomy, but little pyrolysis data had previously existed on these residues and herbaceous grasses that are found to be readily available at our vast agricultural fields. This is because prior efforts had been solely devoted to ethanol or biochemical pathways and thermochemical conversion research tended to focus on wood. Although it has since been modified the first US-DOE "billion-ton" biomass study reported that the amount of biomass available for bioenergy and bioproducts in 2005 was about 194 million dry tons annually; about 16% of the 1.2 billion dry tons of plant material produced on agricultural land at the time. The single largest source of this potential was corn residues or corn stover, totaling to some 75 million dry tons. The figures for crop residues were revised in 2011 to be more than 400 million dry tons with corn stover revised to between 170 and 256 million dry tons. Typical yields of fluidized bed fast pyrolysis products from corn stover and corn cobs are presented in Table 2.1 (Mullen et al., 2010a). That for straw and other residues are presented in Fig. 2.5 accounting for water in the bio-oil separately. These show what is recovered from the operation and what is possible when corrected for accumulation within the reactor system by the assistance of material balance (Mullen et al., 2010b). These numbers are what is expected for agricultural residues although woody biomass have produced higher yields.

Table 2.1 Pyrolysis product yields[a].

	Corn cobs		**Corn stover**	
	Recovery (%)	**Corrected (%)[b]**	**Recovery (%)**	**Corrected (%)[b]**
Bio-oil	40.9	61.0	58.2	61.7
Biochar	18.9	18.9	17.1	17.0
Noncondensable gas	14.7	20.3	5.3	21.9

[a]Average of two pyrolysis experiments for each feedstock.
[b]Corrected yields estimated from mass elemental balance using analysis of the feedstock and products (Mullen et al., 2010a).

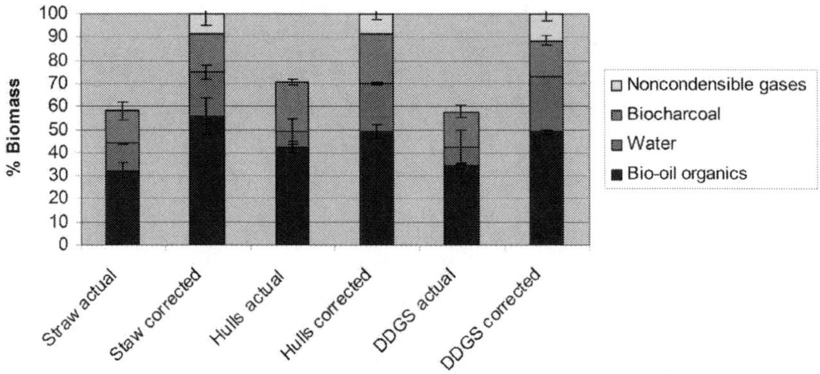

Figure 2.5 Pyrolysis product yields of various biomass feedstocks. Corrected values are based on a mass balance model. Error bars represent one standard deviation (Mullen et al., 2010b).

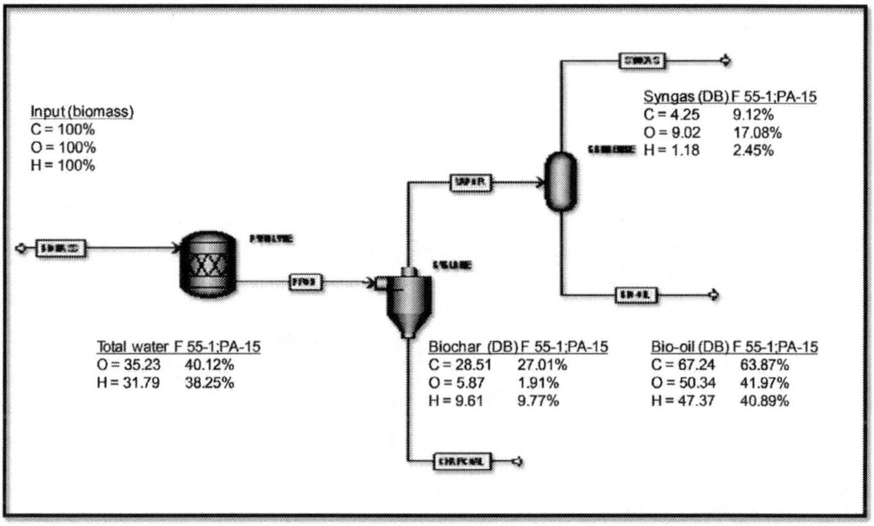

Figure 2.6 Carbon and other elemental conversion by mass balance (Boateng et al., 2010a,b).

Fig. 2.6 depicts what the material balance for the process streams look like on an elemental basis and how the carbon and hydrogen from the biomass are distributed over the pyrolysis products. The data show that in thermal-only pyrolysis as much as 67% of the incoming carbon can be found in the pyrolysis oil, 28% in the biochar, and 4% in the effluent noncondensable gas for agricultural residues such as straw, corn cobs and stover, and hulls.

Energy recovery for producing bio-oil and associated products from corn cob and stover based on the elemental balance can be presented with respect to the bio-oil alone as target energy carrier or with biochar and biochar plus the noncondensable (NCG) gases as equally important energy carriers (Table 2.2).

Energy recovery can be as high as >75% to as low as 50% depending on the target products. The debate as to whether pyrolysis of biomass is energy efficient or not is therefore settled without question since the coproducts such as biochar and the NCG can provide needed heat for the endothermic fast pyrolysis process. However, this is only when the balance is restricted to the reactor boundary. Other energy inputs outside this boundary such as biomass cultivation, harvesting and processing, and transportation are factors that determine the true energy efficiency as well as sustainability of the biorefinery system.

2.4 Product's physical and fuel properties

A representative physical property values for typical bio-oil from fluidized bed fast pyrolysis of biomass are presented in Table 2.3. As shown, and as previously discussed, yields are in the high quantities, about 60–70 wt.% of the biomass depending on the biomass type. However, the bio-oil can contain as high as 30% water and various oxygenates that do not contribute to its fuel value. With an average pH

Table 2.2 Energy recovery[a] (Mullen et al., 2010a).

	Corn cobs (%)	Corn stover (%)
Bio-oil	48.6	54.5
Bio-oil + biochar	71.3	68.7
Bio-oil + biochar + NCG	75.7	74.0

[a]Average of two runs per feedstock.

Table 2.3 Properties of pyrolysis oil.

Physical property	Typical value
Moisture content	15%–30%
pH	2.8–4.0
Specific gravity	1.1–1.2
Elemental analysis	
C	55%–64%
H	5%–8%
O	27%–40%
N	0.05%–1.0%
Ash	0.03%–0.30%
High heat value	16–26 MJ/kg (6878–11,175 Btu/lb)
Viscosity (40°C and 25% water)	25–100 cp

as low as 2–3 bio-oil is acidic. The data also suggest bio-oil is slightly denser than water, highly viscous, and with a heating value of about half that of No-6 diesel fuel oil. These fuel characteristics show that while bio-oil might have an appearance of a crude petroleum oil and is a promising pathway to producing alternative fuels from biomass, it is not fuel per se, but biomass in its liquid form. With oxygen content of up to 40%, bio-oil is an oxygenated liquid mixture comprising a large number of oxygenated organic compounds, many of which are organic acids such as acetic and formic, making it highly unstable and corrosive. Its long-term storage issues must be addressed for it to be commercially viable. Owing to its high oxygen content, it oligomerizes with time forming "gummy" solid with a short shelf life. The rate of viscosity increases is exacerbated with elevated storage temperatures. Refrigeration of thermal-only pyrolysis oil is therefore helpful to reduce oligomerization of the carboxylic acids. Before we discuss bio-oil stability, it is worth mentioning that because of the recent drive for renewable fuels and bio-oil's potential use as biocrude, American Society for Testing of Materials (ASTM) standards have been developed to guide the production and use of bio-oil as combustion fuel or fuel replacement, the purpose of which is to streamline a newly fledged industry.

To produce bio-oil for fuel applications such as home heating and/or industrial furnace combustion, the bio-oil must meet the ASTM standards as specified in the ASTM D7544 (specification for pyrolysis liquid biofuel). The standard (Table 2.4) specifies various test methods that must be followed to determine the liquid fuel properties required for the use of bio-oil in industrial burners.

The reader is referred to the ASTM D7544 document for information on long-term storage, hazards, and microbial contamination. It discusses the fact that although thermal-only pyrolysis oils are unstable when produced following standard procedures, bio-oil has adequate stability properties to withstand normal storage and use without forming excessive amounts of insoluble degradable products. Stability properties are however not well understood and depend on the biomass feedstock type, its cultivation and harvest practices as well as the pyrolysis reaction

Table 2.4 ASTM D7544-9—requirements for pyrolysis liquid biofuels (ASTM International, 2009).

Property	Test method	Specification	Units
Gross heat of combustion	D240	15 min	MJ/kg
Water content	E203	30 max	Mass%
Pyrolysis solids content	Annex A1	2.5 max	Mass%
Kinematic viscosity at 40°C	D445A	125 max	mm^2/s
Density at 20°C	D4052	1.1–1.3	kg/dm^3
Sulfur content	D4294	0.05 max	Mass %
Ash content	D482	0.25 max	Mass %
pH	E70	Report	—
Flash point	D93B	45 min	°C
Pour point	D97	−9 max	°C

conditions. To safeguard the product the ASTM D7544 Standard recommends that bio-oils must not be stored above 30°C as storing at higher temperatures may accelerate fuel degradation (oligomerization). It also suggests that storing at a minimum temperature of 10°C bio-oils can maintain adequate fluidity. Recommended storage temperature is therefore established between 15°C and 20°C, and it must be kept in corrosion-resistant containers. Hydrocarbon fuels are not miscible in pyrolysis liquids; however, polar solvents such as methanol and ethanol may be used as fuel additives and may be employed to improve the long-term storage of pyrolysis bio-oils. Most of the instability characteristics of bio-oils have been correlated to the oxygen content (c. 30 wt.%).

2.5 Product stability

Stability of pyrolysis oils can be established by the accelerated aging tests following Oasmaa and Kuoppala (2003) done by storing bio-oil in a sealed vial in an oven at 90°C followed by gel permeation chromatography (GPC) to determine its molecular weight changes at 0, 8, and 24 hours. Fig. 2.7 along with Table 2.5 presents aging and acidity (pH and TAN) data for the mustard family biomass pyrolysis oils compared with that produced from a perennial grass (switchgrass), woody biomass (oak), legumes (alfalfa), and high-protein feedstock, including distillers dry grains with solubles (DDGS) from barley fermentation (Boateng et al., 2010a,b). The higher the rate of molecular weight increases, the more unstable the bio-oil is, thereby indicating that pyrolysis oil made from different feedstocks could possibly have different stability issues. Specific USDA data show how woody biomass, for example, oak and alfalfa stems, and herbaceous grasses such as switchgrass, an

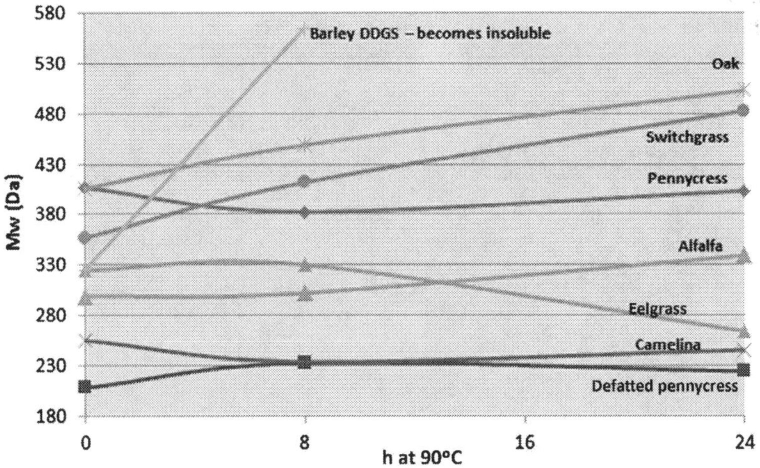

Figure 2.7 Comparison of bio-oil stability as measured by changes in average molecular weight (MW) with storage at 90°C, over 24 h (Boateng et al., 2010a,b).

Table 2.5 Acidity of bio-oils (Boateng et al., 2010a,b).

Bio-oil produced	pH (aqueous phase)	TAN (mg KOH/g)
Defatted pennycress presscake (February 8, 2010)	8.6	62.91
Defatted pennycress presscake (February 25, 2010)	8.9	64.44
Pennycress presscake (January 27, 2010)	8.3	83.23
Pennycress presscake (January 29, 2010)	7.5	84.74
Pennycress seed (February 16, 2010)	6.4	64.15
Pennycress seed (February 18, 2010)	6.4	96.86
Camelina (December 1, 2010)	7.5	70.70
Camelina (April 16, 2010)	7.5	64.55
Perennial grasses (switchgrass, multi-year; 2007–2012)	2.3	157
Wood (oak, 2010–2011)	2.2–2.5	130
Legumes (alfalfa, 2008)	5.6	70.13
Barley DDGS, 2008	6.5	59.28

energy crop, do age gradually and also at similar rates. Proteinaceous organic materials such as biomass from camelina and pennycress residues barely age and are therefore stable over prolonged storage. Others such as distillers dried grains with soluble (DDGS), a residue of barley grain-to-ethanol processing, age unusually strongly and might pose shelf life problems. The chemistry behind such occurrences is complex and we will return to some suggestions leading to this phenomenon. Suffice it to say that biomass selection could be one way of mitigating pyrolysis oil instability issues although influencing the chemistry of the molecular fragments that are depolymerizing from the biomass substrate during the incipient pyrolysis process step or manipulating the chemistry of the pyrolysis oil post pyrolysis is preferable. Some of these mitigating efforts will be discussed in the upcoming chapters.

As shown in Fig. 2.7, most proteinaceous biomass exhibit neutral or reverse aging. For example, the slopes of molecular weight curves for the pyrolysis liquids produced from the mustard family seed value chain were essentially zero, indicating minimal or no potential oligomerization reactions taking place and, as a result, no viscosity increases. However, this is not the case for the barley DDGS, although a comparable TAN value is reported. The conventional cellulosic feedstock pyrolysis liquids show dramatic rate increases in molecular weight over time as expected from their high oxygen content and higher TAN values. The unique stability characteristics of the bio-oils produced from pennycress can be attributed to their ability to partially deoxygenate because of the greatly decreased concentration of the oxygen-containing reactive functional groups.

Chemical analysis shows that nitrogen is contained in cyclic amines, such as pyridines, pyrroles, indoles, piperidines, and their derivatives, a result of feedstock protein decomposition. The higher than usual concentration of these basic compounds has the beneficial effect of neutralizing the organic acids (formic, acetic, propanoic, and others), which would normally reduce pH and cause corrosiveness and contribute to instability. While it is possible that some aging reactions can be catalyzed by acidic conditions, previous reports have demonstrated that neutral pH alone does not produce stable pyrolysis oils, as is the case with DDGS. There is evidence that the proteins and other basic N-functionalized compounds in pyrolysis oil can react with aldehydes or multifunctional compounds, such as hydroxyacetaldyde, to effectively dimerize protein fragments (and also deoxygenate the hydroxyacetaldyde) as shown in Scheme 2.1 and, thereby, increase the average molecular weight and viscosity.

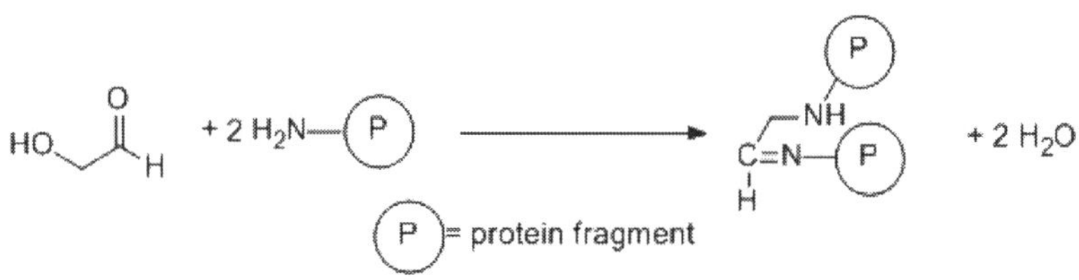

Scheme 2.1

The puzzle therefore lies in why some of the proteinaceous feedstocks such as the presscake from the mustard family of oil seeds produced stable pyrolysis oils. Table 2.6 showing key compositional analysis of proteinaceous and nonproteinaceous feedstock bio-oils indicates that bio-oil derived from the presscakes does not contain detectable levels of furfural, hydroxyacetaldhyde, or acetol, all of which are among the most abundant carbonyl and multifunctional oxygenates that are usually found in pyrolysis oil produced from lignocellulosic feedstocks, such as switchgrass or oak. As mentioned earlier, these compounds are some of the most active species that contribute to aging reactions. Their absence, therefore, is likely a contributing factor to the observed stability of the produced seedcake pyrolysis oils.

Aside from some specific chemical compounds in proteinaceous feedstocks that affect their stability, most lignocellulosic biomass feedstocks yield oxygenated compounds during thermal pyrolysis.

2.6 Product chemical composition and distribution

As we have seen by now, the chemical compositions of bio-oils are determined by many factors, such as biomass type; it is also affected by feedstock pretreatment that includes particle size and shape, moisture, and ash contents; pyrolysis conditions, that is, temperature, heating rate, residence time, pressure, gaseous environment, as well as vapor filtration and condensation filter type, condensing method and medium; cooling rate also plays a major role in determining bio-oil quality. Therefore bio-oils produced from different materials and by different pyrolysis reactors may differ greatly from one another. As a result, bio-oil composition and its fuel properties will almost always vary in wide ranges (Mullen et al., 2009). USDA data on the quantities of the key components of pyrolysis oils from corn cob and stover, agricultural residues, at the various points of collection in a fluidized bed pyrolysis reactor are presented in Table 2.7. As the table indicates, bio-oils are

Table 2.6 Pertinent pyrolysis oil components via gas chromatography and mass spectroscopy (wt.%) for proteinaceous feedstocks (Boateng et al., 2010a,b).

Compound (wt.%)	PC[a] cake	Defatted PC[a] cake	PC[a] seed	Camelina cake	Barley DDGS	Oak	Switchgrass
Acetic acid	6.00	6.22	6.11	6.29	0.88	11.1	2.94
Furfural	0	0	0	0	0	0.87	0.62
Furfuryl alcohol	0.12	0.25	0.19	0.39	0	0	0
Hydroxyacetaldehyde	0	0	0	0	0	1.3	2.40
Acetol	0	0	0	0	<0.2	4.95	0.78
Levoglucosan	2.06	2.44	0.81	1.48	2.75	3.00	6.38
Phenol	0.33	0.44	0.30	0.34	0.13	0.12	1.14
Guaiacol	0.13	0.25	0.21	0.15	0.05	0.25	0.51
Syringol	0.05	0.19	0.17	0.17	0	0.47	0.37
Indole	0.13	0.06	0.06	0.07	0.06	0	Trace
Pyrrole	0.11	0.12	0.08	0.15	0	0	0

[a]*PC*, Penncyress.

Table 2.7 Quantification of some bio-oil components[a] (Mullen et al., 2010a,b).

Method	Compound	Corn cob bio-oil			Corn stover bio-oil		
		Cond 1–4	ESP	Whole oil[b]	Cond 1–4	ESP	Whole oil[b]
K-F[c]	Water (wt.% of bio-oil)	37.68	11.27	24.83	15.54	5.94	9.15
Cellulose/hemicellulose-derived compounds (wt.% of bio-oil)							
HPLC	Levoglucosan	6.91	3.63	5.31	14.75	11.33	12.36
HPLC	Hydroxyacetaldehyde	0.86	2.05	1.44	4.03	3.99	4.00
HPLC	Acetic acid	6.44	6.42	6.43	7.56	6.20	6.26
HPLC	Acetol	10.38	9.89	10.14	8.58	6.40	7.08
GC	Furfural	0.89	1.53	1.20	0.60	0.79	0.71
GC	Furfuryl alcohol	0.24	0.36	0.30	0.14	–	0.04
GC	3-Methyl-2-cyclopenten-1-one	0.26	0.51	0.38	0.08	0.09	0.09
GC	4-Hydroxy-4-methyl-2-pentanone	0.08	0.17	0.12	0.07	0.13	0.11
Lignin-derived compounds (wt.% of bio-oil)							
GC	Phenol	0.64	0.96	0.79	0.13	0.33	0.30
GC	*o*-Cresol	0.15	0.22	0.17	0.08	0.12	0.10
GC	*p*-Cresol	0.15	0.22	0.19	0.05	0.13	0.12
GC	*m*-Cresol	0.13	0.15	0.17	0.03	0.03	0.03
GC	2,4-Dimethyl phenol	0.05	0.04	0.05	–	–	–
GC	3,5-Dimethylphenol	0.03	0.38	0.02	–	–	–
GC	4-Ethyl phenol	0.37	0.33	0.54	0.17	0.25	0.22
GC	3-Ethylphenol	0.03	–	0.02	–	–	–
GC	2-Ethylphenol	0.02	–	0.01	–	–	–
GC	Guaiacol	0.44	0.78	0.61	0.19	0.29	0.25

(Continued)

Table 2.7 (Continued)

Method	Compound	Corn cob bio-oil			Corn stover bio-oil		
		Cond 1–4	ESP	Whole oil[b]	Cond 1–4	ESP	Whole oil[b]
GC	2-Methoxy-4-methyl phenol	0.37	0.25	0.31	0.22	0.15	0.17
GC	Isoeugenol	0.22	0.35	0.28	0.14	1.26	0.82
GC	2,6-Dimethoxyphenol	0.41	0.84	0.63	0.24	0.46	0.38

ESP, Electrostatic precipitator; *GC*, gas chromatography.
[a]Average from two pyrolysis runs for each feedstock.
[b]Whole bio-oil percentages based on a weighted sum of the condenser and ESP bio-oils.
[c]K-F, Karl-Fischer titration.

highly oxygenated compounds derived from the depolymerization of the primary biomass polymers. The main oxygenates are typically derived from carbohydrate (cellulose and hemicellulose) polymer. Lignin depolymerization yields oligomers such as phenol, creosol, guaiacol, and related compounds, but these are only those detectable by gas chromatography and mass spectroscopy (GC/MS) standards. As is seen here, no hydrocarbons, that is, nonoxygenated compounds are observed in any appreciable amounts in the products of thermal-only fast pyrolysis.

It is evident from Table 2.7 that the compounds most abundant in the bio-oils (other than water) are small water-soluble oxygenated molecules derived from the polysaccharides in the corn crop residues. Same can be said about the biomass of other agricultural residues. The most abundant compounds in this class are levoglucosan, hydroxyacetaldehyde, acetol, and acetic acid. Most of these compounds are found in slightly higher concentrations in the bio-oils collected at the condensers than at the ESPs. Compounds derived directly from cellulose, including levoglucosan and hydroxyacetaldehyde, are generally found in higher concentration in bio-oils from corn stover, reflective of the higher concentration of cellulose in the feedstock. Decomposition of the lignin found in biomass results in the production of numerous water-insoluble phenolic compounds, both GC observable lignin monomer derivatives and higher molecular weight lignin oligomers. The most abundant lignin-derived small molecule in the corn cobs bio-oil is phenol (0.8 wt.%) and the most abundant small phenolic compound in the bio-oils from corn stover is isoeugenol (0.8 wt.%). The phenolic compounds are found in higher concentrations in bio-oils collected at the ESP than the bio-oils collected at the condensers.

As is also seen, the highly oxygenated carbohydrate-derived compounds, that is, anhydrous sugars such as levoglucosan and the carboxylic acids are high in concentration in bio-oils created by thermal-only pyrolysis (noncatalytic). These are what we call "the bad actors" that greatly contribute to the instability of the pyrolysis oil. As oxygenates they also have little fuel value. The stability of thermal-only pyrolysis is problematic and prevents larger utility of the pyrolysis oil as fuel intermediate. In order to make pyrolysis oil that is commercially viable as fuel substitute or energy carrier that meets advanced biofuels quality, much research efforts and financial resources have been devoted to improving the stability of pyrolysis oils by not only searching for the right feedstock but also reducing the carboxylic acids through upgrading methods that will reject the oxygen as either CO_2 or water. These include influencing the process at the devolatilization steps or postprocess as mentioned earlier. Nonetheless, the characteristic of thermal-only pyrolysis that qualifies it as a fuel intermediate or an energy carrier cannot be underestimated.

NMR spectroscopy, including ^{1}H, ^{13}C, and DEPT spectra, have been useful in characterizing fast pyrolysis oils from numerous energy crops and other agricultural feedstocks at the USDA (Mullen et al., 2009; Strahan et al., 2011). Herein are some NMR results and their correlations that characterize bio-oil as fuels and chemicals. NMR spectroscopy data of bio-oils created from switchgrass, alfalfa stems, corn stover, guayule (whole plant and latex-extracted bagasse), and chicken litter are presented. The ^{1}H and ^{13}C NMR spectra are successful in integrating over spectral regions to quantify the classes of carbon and hydrogen atoms in each bio-oil sample. DEPT spectra can be used to quantify the constituents by carbon protonation number, and the distribution of the carbon atoms in each of those classes used to provide further information on the types of molecules that are found in the bio-oil. The results (Tables 2.8 and 2.9) provide information on the concentrations of chemical functionalities that are potentially useful for synthetic modifications and hence may help guide those interested in using bio-oils as chemical feedstock. Bio-oil fuel and chemical potentials and their similarities among the various biomass types in the study can be summarized as follows:

Fuel quality. The relative number of aliphatic carbons not adjacent to heteroatoms, represented by those carbon atoms and protons that resonate in the upfield regions of the NMR spectra, track with the dry-basis energy content of the pyrolysis oils. For example, it is found that bio-oil from chicken litter has some of the lowest overall amount of methyl groups among the biomass studied. Bio-oils derived from corn stover and switchgrass have the fewest aliphatic carbons. The switchgrass bio-oil has more straight-chain aliphatics, while that of corn stover are branched. The carbohydrate content of corn stover was also found to be high.

Chemical quality. Since unsubstituted aromatic carbons (CH) are potentially important reactive sites for synthetic modification, the aromatic CH0:CH1 ratio may be an important consideration when evaluating the suitability of a feedstock for product derivitization. The chemical shifts indicate that the aromatic CH0:CH1 ratio for the bio-oil from alfalfa stems for example is among the lowest (1:1) of the biomass studied, bio-oils from guayule bagasse and switchgrass had similar aromatic ratios ($\sim$2.1–2.3) with corn stover, chicken litter, and whole plant guayule having the highest ratios ($\sim$2.7–2.8). With regard to aldehydes and ketones, switchgrass, corn stover, and guayule bagasse, bio-oils had only detectable levels of aldehydes, but bio-oil from chicken litter had the highest ketone content. Bio-oil from alfalfa stem had mostly esters or acids.

To further elucidate the discrimination among bio-oils and their parent biomass type, the in-depth NMR analysis of intact pyrolysis oils was expanded by, including additional feedstocks; their chemical components based on functional groups, which were in turn, quantified and compared with fossil fuels, diesel and gasoline. The discrimination methodology employed the use of unsupervised principle component analysis (PCA) of the ^{13}C NMR spectra (Strahan et al., 2011). By using PCA, bio-oil samples produced by thermal-only pyrolysis could be quickly analyzed, compared to each other, and tested against an existing model to evaluate their salient features and probable energy contents of the energy carriers, without the need of purification, extraction, separation, or any other special handling.

The score plot of PC1 versus PC2 in (Fig. 2.8) reveals a distinct clustering of the fossil fuels (cluster E) compared with the pyrolysis oils regardless of biomass type although some are closer to fossil fuels than others. Comparing the score plots, one can annotate the major contributing chemical components (cluster groups A–E) and interpret their fuel quality due to chemical moieties. The analysis reveals, as expected, that the fossil fuels in cluster E are discriminated from pyrolysis oils based on only a few significant resonances arising from methyls, hydrocarbon CH2 and limited range of aromatics. Biomass cluster D, which is closest to the fossil

Table 2.8 Percentage of hydrogen based on ^{1}H NMR analysis of bio-oil from fast pyrolysis of various feedstocks, grouped according to chemical shift range (Mullen et al., 2009).

Chemical shifts (ppm)	Proton assignment	Switchgrass (%)	Corn stover (%)	Alfalfa stems (%)	Guayule—whole (%)	Guayule bagasse (%)	Chicken litter (%)
0.15–1.5	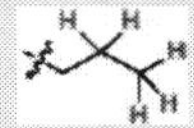 Alkanes	9.8	11.8	20.9	29.4	28.7	34.6
1.5–3.0	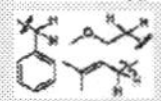 Aliphatics α to heteroatom or unsaturation	24.3	18.3	54.0	42.0	34.5	45.9
3.0–4.4	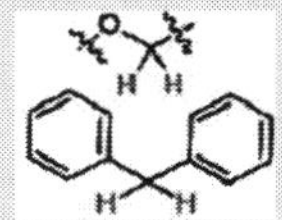 Alcohols, methylene-dibenzene	21.3	20.5	7.2	10.4	12.5	9.8
44–6.0	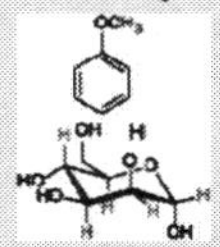 Methoxy, carbohydrated	25.7	30.3	2.3	6.8	9.7	1.8
6.0–8.5	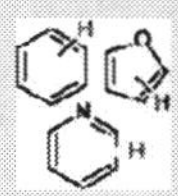 (Hetero-) aromatics	17.5	15.1	15.1	11.2	15.6	7.9
9.5–10.1	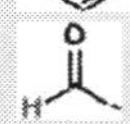 Aldehydes	1.3	1.7		0.2	0.5	

Table 2.9 Percentage of carbon based on ^{13}C NMR analysis of bio-oil from fast pyrolysis of various feedstocks, grouped according to chemical shift range (Mullen et al., 2009).

Chemical shifts (ppm)	Carbon assignments	Switchgrass (%)	Corn stover (%)	Alfalfa stems (%)	Guayule—whole (%)	Guayule bagasse (%)	Chicken litter (%)
0–28	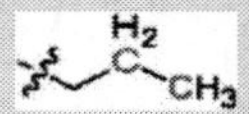 Short aliphatics	13.8	13.8	17.2	28.5	19.1	25.8
28–55	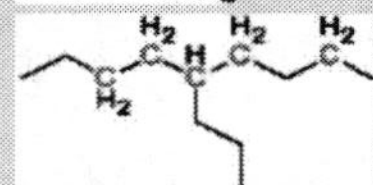 Long and branched aliphatics	7.3	10.3	12.2	24.4	29.0	21.8
0–55 Total	All of above	21.1	24.1	29.4	52.9	48.1	47.6
55–95	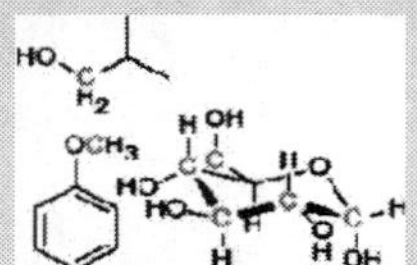 Alcohols, ethers, phenolic-methoxys, carbohydrates sugars	24.7	30.8	16.1	6.7	7.7	13.6
95–165	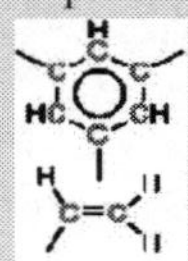 Aromatics, olefins	53.0	36.0	51.9	39.5	43.5	36.2
165–180	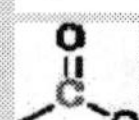 Esters, carboxylic acids		3.8	2.6	0.8	0.4	
180–215	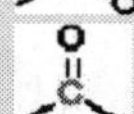 Ketones, aldehydes	1.2	1.5		0.1	0.2	2.6

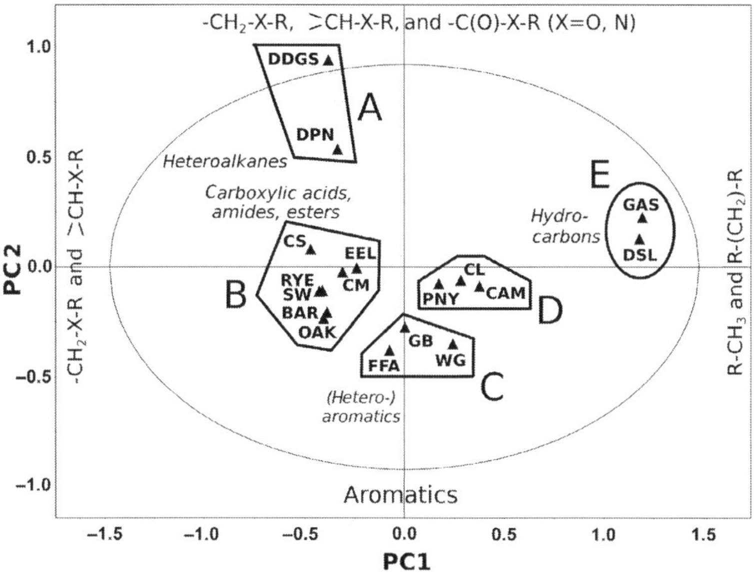

Figure 2.8 Principal component analysis (PCA) score plot of Model 1, PC2 vs PC1 for biomass pyrolysis oils and fossil fuels. Here, X may be either heteroatom N or O; and R can be H or any organic moiety that does not have a heteroatom directly attached. General descriptions of the chemical moieties that were determined to be most important in discriminating the clusters are listed as annotations (in italics) based on the analysis of the loadings plots in terms of chemical shifts and the carbon proton substitution number. Samples are *GAS*, gasoline; *DSL*, diesel; *CL*, chicken litter; *CAM*, camelina presscake; *PNY*, pennycress presscake; *GB*, guayule bagasse; *WG*, whole guayule; *FFA*, alfalfa stems; *CS*, corn stover; *EEL*, eelgrass; *CM*, cow manure; *OAK*, oak; *BAR*, barley straw; *SWG*, switchgrass; *RYE*, ryegrass; *DPN*, defatted pennycress presscake; *DDGS*, barley DDGS.

fuels, consists of the pyrolysis oils derived from camelina and pennycress presscakes, as well as that from chicken litter, that is, lipid-based feedstock. This cluster is also dominated by primary and secondary hydrocarbons as they originate from feedstocks containing residual triglycerides (vegetable oils), the likely cause for their relative proximity to the petroleum cluster. The triglycerides enable a higher proportion of aliphatic carbon chains, compared to those derived from lignocellulose alone (found in cluster B). The presscakes and the chicken litter also have a higher protein or protein-derivative content in their feedstocks, compared to those for pyrolysis oils in cluster D. This results in a higher nitrogen content in both their feedstocks and their pyrolysis oils. As mentioned earlier, the presence of nucleophilic organonitrogen compounds resulting from pyrolytic breakdown of proteins can have a deoxygenating effect, that is, may result in fewer alkyl oxygenates (alcohols, carbohydrates). Similar arguments can be made for cluster C, which consists of pyrolysis oils from whole guayule, guayule bagasse, and full flower alfalfa plant biomass. Guayule, which is a natural latex producing plant, is different from those

Table 2.10 Higher heating values (HHVs) of pyrolysis oils (Strahan et al., 2011).

^{13}C NMR PCA cluster (Fig. 2.8)	Pyrolysis oil feedstock	Pyrolysis oil HHV (MJ/kg, DB)
A	Defatted pennycress presscake	29.1
A	Barley DDGS	27.7
B	Corn stover	24.3
B	Eelgrass	26.1
B	Cow manure	28.8
B	Ryegrass	25.7
B	Switchgrass	23.1
B	Barley straw	26.3
B	Oak	22.5
C	Alfalfa stems	30.6
C	Guayule (whole)	30.4
C	Guayule bagasse	30.5
D	Chicken litter	31.2
D	Pennycress presscake	31.9
D	Camelina presscake	32.0
E	Gasoline (E10)	45.3
E	Diesel	45.0

in cluster B because of the presence of hydrocarbons derived from residual latex and plant resins.

In summary, the results of NMR studies show that the PCA clusters based on the binned ^{13}C NMR intensities (as a function of chemical shifts) correlate with their energy contents, even though their high heating values (HHVs) are not used as input for the analysis. The ranking based on measured energy content on the HHV scale is presented in Table 2.10. As expected, pyrolysis oils in cluster B (e.g., Fig. 2.8), which were created from the mostly lignocellulosic feedstocks, have the lowest HHV of all the samples studied. The clustering of the pyrolysis oils into these groups correlates chemical structures and physical properties as characterized using the ^{13}C NMR and PCA analysis. This indicates that the higher the energy content of the biomass the higher the energy content of the bio-oil produced by thermal-only pyrolysis and that thermal-only pyrolysis yields biomass in its liquid form. Biofuels similar to petroleum product cannot be produced without modifying the process (CFP) or without upgrading the product.

2.7 Effect of temperature

The compounds analyzed are based on bio-oil production at the 450°C–550°C optimum for the highest yield of the pyrolysis oil as we saw in Fig. 2.2. This is the condition at which the product distribution has been mostly analyzed as it gives the highest product yield. However, while higher temperatures may reduce yield, their product distribution analyzed by NMR may be different. Recent work using diffusion-ordered NMR spectroscopy (DOSY) (Mullen et al., 2019) reveals some stark differences not only in the chemical makeup of the bio-oil at temperature but also the way the various functional groups or chemical classes are distributed over the molecular weight ranges. The DOSY performed of on bio-oil samples derived from switchgrass by fast pyrolysis over 500°C, 600°C, and 700°C reveals that carbohydrates are present mostly as monosaccharides, disaccharides, and trisaccharides, but their relative proportions are temperature dependent. It shows that the aromatic compounds found in the 500°C and 600°C samples primarily consist of pyrolytic lignin, most of which exist as dimeric or trimeric phenolic units. In contrast, the 700°C sample tends to consist mostly of polyaromatic hydrocarbons (PAHs) containing two or three rings, but PAHs with up to eight rings may be present. Specifically, this work makes some specific conclusions on the temperature effect on thermolysis product quality. In applying the DOSY characterization to a set of pyrolysis oils produced at the three different temperatures, each having different oxygen contents, enabled the extraction of significantly more information about the chemical composition of the samples and their approximate molecular weights, than can be derived from a typical pyrolysis oil characterization protocol beyond elemental analysis, GC/MS, GPC, and conventional ^{1}H and ^{13}C NMR presented in Tables 2.8 and 2.9. Significant observations from the diffusion NMR include the following specific conclusions:

- In the carbohydrate region of the spectra of the bio-oils produced at 500°C and 600°C, the diffusion plots and spectral intensity data indicate that mono- and disaccharides are the most prevalent carbohydrates. By comparison, the 700°C sample has a lower overall concentration of carbohydrates than the other samples, but those that remain have higher molecular weights, especially in the trisaccharride molecular weight range. Diffusion NMR also suggests that even though GC analysis indicates that levoglucosan has been largely eliminated at 700°C temperature, small amounts of some other monosaccharides remain.
- In the aromatic region the bio-oils produced at 500°C and 600°C tend to contain phenolic oligomers (pyrolytic lignin). The DOSY NMR data suggests that the bulk of this material is in a molecular weight range consistent with phenolic dimers, trimers, or tetramers.
- Of all the bio-oils, that produced at 700°C has significantly less pyrolytic lignin and more PAHs as determined by their characteristic chemical shifts. Most of the observed PAHs have estimated molecular weights that correspond to two and three rings, but signals for PAHs with up to a possible eight aromatic rings may be detected. However, given that PAHs have a higher propensity for self-aggregation, this is a realistic possibility that must not be ignored.
- Most of the alkyl resonances over the entire molecular weight range are associated with aromatics, including both phenolics and hydrocarbons.
- Some of the small, fast diffusing molecules, including acetic acid and acetol, are easily identified and quantitated using DOSY NMR, as their signals are well resolved from the bulk of the peaks.
- While the water (HOD) and methanol solvent peaks interfere in the analyses of $^{1}D-^{1}H$ spectra of bio-oils, they are easily identified and separated from nearby resonances through the use of diffusion NMR.

2.8 Sustainability

Increased use of biomass for energy will provide a wider range of renewable fuel resource. Although the heating value of pyrolysis oils is not too different from that of the parent biomass if pyrolysis liquids are produced and used as combustion fuel "as is," then one of the several benefits include ease of transport of liquid as against transporting the bulky biomass to the combustion site or refinery sites. This concept lends itself to satellite or distributed processing whereby the bio-oil is produced at several biomass source locations and refining/upgrading done at a centralized location at an economic cost especially if the field units are compact fluidized bed designs. Another important advantage is that this pathway to fuel production is considered a carbon negative process since the coproduct biochar can be sequestered for a millennium and retained to amend the soil as illustrated by the experience of the Terra Preta, an anthrosol limited to the Amazonia regions of South America (Verheijen et al., 2010). Sequestration of the 20%–28% carbon associated with the biochar is the same amount that is not released to the atmosphere as CO_2 will be deficient. With the one-to-one stoichiometry for CO_2 products of biomass combustion and photosynthesis reaction, the replacement plants would need the 20%–28% CO_2 equivalent carbon that will be extracted from the atmosphere.

However, combustion of pyrolysis oils in general has been problematic for all the stability reasons presented before. Some of the combustion applications will be presented in the latter chapters, but it is imperative to state that the goal of the renewable fuel standards is to produce fungible transportation fuels. The Bioeconomy Initiative Implementation Framework of 2019 requires the production of cellulosic biofuels produced from structural fibers of plants. The fact that there are limited commercial facilities operating in the United States indicates the advanced biofuel quotas of the RFS2 are not being fulfilled as one would like and that there are several technology barriers to overcome, some of which have led to several false starts and closure of pyrolysis facilities. The task is daunting and has gone begging for research and development that can reduce the uncertainties and scale-up risks. Pyrolysis shows promise since it is feedstock-neutral and uses the whole biomass compared with biochemical conversion processes that use only the carbohydrate content of the biomass and is challenged by the recalcitrance of lignin. As we have seen from the foregoing, the major barrier is the high oxygen that needs to be reduced, so drop-in fuels can be produced that can be inserted into existing petroleum refinery or quality refinery feedstock can be produced via thermolysis. While feedstock selection choice can help, that alone cannot be a sustainable solution as all biomass resources must be exploited and utilized. The only option is to develop technologies that will target low-oxygen pyrolysis liquids by influencing the pyrolysis process and/or condensed-phase upgrading done economically and sustainably. The next chapters will address some of these technologies and their state of the art as well as their product life cycles.

References

ASTM International, 2009. 100 Barr Harbor Dr., P.O. box C-700 West Conshohocken, Pennsylvania 19428-2959, United States of America.

Boateng, A.A., Mtui, P.L., 2012. CFD modeling of space-time evolution of fast pyrolysis products in a bench-scale fluidized-bed reactor. Appl. Therm. Eng. 33 (34), 190–198.

Boateng, A.A., Daugaard, D.E., Goldberg, N.M., Hicks, K.B., 2007. Pilot-scale fluidized-bed pyrolysis of switchgrass for bio-oil production. Ind. Eng. Chem. Res. 46 (7), 1891–1897.

Boateng, A.A., Mullen, C.A., Goldberg, N.M., Hicks, K.B., Jung, H. G.-J., Lamb, J.F.S., 2008. Production of bio-oil from Alfalfa stems by fluidized-bed fast pyrolysis. Ind. Eng. Chem. Res. 47, 4115–4122.

Boateng, A.A., Mullen, C.A., Goldberg, N.M., Hicks, K.B., Devine, T.E., Lima, I.M., et al., 2010a. Sustainable production of bioenergy and biochar from the straw of high-biomass soybean lines via fast pyrolysis. Environ. Prog. Sustain. Energy 29 (2), 127–130.

Boateng, A.A., Mullen, C.A., Goldberg, N.M., 2010b. Producing stable pyrolysis liquids from the oil-seed presscakes of mustard family plants: pennycress (*Thlaspi arvense* L.) and camelina (*Camelina sativa*). Energy Fuels 24, 6624–6632.

Boateng, A.A., Serapiglia, M.J., Mullen, C.A., Dien, B.S., Hashem, F.M., Dadson, R.B., 2015. Bioenergy crops grown for hyperaccumulation of phosphorous in the delmarva peninsula and their biofuels potential. Environ. Manage. 150 (1), 39–47.

Bridgwater, A.V., 2012. Review of fast pyrolysis of biomass and product upgrading. Biomass Bioenergy 30, 68–94.

Mahmood, A.S.N., Brammer, J.G., Hornung, A., Steele, A., Poulston, S., 2013. The intermediate pyrolysis and catalytic steam reforming of Brewers spent grain. J. Anal. Appl. Pyrolysis 103, 328–342.

Mullen, C.A., Strahan, G.D., Boateng, A.A., 2009. Characterization of various fast pyrolysis bio-oils by NMR spectroscopy. Energy Fuels 23, 2707–2718.

Mullen, C.A., Boateng, A.A., Goldberg, N.M., Hicks, K.B., Moreau, R., 2010a. Analysis and comparison of bio-oil produced by fast pyrolysis from three barley biomass/byproduct streams. Energy Fuels 24, 699–706.

Mullen, C.A., Boateng, A.A., Goldberg, N.M., Lima, I.M., Laird, D.A., Hicks, K.B., 2010b. Bio-oil and bio-char production from corn cobs and stover by fast pyrolysis. Biomass Bioenergy 34, 67–74.

Mullen, C.A., Strahan, G.D., Boateng, A.A., 2019. Characterization of Biomass Pyrolysis Oils by Diffusion Ordered NMR Spectroscopy. ACS Sustain. Chem. Eng. 7, 19951–19960.

Oasmaa, A., Kuoppala, E., 2003. Fast pyrolysis of forestry residue. 3. Storage stability of liquid fuel. Energy Fuels 17, 1075–1084.

Patwardhan, P.R., Satrio, J.A., Brown, R.C., Shanks, B.H., 2010. Influence of inorganic salts on the primary pyrolysis products of cellulose. Bioresour. Technol. 101, 4646e4655.

Perlack, R.D., 2005. Biomass as Feedstock for a Bioenergy and Bioproducts Industry: The Technical Feasibility of a Billion-Ton Annual Supply. A Joint Study Sponsored by U.S. Department of Energy and U.S. Department of Agriculture.

Ringer, M., Putsche, V., Scahill, J., November 2006. Large-Scale Pyrolysis Oil Production: A Technology Assessment and Economic Analysis. Technical Report. NREL/TP-510-37779.

Strahan, G.D., Mullen, C.A., Boateng, A.A., 2011. Characterizing biomass fast pyrolysis oils by ^{13}C-NMR and chemometric analysis. Energy Fuels 25, 5452–5461.

U.S. Department of Energy, 2011. U.S. Billion-Ton Update: biomass supply for a bioenergy and bioproducts industry. In: Perlack, R.D., Stokes, B.J. (Leads), ORNL/TM-2011/224. Oak Ridge National Laboratory, Oak Ridge, TN. 227p.

Verheijen, F., Jeffery, S., Bastos, A.C., van der Velde, M., Diafas, I. 2010. Biochar Application to Soils. A Critical Scientific Review of Effects on Soil Properties, Processes and Functions. EUR 24099 EN – 2010. Joint Research Centre, JRC Report 55799, European Commission.

3 Catalytic fast pyrolysis

3.1 Introduction

We saw in Chapter 2, Thermal Pyrolysis, that under appropriate pyrolysis operating conditions, biomass can be converted to relatively high yields ($\sim$60–70 wt.%) of pyrolysis liquids (bio-oil, biocrude, etc.)—a mixture of organic compounds and water. We also saw that the liquid organics are oxygenated compounds that result from the thermal breakdown of the biomass polymers, that is, cellulose, hemicellulose, and lignin. The product distribution presented for important agricultural residues shows that, collectively, pyrolysis oil comprises a complex mixture of acids, alcohols, aldehydes, esters, ketones, sugars, phenols, furans, and multifunctional compounds such as hydroxyacetaldehyde to mention but a few. The relative amounts of each compound class can vary depending on the biomass feedstock used and the operating conditions employed during pyrolysis. We have also established that these oxygenated compounds are responsible for the stability problems that plague pyrolysis oils. Department of Energy (DOE) characterizes stabilizing bio-oils as reducing the oxygen content within the various organic compounds collectively comprising pyrolysis oil, with a preference toward rejecting the oxygen at an economically optimum balance between carbon oxides and water. A preferred level of stability is to remove oxygen present as carboxylic acid groups such that the total acid number (TAN) of the pyrolysis oil is dramatically reduced to about 5 mg KOH/g. For pyrolysis oil to be compatible with petroleum, it must be stabilized either by selecting the right feedstock or by influencing the depolymerizing process to reduce the oxygen content of the resulting bio-oil. Since all feedstock must be able to participate in the biorefinery for bioeconomy to be viable, the latter approach constitutes a more realistic proposition to the production of fuels and chemicals from pyrolysis biorefining. In fact, part of the pyrolysis biorefining appeal is that some of the technologies required for doing so are often found in existing petroleum refineries. For example, research supported by the US-DOE has shown the promise of upgrading these oxygenated compounds to fungible hydrocarbon fuels, such as gasoline and diesel, by employing conventional petroleum refining techniques, such as hydrotreating and hydrocracking.

Huber and Corma (2007) presented synergies between bio-oil and petroleum oil refineries as depicted in Fig. 3.1. Since thermal degradation of biomass by pyrolysis is agnostic to feedstock type, the opportunity to colocate and use the same processes in the existing petroleum infrastructure for the processing of pyrolysis liquid to fuels and chemicals makes pyrolysis biorefineries very attractive and economically expedient. We will discuss hydrotreating of pyrolysis oils as condensed phase upgrading in

Pyrolysis of Biomass for Fuels and Chemicals. DOI: https://doi.org/10.1016/B978-0-12-818213-0.00003-5

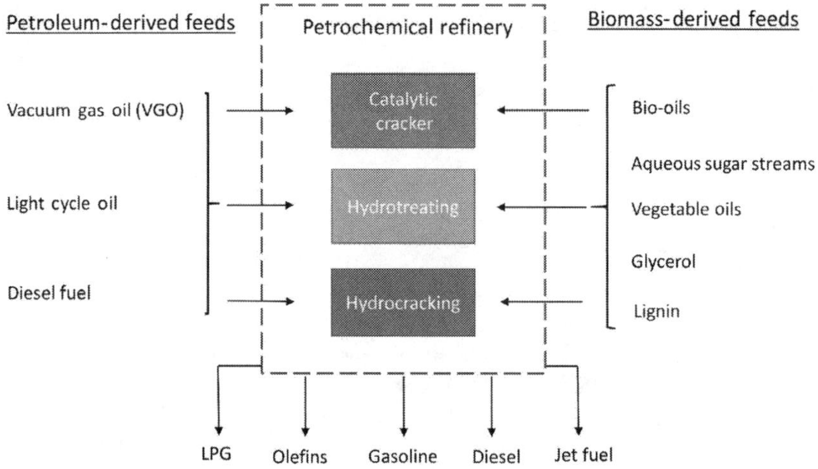

Figure 3.1 Conversion of petrochemical- and biomass-derived feedstocks in a petroleum refinery.
Source: Adapted from Huber, G.W., Corma, A., 2007. Synergies between bio- and oil refineries for the production of fuels from biomass. Angew. Chem. Int. Ed. 46, 7184–7201.

later chapters, but we start in this chapter with catalytic pyrolysis whereby heterogeneous catalysts such as zeolites can be added to the biomass to influence the incipient pyrolysis process and reject oxygen as water, CO, and/or CO_2 leaving hydrocarbon compounds as biofuels. For pyrolysis biorefinery, catalytic cracking is what we will call catalytic fast pyrolysis (CFP). CFP modifies bio-oil chemical composition by incorporating catalysts in the pyrolysis process. The synergy here with petrochemical processing is to use deoxygenating catalysts during the incipient pyrolysis process when at temperatures to allow the deoxygenation of organic species before the bio-oil is formed. The catalyst may be incorporated directly as the fluidizing medium in lieu of sand (in situ), facilitating catalyst recovery and regeneration to reduce cost. Also, the catalyst may be placed after the vapors are formed (ex situ) to catalyze gas-phase reactions before the liquid is collected.

3.2 Fluid catalytic cracking of bio-derived feedstock

If one is to explore the synergy between petroleum and biorefining, particularly thermolysis, one needs to explore the ability for some of the heterogeneous catalysts used in the cracking of petroleum-derived feedstock to perform similar functions for biomass-derived feedstocks in a biorefinery. Zeolite materials have been extensively studied and used in the petrochemical industry for fluid catalytic cracking (FCC). Zeolites are crystalline aluminosilicates with a defined pore structure. They comprise a large variety of pore mouths, channels, and interfaces to be sometimes designated as molecular sieves. Due to their microporosity, zeolites have very high surface areas ($>$ 200 mg/m^2) expanding their applications to, not only as catalysts, but as also absorption, ion exchange and separations (Larsen, 2007). Specific zeolites can be identified by their chemical composition particularly the proportions of silica and alumina (SiO_2/Al_2O_3), pore size, pore morphology, etc. Two commonly used zeolites ZSM-5 with micro-flow imaging (MFI) structure and Y zeolite with faujasite structure are shown in Fig. 3.2. Others include Beta and mordenite. MFI is a zeolite with a 10-membered ring, 3-dimensional pore topology that is well known in the literature as an active catalyst for conversion of oxygenates to aromatics. Zeolites reveal Bronsted and Lewis acidic properties that can be varied in a wide range thereby their designation as "solid acids." Almost all zeolites take advantage of their acidic properties. While they are not very strong acids at low temperatures, at 500°C considered optimum for fast pyrolysis, they are able to protonate paraffins and initiate their cracking and has, therefore, found successful applications in the petroleum industry where feedstock to crackers range from gas oil to residuum.

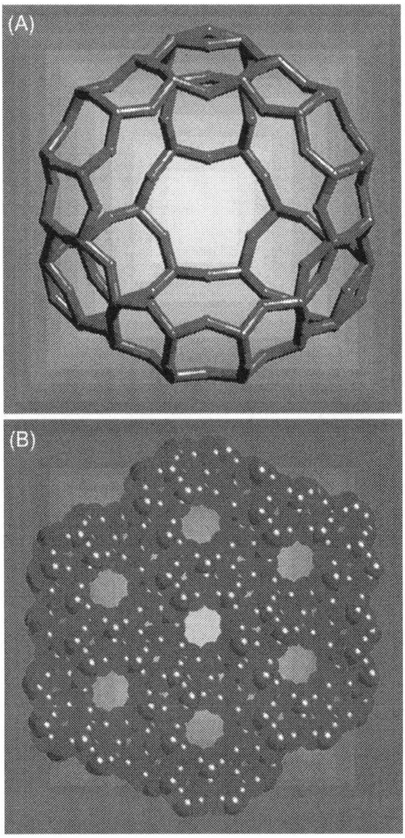

Figure 3.2 Two common zeolite structures: (A) Y with faujasite structure and (B) ZSM-5 with MFI-morphology (Larsen, 2007). *MFI*, Micro-flow imaging.

In hydrocarbon cracking the acid catalysts work by forming carbocations either by protonation of olefins (Scheme 3.1) in the feedstock or hydride abstraction from saturated hydrocarbons. This leads to isomerization, cyclization, and aromatization of the hydrocarbons opening up a slew of petrochemicals.

The addition of zeolitic and other inorganic catalysts to the pyrolysis reactor has been shown to have a major impact on pyrolysis oil composition (Williams and Nugranad, 2000; Adam et al., 2005). Reduction in oxygen content and increased formation of aromatic hydrocarbons have been reported along with changes to the amounts of low molecular weight carboxylic acids when compared with thermal-only pyrolysis done over sand. Unlike petroleum feedstocks, however, pyrolysis vapors are highly oxygenated and oxygen groups are more reactive toward both Brønsted and Lewis acid activation. This makes the acid catalyst idea well suited for the cracking of biomass pyrolysis vapors also. However, the oxygen groups also make them more likely to strongly bind to an acid catalyst and, therefore, tend to deactivate it. Furthermore, protonation of hydroxyl groups followed by dehydration to form carbocations removes hydrogen from the already hydrogen-deficient biomass molecules. With the ability to dehydrogenate, coke formation becomes prevalent due to a lack of available hydrogen to support hydrocarbon formation. In addition, biomass pyrolysis vapors do not tend to have long carbon chains as petroleum feedstocks make the formation of small olefins and intermolecular recombination a likely more prevalent mechanism in cracking that would lead to production of aromatic hydrocarbons from these oxygenates (Scheme 3.2).

Some of the observations identified in the literature on zeolite catalyzed fast pyrolysis of bio-derived feedstocks are specific to the zeolite type and morphology. The first work on bio-feedstock conversion over zeolite catalysts was performed by researchers at Mobil who showed that the molecular shape selective ZSM-5 catalyst

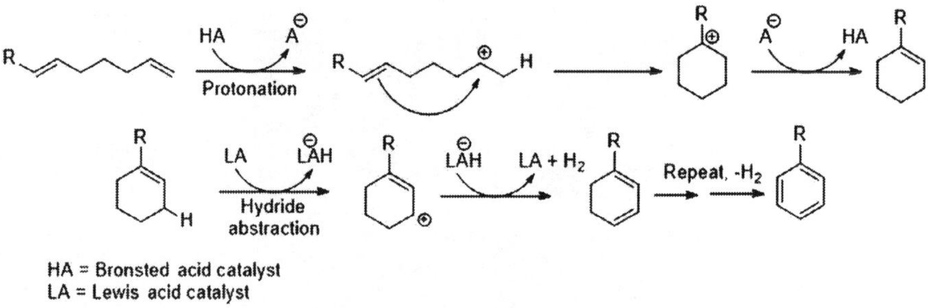

Scheme 3.1

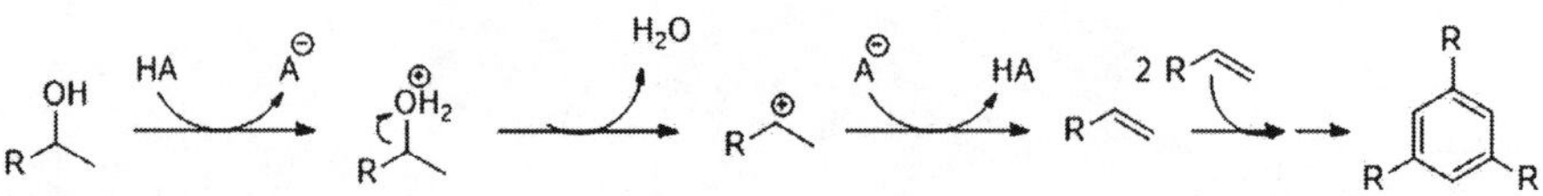

Scheme 3.2

could be used to convert feedstock such as latex, corn oil, and peanut oil to hydrocarbons (Weisz et al., 1979). Their work was motivated by the understanding that the efficiency of conversion to hydrocarbon fuels increases for organic compounds with higher hydrogen-to-carbon ratios and lower heteroatom compounds. They determined that the constraining size and shape of the pores of the zeolite could lead to the selective production of hydrocarbon units with up to ~ 10 carbon atoms per molecule. As a result, ZSM-5 was applied to biomass molecules in hopes of generating gasoline type hydrocarbons. Surprisingly, a high degree of conversion from biomass to aromatics such as benzene, toluene, and xylene was observed with conversions of up to 74% defying the expectation that the large molecular dimensions of the biomass polymers compared to the much smaller dimensions of the catalyst pores would not work. The logical explanation offered was that the shape modification allows for entrance of the cellulosic biomass fragments into the pores of the shape-selectively controlled zeolite. To further explore the synergy between petrochemical processes such as the FCC and CFP, several groups have explored addition of various modifications of ZSM-5 type catalysts at varying Si/Al ratios to fluidized-bed reactors so as to obtain information on the conversion of woody biomass, such as sawdust, to aromatics (Olazar et al., 2000; Horne et al., 1995; French and Czernik, 2010; Lappas et al., 2002; Aho et al., 2007; Carlson et al., 2009; etc.). Results have been shown to vary depending on the Si/Al, that is, acidity and pore sizes. At the optimum temperature conditions typical of fluidized-bed pyrolyzers ($\sim 500°C$), it has been shown that during CFP, biomass-derived vapors undergo various deoxygenation reactions, including dehydration, decarboxylation, and decarbonylation similar to petroleum FCC. The larger pore zeolites tend to be less efficient in the deoxygenation of pyrolysis vapors derived from biomass feedstocks. With respect to gas formation, it happens that the presence of catalyst does increase the production of all gases with the loss of oxygen to CO, CO_2, and/or water and as a result bio-oil yield is vastly reduced compared with noncatalytic pyrolysis but quality vastly improves. Corma et al. (2008) suggest that the reaction chemistry pathways for the CFP with biomass-derived oxygenates involves H_2-producing reactions including steam reforming, water gas shift (WGS) and dehydrogenation, with some reactions producing more H_2 per carbon than others (Fig. 3.3). For this reason a typical biomass with a low H/C ratio will tend to produce aromatics as the product, with prevalent coke formation that will quickly foul the active sites and deactivate the catalysts unless the catalyst is regenerated. In the next sections, we will explore our own findings on the addition of zeolite catalysts in CFP as a viable pathway to stabilizing pyrolysis oil toward the production of biofuels compatible with petrochemicals.

3.3 Screening catalysts for biomass catalytic fast pyrolysis

In a collaborative research, Universal Oil Products (UOP), LLC, a Honeywell company, the inventor of the FCC and leading provider of zeolite catalysts to the

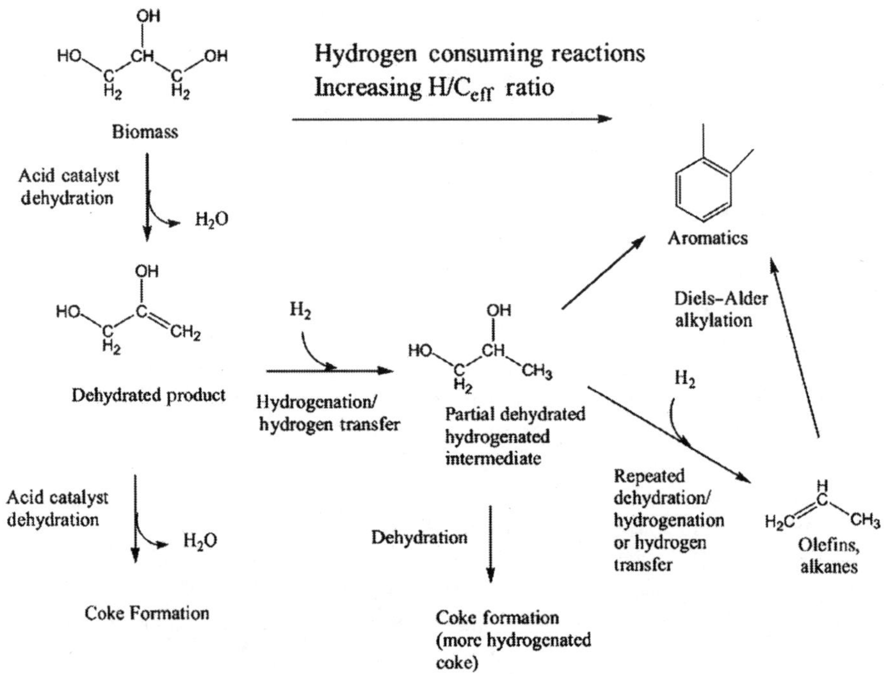

Figure 3.3 Hydrogen-consuming reactions for catalytic cracking of biomass.

petrochemical refinery industry teamed up with USDA's Agricultural Research Service (ARS) to evaluate various zeolite catalysts used in the petroleum for their efficacy on pyrolysis oil stabilization through in situ catalytic upgrading (Mihalcik et al., 2011a,b). Several heterogeneous catalysts were screened at USDA-ARS Eastern Regional Research Center (ERRC) lab near Philadelphia to confirm some of the information reported above and downselect a few for detailed studies so that compositional changes resulting from catalyst incorporation could be correlated with pyrolysis oil stability. The first step in the approach was to identify the heterogeneous catalysts that would perform the functions of oxygen reduction and thereby stabilize the bio-oil as per the DOE stability target goals. About 20 catalysts were tested, including basic catalysts (e.g., hydrotalcite), acidic catalyst (e.g., FCC or HC), metal hydrogenation catalysts, e.g., the hydrotreating type, and other zeolitic catalyst types ranging from mild to medium to strong acid catalysts. While a set was not fully disclosed, the partial listing presented in Table 3.1 places these catalysts in various sets of alpha-numeric categories, including substituted Y-zeolites and hydrotreating catalysts (1–5), high and moderate temperature WGS catalysts (A–E), substituted and hydroprocessing zeolite catalysts of various acidity (F–J), and various β-zeolite and hydrotalcite catalysts (L–P). Screening was done using the individual catalysts "as supplied" and/or a mixture of any two or three from the subgroups to meet certain desired potency. The experimental approach included the use of various reactor scales ranging from purely analytical pyrolysis gas

Table 3.1 Partial list of commercial heterogeneous catalyst sourced for screening.

Code	Description
1	Ba/Y-54
2	Ca/Y-54
3	HC-1(S) DHC-8 sulfided
4	HT
5	HC-2(S)
A	HT
B	Zeolite
C	HT-S (sulfided hydrotreating)
D	HT-WGS A
E	MT WGS B
F	Ca-zeolite
G	HC low acidity
H	HC moderate acidity
I	HC high acidity
J	NM moderate acidity
L	Zeolite
M	Zeolite (beta)
N	Zeolite
O	Zeolite
P	Zeolite

HC, Hydroprocessing catalysts; *HT*, hydrotreating; *HT-S*, high temperature shift catalyst; *HT-WGS*, high temperature water gas shift catalysts.

chromatograph/mass spectrometer (py-GC/MS) requiring only milligram size samples connected to a GC/MS for direct analysis of the eluting compounds through a small-scale fixed-bed batch reactor (at USDA-ARS/NCAUR, Peoria, IL) that allowed small samples of bio-oil to be collected for analysis. In the end two catalysts a Ca-exchanged zeolite (Ca-Y54, Catalyst 2) and a β-zeolite catalyst (M in this study) were downselected for scale-up to bench scale ($\sim$2 kg biomass/h) fluidized-bed pyrolysis system at the ARS/ERRC, the "Kwesinator" for which more will be said later. Stability was characterized by TAN and accelerated aging/viscosity tests. Biomass used was oak, a woody biomass, with low ash content to reduce catalytic effect of minerals in the ash.

The use of microgram quantities of biomass in the pyroprobe reactions does not allow for direct collection and quantification of bulk fractions but the gram-scale, packed-bed reactor does where the three product fractions from the pyrolysis of oak sawdust, that is, char, liquid, and gas could be collected. Sand medium was used as a control for the catalytic pyrolysis reactions, so differences between thermal-only pyrolysis yields can be compared.

Table 3.2 summarizes the effect of the addition of zeolitic catalysts (G–P) on the pyrolysis product fractions, that is, the bulk char, liquid, and noncondensable gas from the packed-bed screening reactions. The weight percentages of the bio-oil fractions are corrected for water entrained in the catalysts that is carried over into the

Table 3.2 Product fractions from the packed-bed pyrolysis of oak sawdust (Compton et al., 2011).

Catalyst	Char (wt.%)[a]	Liquid (wt.%)[b]	Gas (wt.%)
Sand	26.0 (0.4)	49.8 (1.8)	10.2 (1.6)
G	48.0 (4.4)	44.2 (9.6)	11.8 (0.1)
H	37.9 (0.0)	58.9 (6.7)	5.6 (0.6)
I	46.0 (5.9)	49.7 (6.2)	13.6 (1.9)
L	47.7 (6.4)	48.8 (6.00)	13.4 (1.0)
M	43.8 (3.6)	39.9 (3.3)	15.7 (2.4)
N	37.1 (2.3)	58.4 (5.0)	16.8 (1.6)
O	37.7 (0.2)	42.2 (6.4)	12.3 (0.38)
P	43.9 (2.6)	40.2 (0.4)	16.8 (1.9)

[a]Char, liquid, and gas fractions are reported as a wt.% relative to the initial mass of the oak sawdust. Data are reported as the average of $n \geq 2$ trials with standard deviations in parentheses.
[b]The liquid fractions have been corrected for extraneous water that was entrained in the catalysts. Standard deviation in parenthesis.

product stream during the reactions. In general, all the zeolitic catalysts display the same general trends when compared to the control reaction over sand bed. Like most studies reported earlier, the screening exercise shows that the use of zeolite catalysts increases char yields and slightly increases the gas yields while reducing the pyrolysis liquid yields for all the catalysts studied except for a couple. Although the feedstock is oak, a hardwood, this observation was in agreement with the catalytic pyrolysis of pine sawdust, a softwood, carried out at 500°C where the use of a zeolite, hydrotalcite, and mordenite (comparable to catalyst M, N, and O, respectively) yielded some 37%–59% char, 14%–31% liquid, and 9%–13% gas (Ingram et al., 2008).

The major compounds formed during CFP over the heterogeneous catalysts are summarized in Table 3.3 and Figs. 3.4–3.6 for the oak sawdust used as pyrolysis feedstock. Again, the control reaction for the packed-bed reactions was performed over sand (Jackson et al., 2009; Compton et al., 2011) while that for the pyroprobe reactions was conducted neat (Mihalcik et al., 2011a,b). Noncatalytic pyrolysis (the controls) formed large amounts of oxygenated compounds such as acetic acid and significant quantities of furfural and levoglucosan as measured by GC–MS and confirmed by high performance liquid chromatography.

The screening shows that a consequence of catalytic pyrolysis is the cracking of the lignin-derived components to pyrolysis products such as syringol, guaiacol, and methylmethoxy phenol; the cellulose-derived components such as levoglucosan, and the hemicellulose-derived components such as furfural (Demirbas, 2000). All the catalysts tested using the pyroprobe and the packed-bed methods reduced the amount of the oxygenates, including syringol, guaiacol, and methyl guaiacol as well as the levoglucosan, furfural, and acetic acid (Figs. 3.4 and 3.5). The cracking of the lignin, cellulose, and hemicelluloses-derived components resulted in an increase of lower molecular weight aromatics such as benzene, toluene, and xylenes (Fig. 3.6). While the use of pyroprobe and packed bed as screening reactors showed the same general

Table 3.3 The most prevalent compounds found in the liquid fraction from the pyrolysis of oak sawdust using the packed bed and pyroprobe reactors (Compton et al., 2011).

		Catalyst[c]								
Compound (wt.%)[a]	**Reactor[b]**	**None[d]**	**G**	**H**	**I**	**L**	**M**	**N**	**O**	**P**
Acetic acid	PB	4.10 (0.25)	2.84 (0.72)	2.14 (0.10)	1.64 (0.07)	1.56 (0.87)	1.30 (0.65)	2.63 (0.82)	1.67 (0.02)	1.28 (0.27)
	PP	4.22 (0.92)	0.76 (0.27)	3.40 (2.76)	1.54 (1.10)	0.63 (0.30)	0.88 (0.46)	0.44 (0.12)	0.57 (0.05)	0.94 (0.18)
Benzene	PB	–	0.04 (0.00)	0.06 (0.00)	0.06 (0.01)	0.10 (0.07)	0.04 (0.02)	–	–	0.06 (0.00)
	PP	–	0.07 (0.07)	0.35 (0.27)	0.26 (0.08)	0.23 (0.06)	0.34 (0.03)	0.07 (0.01)	0.24 (0.05)	0.13 (0.01)
Toluene	PB	–	0.06 (0.00)	0.08 (0.00)	0.08 (0.00)	0.18 (0.08)	0.15 (0.02)	–	–	0.18 (0.00)
	PP	0.07 (0.05)	0.15 (0.14)	0.45 (0.11)	0.61 (0.13)	0.61 (0.07)	0.91 (0.09)	0.12 (0.03)	0.54 (0.10)	0.45 (0.00)
Xylenes	PB	–	0.09 (0.01)	0.06 (0.00)	0.06 (0.00)	0.19 (0.07)	0.16 (0.02)	–	–	0.25 (0.01)
	PP	–	0.09 (0.09)	0.34 (0.10)	0.51 (0.15)	0.42 (0.08)	0.66 (0.08)	tr	0.28 (0.05)	0.39 (0.01)
Benzene, trimethyl	PB	–	0.05 (0.01)	tr	0.02 (0.00)	0.12 (0.04)	0.09 (0.02)	–	–	0.09(0.001)
	PP	–	dt	dt	dt	dt	dt	dt	dt	dt
Benzene, tetramethyl	PB	–	–	–	–	0.03 (0.02)	0.07 (0.02)	–	–	tr
	PP	–	–	dt	dt	dt	dt	–	dt	dt
Benzene, pentamethyl	PB	–	–	–	–	–	0.08 (0.02)	–	–	–
	PP	–	–	–	–	–	–	–	–	–
Benzene. hexamethyl	PB	–	–	–	–	–	0.07 (0.02)	–	–	–
	PP	–	–	–	–	–	dt	–	–	–
Benzene, ethyl	PB	–	tr	–	tr	tr	tr	–	–	0.03 (0.01)
	PP	–	tr	0.08 (0.05)	0.08 (0.05)	–	0.13 (0.03)	tr	0.05 (0.02)	0.11 (0.01)
o-Cresol	PB	0.02 (0.01)	0.04 (0.01)	0.03 (0.00)	0.03 (0.02)	tr	tr	0.04 (0.01)	–	0.03 (0.00)
	PP	tr	–	0.07 (0.04)	0.04 (0.00)	–	–	–	–	–
p-Cresol	PB	0.03 (0.03)	0.03 (0.01)	0.04 (0.01)	0.02 (0.01)	0.02 (0.01)	tr	0.13 (0.06)	0.04 (0.00)	0.03 (0.00)
	PP	tr	–	0.17 (0.03)	0.11 (0.02)	0.06 (0.00)	0.06 (0.01)	0.06 (0.01)	0.06 (0.01)	0.06 (0.00)
Cyclopentenone	PB	0.07 (0.01)	0.11 (0.02)	0.10 (0.01)	0.06 (0.00)	tr	0.03 (0.00)	0.11 (0.07)	0.13 (0.00)	0.02 (0.01)
	PP	dt	dt	dt	dt	–	dt	dt	–	dt
Furfural	PB	0.78 (0.06)	0.43 (0.12)	0.30 (0.03)	0.16 (0.03)	0.04 (0.03)	0.07 (0.07)	0.40 (0.21)	0.40 (0.00)	0.10 (0.03)
	PP	0.88 (0.19)	0.62 (0.34)	1.24 (0.60)	0.64 (0.20)	0.10 (0.05)	0.33 (0.15)	0.08 (0.01)	0.36 (0.01)	0.43 (0.08)

(Continued)

Table 3.3 (Continued)

		Catalyst[c]								
Compound (wt.%)[a]	**Reactor[b]**	**None[d]**	**G**	**H**	**I**	**L**	**M**	**N**	**O**	**P**
Furfural, methyl	PB	0.27 (0.02)	0.10 (0.05)	0.05 (0.01)	0.06 (0.00)	tr	–	–	0.20 (0.00)	–
	PP	dt	dt	dt	dt	–	dt	dt	dt	dt
Levoglucosan	PB	1.21 (0.30)	–	–	–	–	–	0.15 (0.06)	0.22 (0.00)	–
	PP	1.03 (0.86)	–	–	–	–	–	–	–	–
Naphthalene	PB	–	0.03 (0.01)	tr	tr	tr	0.03 (0.01)	–	0.03 (0.00)	0.04 (0.01)
	PP	–	–	–	–	–	0.06 (0.03)	–	tr	–
Naphthalene, methyl	PB	–	0.05 (0.01)	0.02 (0.00)	0.02 (0.00)	tr	0.08 (0.01)	–	–	0.07 (0.01)
	PP	–	0.06 (0.03)	0.12 (0.05)	0.13 (0.04)	0.09 (0.02)	0.29 (0.04)	tr	0.14 (0.01)	0.13 (0.01)
Naphthalene, dimethyl	PB	–	0.02 (0.00)	–	–	–	0.05 (0.02)	–	–	0.04 (0.00)
	PP	–	–	–	–	–	dt	–	dt	dt
Naphthalene, trimethyl	PB	–	–	–	–	–	0.03 (0.02)	–	–	–
	PP	–	–	–	–	–	dt	–		
Naphthalene, tetramethyl	PB	–	–	–	–	–	tr	–	–	–
	PP	–	–	–	–	–	dt	–	–	–
Phenol	PB	0.04 (0.01)	0.08 (0.02)	0.06 (0.01)	0.06 (0.00)	tr	tr	0.05 (0.01)	0.05 (0.00)	0.04 (0.02)
	PP	0.11 (0.02)	0.08 (0.04)	0.23 (0.07)	0.05 (0.03)	tr	0.05 (0.02)	tr	0.05 (0.01)	0.07 (0.01)
Phenol, 2-methoxy (guaiacol)	PB	0.06 (0.04)	0.03 (0.01)	0.02 (0.00)	0.02 (0.00)	tr	tr	0.17 (0.07)	–	0.02 (0.01)
	PP	–	0.10 (0.05)	0.19 (0.10)	0.13 (0.03)	0.06 (0.02)	0.11 (0.04)	0.08 (0.01)	0.06 (0.00)	0.12 (0.01)
Phenol, dimethoxy (syringol)	PB	0.37 (0.09)	0.03(0.01)	tr	–	tr	–	0.51 (0.24)	–	–
	PP	0.52 (0.03)	0.11 (0.05)	0.17 (0.10)	0.17 (0.03)	–	–	0.10 (0.01)	0.09 (0.02)	0.12 (0.02)
Phenol, methylmethoxy	PB	0.19 (0.05)	tr	–	–	–	–	0.16 (0.06)	–	–
	PP	0.29 (0.03)	0.08 (0.03)	0.12 (0.06)	0.10 (0.01)	0.06 (0.01)	0.09 (0.03)	0.07 (0.01)	0.07 (0.01)	0.09 (0.01)

[a]Weight % of the liquid fraction compounds is relative to the initial mass of the oak sawdust and was determined by GC versus A dash indicates that the compound was not detected and "tr" indicates the compound was detected at <0.02 wt.%, "dt" indicates that the compound was detected but not quantified. Data are reported as the average of $n \geq 2$ trials with standard deviations in parentheses.
[b]*PB*, Packed-bed reactor; *PP*, pyroprobe reactor.
[c]Catalyst defined in Table 3.1.
[d]Sand was used for the control reaction in the packed-bed reactor and the oak sawdust was pyrolyzed neat for the control reaction in the pyroprobe reactor.

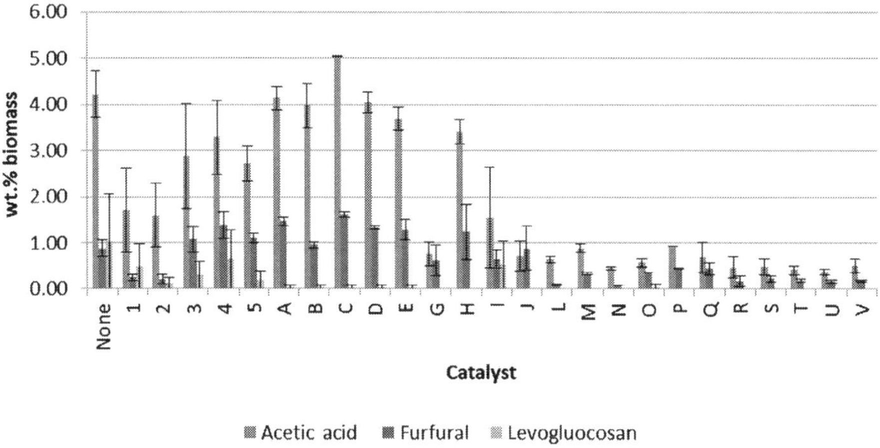

Figure 3.4 Results of pyroprobe screening of catalysts for pyrolysis of oak wood: production of selected nonphenolic oxygenates (acetic acid, furfural and levoglucosan, based on mass of input biomass) (Mihalcik et al., 2011a,b).

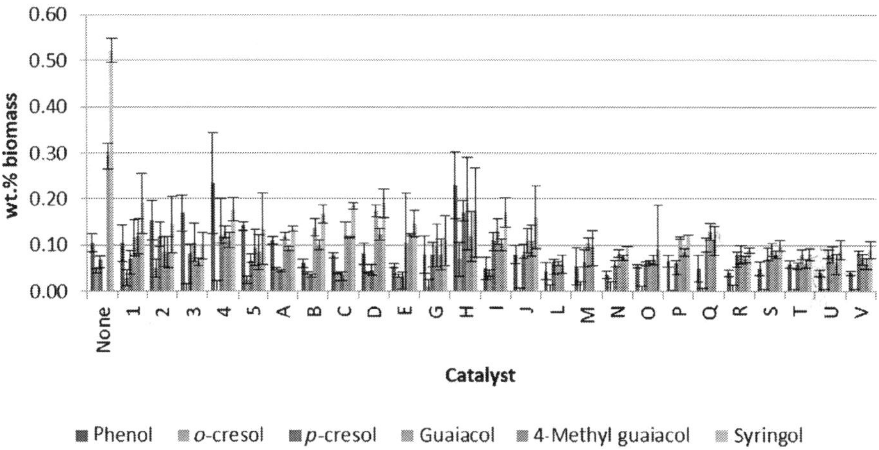

Figure 3.5 Results of pyroprobe screening of catalysts for pyrolysis of oak wood: production of selected phenolics (phenol, cresols, guaiacol, 4-methyl guaiacol, and syringol) (Mihalcik et al., 2011a,b).

trend, the quantities of specific compounds obtained with specific catalysts were sometimes disparate. For example, some of the catalyst such as L, resulted in some of the lowest furfural yields for both screening methods, but the pyroprobe screening reactions produced twice as much furfural on a weight percent basis. The most disparate results, however, were the quantities of aromatic hydrocarbon yields, including benzene, toluene, and xylenes produced over the catalysts screened. Screening via

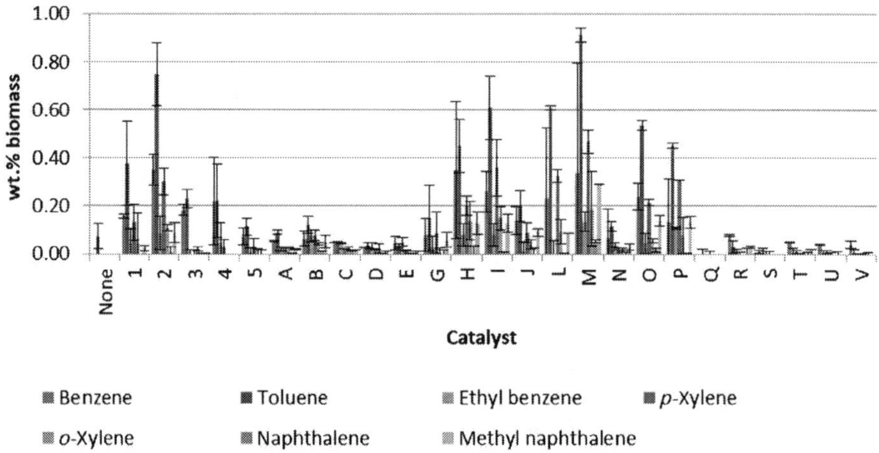

Figure 3.6 Results of pyroprobe screening of catalysts for pyrolysis of oak wood: production of selected aromatic hydrocarbons (benzene, toluene, xylenes, naphthalene, and 1-methyl naphthalene) (Mihalcik et al., 2011a,b).

pyroprobe resulted in four to eight times abundance of these lower molecular weight aromatics than observed for the packed-bed screening reactions.

The screening shows the uniqueness of catalyst M (β-zeolite) in forming lower molecular weight aromatics, including tetra- through hexamethylbenzenes as well as the methyl substituted naphthalenes, while catalyst N in the same family was least effective in cracking the lignin, cellulose, and hemicelluloses-derived components producing large amounts of acetic acid, furfural, guaiacol, and syringol and produced no benzene, toluene, or xylenes (Table 3.3). With regard to gas yields (Table 3.4) the catalytic reactions in both screening methods increased the $CO:CO_2$ ratios relative to their controls, although the pyroprobe reactions consistently produced two to five times the CO and CO_2 of the packed-bed reactions. Obviously not all the zeolite catalysts exhibited the same yield amounts of permanent gases as some catalysts affected increases in the minor gas components. For example, catalysts I and O increased H_2 formation ~15-fold and more than doubled C_2H_4, C_2H_6 formation, relative to the control reaction, while lowering CO_2 yields in the packed-bed reactor. Catalyst L produced the most C_2H_4, a 28-fold increase relative to the control and increased CH_4 and H_2 formation while lowering CO_2 yields. Some catalysts (G and M) had a relatively insignificant effect on the gas yields.

The general trend observed for the zeolitic catalysts in reducing oxygenated compounds such as acetic acid yields while increasing aromatic hydrocarbon yields translates to the ability to effect stability of the bio-oil fractions. For example, catalyst M could lower the acetic acid yield by ~70% while forming aromatic hydrocarbons ranging from benzene to pentamethyl benzene as well as ethyl benzene and naphthalene through tetramethyl naphthalene and concurrently increasing the CO and CO_2 yields. This demonstrated the potential for using cracking catalysts to lower the oxygen-containing acid content of the liquid fraction and increasing the

Table 3.4 Gas species formed during the pyrolysis of oak sawdust in the packed bed and pyroprobe reactors (Compton et al., 2011).

Catalyst[a]	Reactor[b]	H_2 (wt.%)[c]	CO (wt.%)	CH_4 (wt.%)	CO_2 (wt.%)	C_2H_4 (wt.%)	C_2H_6 (wt.%)	Total (wt.%)
Sand	PB	tr	3.29 (0.53)	0.67 (0.09)	6.02 (1.02)	0.09 (0.02)	0.11 (0.01)	10.20 (1.65)
	PP	–	2.62 (1.44)	tr	9.85 (1.52)	–	–	12.47 (3.31)
G	PB	tr	5.29 (0.98)	0.54 (0.14)	5.71 (1.10)	0.18 (0.03)	0.08 (0.02)	11.83 (2.28)
	PP	–	12.19 (0.78)	tr	14.76 (1.19)	tr	–	26.95 (2.88)
H	PB	0.07 (0.04)	2.51 (0.14)	0.35 (0.08)	2.58 (0.31)	0.06 (0.03)	0.06 (0.04)	5.63 (0.55)
	PP	–	8.02 (1.26)	tr	10.07 (1.16)	tr	–	18.08 (2.83)
I	PB	0.15 (0.03)	6.15 (0.84)	0.91 (0.22)	5.95 (0.65)	0.21 (0.06)	0.24 (0.09)	13.62 (1.89)
	PP	–	12.43 (0.28)	tr	12.23 (2.01)	tr	–	24.66 (2.56)
L	PB	0.08 (0.05)	4.20 (0.40)	1.43 (0.32)	4.45 (0.78)	2.54 (0.25)	0.72 (0.17)	13.40 (1.00)
	PP	–	10.73 (0.58)	tr	21.72 (0.79)	tr	tr	32.46 (1.68)
M	PB	tr	6.61 (0.98)	0.58 (0.14)	7.85 (1.19)	0.41 (0.11)	0.18 (0.01)	15.65 (2.37)
	PP	–	12.74 (3.40)	tr	14.54 (2.23)	tr	tr	27.28 (6.11)
N	PB	0.05 (0.01)	3.54 (0.16)	1.14 (0.15)	11.58 (1.27)	0.20 (0.05)	0.28 (0.08)	16.79 (1.61)
	PP	–	3.59 (0.83)	tr	26.10 (1.98)	tr	tr	29.69 (3.63)
O	PB	0.14 (0.00)	5.49 (1.03)	0.71 (0.12)	5.22 (1.05)	0.39 (0.11)	0.30 (0.16)	12.26 (0.38)
	PP	–	11.86 (0.49)	tr	18.30 (0.28)	tr	tr	30.16 (1.14)
P	PB	tr	7.29 (0.89)	0.82 (0.28)	7.94 (0.82)	0.45 (0.03)	0.29 (0.09)	16.80 (1.94)
	PP	–	12.07 (0.80)	tr	14.78 (0.90)	tr	tr	26.85 (2.16)

[a]Catalyst abbreviations are as in Table 3.1.

[b]*PB*, Packed-bed reactor; *PP*, pyroprobe reactor.

[c]The gasses are reported as a wt.% relative to the initial mass of the oak sawdust. Data are reported as the average of $n \geq 2$ trials with standard deviations in parentheses. A dash indicates that the gas was not detected and "tr" indicates that the gas was detected at <0.05 wt.%.

Table 3.5 Total acid number (TAN, mg KOH/g) and pH of the liquid fractions from the packed-bed pyrolysis of oak sawdust (Compton et al., 2011).

Catalyst	TAN	pH
Sand	133.0	1.63
G	61.9	2.24
H	53.6	1.71
I	59.5	1.96
L	40.3	2.79
M	36.7	2.63
N	11.1	3.08
O	–	2.47
P	48.0	2.60

hydrocarbon pool. As Table 3.4 demonstrates, the stability measure such as the TAN of the liquid fractions could be lowered from 133 mg KOH/g for the control to substantially low values. The lower acidity of the liquid fractions should theoretically stabilize the pyro-oil (Table 3.5).

3.4 Scale-up into continuous process with downselect catalysts

To see how some of the best performing catalysts in the screening exercise fare in a continuous process synergetic to the FCC in the petroleum refinery system further experiments were conducted in situ and ex situ CFP with two downselect catalysts, that is, Ca-Y zeolite (catalyst 2) and β-zeolite (catalyst M). These two showed the most promise to achieve the goal of reducing oxygen content of pyrolysis liquid and hitherto for to achieve bio-oil stability. Extrudates of the downselect catalysts were used as the bed material in lieu of sand in the USDA-ARS fluidized-bed reactor (Fig. 3.7). Samples of bio-oil were collected from each of the collection points at timed intervals to study any deactivation of the catalyst over time. It was expected that the results would be somewhat different at this scale than at the screening scale. Table 3.6 (Mullen et al., 2011) shows that only up to 30 wt.% yield, but good quality pyrolysis oil was produced overall on mass-averaged basis not accounting for unrecovered bio-oil stuck in the reactor system. As much as 40 wt.% is the expected liquid yield based on previous experiences (Mullen et al., 2011; Park et al., 2007; Aho et al., 2007) compared with noncatalytic pyrolysis; about 20% reduction in yield. The bio-oil produced via CFP with these two downselect catalysts does contain a much higher proportion of water than a non-CFP process, with water constituting 43% of the liquid collected at the first collection time (15 minutes into the run) as against only 15% in the case of sand bed. It is fair to ascertain that some of this excess water originates from deoxygenation of the

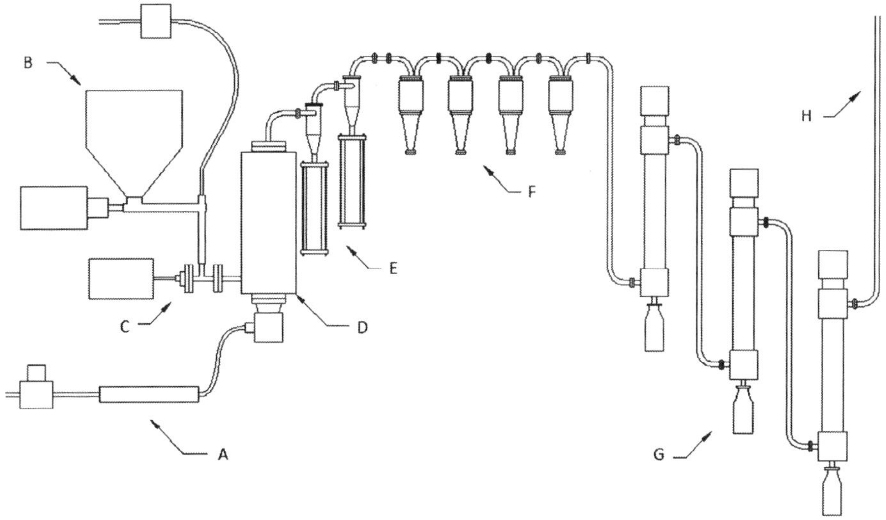

Figure 3.7 USDA-ARS fluidized-bed fast pyrolysis unit. Condenser's (1−4) and ESP's are depicted as letters F and G, respectively (Mihalcik et al., 2011a,b). *ARS*, Agricultural Research Service; *ESP*, electrostatic precipitator; *USDA*, United States Department of Agriculture.

Table 3.6 Pyrolysis product recoveries (wt.%) for catalytic pyrolysis experiments without regeneration (Mullen et al., 2011).

Catalyst	None	CaY	Catalyst M
Total pyrolysis oil (%)	62	30	29
Water (%)	7.8	13	18
Organics (%)	54	17	11
Char (%)	18	13	24
Total noncondensable gases (%)	13	27	14
CO_2 (%)	5.9	5.0	3.5
CO (%)	6.0	16.5	8.3
CH_4 (%)	1.0	4.5	1.5
H_2 (%)	0.2	0.7	0.7
Total mass recovery[a] (%)	93	70	67

[a]Not including deposits on catalysts or material stuck in the system.

pyrolysis vapors over the solid catalyst via dehydration. The bio-oils of the continuous FCC-like cracking of bio-feedstocks for the downselect catalysts are slightly deoxygenated. The C/O molar ratio increases from 1.7/1 for reactions over sand to 2.6/1 for the bio-oil collected at the first-time interval over the Ca-Y zeolite (Table 3.7) reaction. This constitutes to 17% increase in carbon content and 22% decrease in oxygen content. GC/MS analysis (Table 3.8) indicates that the

Table 3.7 Elemental analysis of pyrolysis oil produced over catalysts without regeneration (Mullen et al., 2011).

Catalyst	None as is	None dry basis	Ca Y as is[a]	CaY dry basis[a]	M as is[a]	M dry basis[a]
Water (wt.%)	15.1		43.8		57.6	
Carbon (wt.%)	44.33	52.46	34.72	61.75	23.0	54.40
Hydrogen (wt.%)[b]	4.65	5.48	3.17	5.62	2.31	5.45
Nitrogen (wt.%)	0.41	0.47	0.455	0.80	0.07	0.16
Sulfur (wt.%)	0.00	0.00	0.00	0.00	0.00	0.00
Oxygen (wt.%)[b]	33.79	41.01	17.9	31.84	17.0	40.0
C:O (mol)[b]	1.7:1	1.7:1	2.6:1	2.6:1	1.8:1	1.8:1
H:C (mol)[b]	1.3:1	1.3:1	1.1:1	1.1:1	1.2:1	1.2:1
HHV (MJ/kg)	19.4	22.9	14.2	25.1	10.0	23.6
TAN (mg KOH/g)	126	–	88–112	–	124	
pH	2.5		2.9		2.6	

HHV, High heating value.
[a]Combined fractions of pyrolysis oil collected during first collection of run.
[b]Hydrogen and oxygen values are organic only.

Table 3.8 Pertinent pyrolysis oil components (wt.%) in pyrolysis oil produced over catalyst without regeneration[a] (Mullen et al., 2011).

Catalyst	None	CaY	M
Acetic acid	9.6 (12.4)	7.2 (10.2)	5.5 (12.9)
Furfural	0.6 (0.8)	0.6 (0.9)	0.3 (0.8)
Acetol	4.6 (6.0)	2.3 (3.2)	1.5 (3.5)
Levoglucosan	9.3 (12.0)	4.2 (8.0)	2.6 (6.3)
Phenol	0.2 (0.2)	0.4 (0.5)	0.2 (0.5)
Cresols	0.3 (0.4)	0.6 (0.8)	0.3 (0.6)
Guaiacol	0.3 (0.4)	0.2 (0.3)	0.1 (0.2)
Syringol	0.7 (0.9)	0.2 (0.3)	0.2 (0.4)
Toluene	Trace	Trace	Trace
Xylenes	Trace	Trace	Trace
Naphthalene	Trace	Trace	Trace

[a]Dry basis values in parenthesis.

deoxygenation is a result of decreased concentration of highly oxygenated species such as acetic acid, levoglucosan, and acetol, consistent with the results we saw for the py-GC/MS screening study. The overall presence of acids decreases as evidenced by a drop in the TAN from 128 for bio-oil produced over sand to a lower range of 88–112 for bio-oil at first collection over the CaY zeolite (Table 3.7). Concurrent with the decrease in oxygenates is an increase in noncondensable gas (NCG) production overall. Consistent with results from the small-scale fixed-bed

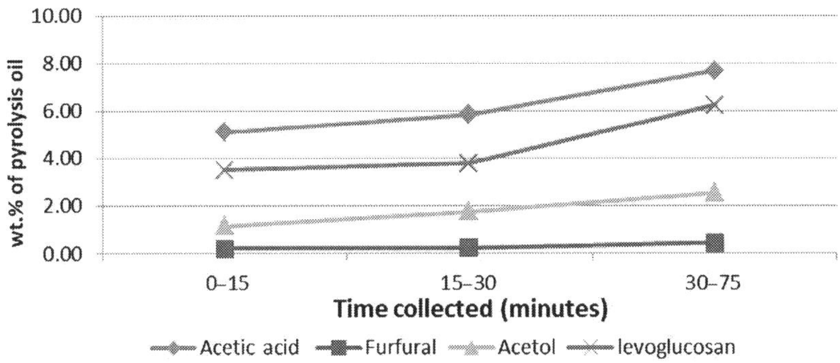

Figure 3.8 Concentration of oxygenates in pyrolysis oil produced over CaY zeolite at different collection times (Mullen et al., 2011).

batch-screening reactor, the noncondensable gas produced over the catalysts in a continuous system is richer in CO, CH_4, and H_2 than that produced over sand, with the latter containing a higher proportion of CO_2 (Table 3.6). The increased production of H_2 in the large-scale continuous fluidized-bed system suggests the dehydrogenation mechanisms are most responsible for the presence of aromatics, rather than direct deoxygenation of lignin-derived phenolics. The high level of CO produced suggests that in addition to decarboxylation of acids to generate CO_2, decarbonylation of other carbonyl groups such as ketones and aldehydes could be an active pathway (Carlson et al., 2009) to deoxygenation at the industrial scale process. This is further supported by the observed reduction in the acetol content compared with bio-oil produced over sand. It is worth noting that there was no decrease in the aldehyde (furfural), but that could be due to its concurrent formation via a dehydration reaction of carbohydrate derivatives such as levoglucosan which, as observed, is also reduced in meaningful concentrations compared with the noncatalytically produced bio-oil.

Fig. 3.8 illustrates how, over time deoxygenation decreases as the collection of the liquid product continues. The characteristics of the bio-oil continually become more like that produced over sand indicating catalyst poisoning or deactivation. Oxygenated compounds such as acetic acid, acetol, and levoglucosan increase as the catalyst is exposed to greater amounts of biomass. The short-term mechanism for the deactivation is coke-like carbon deposits that block the active sites of the catalysts as evidenced by reduction in its surface area. The Brunauer–Emmett–Teller (BET) surface area decreased from 627 to 118 m^2/g over the course of the experiment reported herein.

3.5 Catalysts deactivation and regeneration

We saw that the β-zeolite catalyst (M) was most effective on stability with an ability to promote levoglucosan conversion. However, prior studies have shown that levoglucosan yield is sensitive to the presence of any inorganic material including alkali and alkaline earth minerals sequestered in the biomass ash (Fahmi et al., 2008). Because the pyroprobe screening showed H-exchanged zeolite catalyst set (e.g., catalyst M) to be effective in producing deoxygenated pyrolysis vapors but does not show the same response in the continuous fluidized-bed process such as FCC, it is not farfetched to assume that rapid deactivation by coke deposits is the culprit. In comparison the calcium-exchanged catalyst, for example, Ca-Y zeolite was less effective in producing deoxygenated vapors in the screening but deactivated less quickly. One could attribute the discrepancy to the attenuation of the Brønsted acid sites on the Y zeolite because of the Ca^{2+} leading to fewer dehydrogenation reactions and therefore less coke formation than is the case for the strongly acidic zeolite catalysts. The foregoing study strongly singles out zeolite catalysts, for example, catalyst M as the catalyst of choice for catalytic pyrolysis of biomass for the production of stable pyrolysis oil, but like petroleum refining, rapid deactivation is a problem that needs to be addressed. In order to do so, it is necessary to explore the rate of catalyst deactivation and the extent of regeneration required to mitigate the fouling process and inform their engineering and expected catalyst lifetime in CFP biorefineries.

An experiment to regenerate spent catalyst in situ by intermittently introducing air into the system over a set time could prolong catalyst lifetime. Time intervals of 5 and 10 minutes, corresponding to ~175 and 350 g of biomass per cycle, were tested in the fluidized-bed process development unit (PDU) processing oak as feedstock. As Fig. 3.9 shows, the bed temperature rises above the pyrolysis temperature set-point upon an introduction of air due to the combustion exotherm of the carbon deposits which, upon completion, drops and at which point biomass feeding is resumed. This procedure shows the need for shorter time cycles that is perhaps not practical in intermittent air cycles. What is needed is a continuous regeneration system such as the commercial FCC systems, where catalyst is cycled through a separate and dedicated secondary combustion chamber to be regenerated in situ.

Nevertheless, the results of using the intermittent in situ regeneration process, in both 5- and 10-minute pyrolysis cycles, could be informative. The bio-oil (total liquid) yield with the in situ regeneration of β-zeolite catalyst (catalyst M) ranged

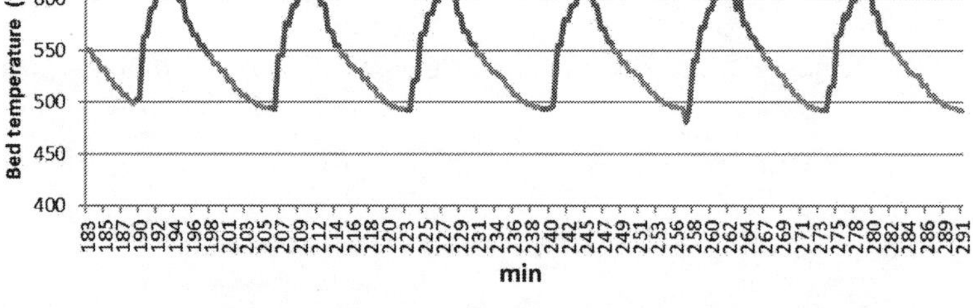

Figure 3.9 Example fluidized-bed temperature profile for pyrolysis (blue)/catalyst regeneration (red) cycles (Mullen et al., 2011).

from 25% to 28% of the input biomass; a large amount of the bio-oil was water (Table 3.8). Since the liquid is collected via condensers at various locations downstream of the fluidized-bed reactor, the fractional condensation configuration allows for separation of the aqueous phase with a large amount of water ($>$65 wt.% H_2O) in the condenser fraction allowing for the collection of mostly the organic fraction at the ESP. At ~350 g oak wood per cycle the bio-oil at the ESP exhibited a C/O ratio of 3.1/1 compared with 1.8/1 for noncatalytic pyrolysis (Table 3.9). When the cycle rate was reduced to ~175 g, the C/O ratio further increased to 5.9/1, representing an overall increase in carbon content of 40% and a decrease in oxygen content of 55% (Table 3.9). GC/MS analysis of the samples (Table 3.10) shows that major oxygenates such as acetic acid, acetol, and levoglucosan decreased with greater exposure to the regenerated/active catalyst, thereby reducing bio-oil acidity as measured by TAN and improving the heating value (Table 3.9).

While the regeneration of the spent catalyst in situ successfully improved bio-oil quality (Table 3.11), it did not change the yield of the noncondensable gas produced compared with noncatalytic pyrolysis; rather, the gas was richer in H_2, CO, and CH_4 with lesser CO_2 concentration (Table 3.9). This suggests that dehydrogenation and decarbonylation of carbonyls are also possible operative mechanisms over the acid catalyst. Dehydration and dehydrogenation contribute to the loss of hydrogen; hence the molar H/C ratios of the pyrolysis oil reduced from 1.3/1 (noncatalytically produced liquid) to 1.1/1 for the ~350 g/cycle and 0.9/1 for the ~175 g/cycle operations, respectively (Table 3.9). The hydrogen loss due to dehydration results in higher concentration of aromatic compounds, including very low H/C compounds such naphthalenes (H/C ratio 0.8/1) (Table 3.10). GC/MS analysis also reveals that non-methoxylated phenolic compounds, phenol and alkyl phenols, increased in concentration with the shorter regeneration cycle run producing more quantities. Phenols are known to have a low deoxygenation reactivity on acidic zeolite type catalysts;

Table 3.9 Pyrolysis product recoveries for catalytic pyrolysis experiments with in situ regeneration (Mullen et al., 2011).

Catalyst	None	M (350 g cycles)	M (175 g cycles)
Total pyrolysis oil (%)	62	28	25
Water (%)	7.8	14.5	16.5
Organics (%)	54	13.5	8.5
Char (%)	18	21	16
Total noncondensable gases (%)	13	14	16
CO_2 (%)	5.9	2.9	2.6
CO (%)	6.0	9.1	12.0
CH_4 (%)	1.0	1.3	1.0
H_2 (%)	0.2	0.6	0.4
Total mass recovery[a] (%)	93	63	57
Coke-like deposits on catalyst	N/A	nd	25%[b]

[a]Not including deposits on catalysts or material stuck in the system.
[b]Elemental Analysis of coked catalyst (wt.%): C 5.59H 0.465 (H:C 0.6:1).

Table 3.10 Elemental analysis of electrostatic precipitator (ESP) pyrolysis oil[a] produced over catalyst M with in situ regeneration (Mullen et al., 2011).

	No catalyst	No catalyst dry basis	350 g cycles	350 g cycles dry basis	175 g cycles	175 g cycles dry basis
Water (wt.%)	11.7		8.6		7.7	
Carbon (wt.%)	48.21	54.59	60.19	65.8	70.59	76.5
Hydrogen (wt.%)[b]	5.10	5.78	5.28	5.78	5.28	5.63
Nitrogen (wt.%)	0.14	0.16	0.203	0.22	0.350	0.38
Oxygen (wt.%)[b]	34.85	39.48	25.83	28.22	16.08	17.4
C:O (mol)	1.8:1	1.8:1	3.1:1	3.1:1	5.9:1	5.9:1
H:C (mol)	1.3:1	1.3:1	1.1:1	1.1:1	0.9:1	0.9:1
HHV (MJ/kg)	19.9	22.5	22.9	25.0	29.7	32.3
TAN (mg KOH/g)	130		112		68	
pH	2.5		2.5		2.8	

HHV, High heating value.
[a]ESP Pyrolysis oil collected during first collection of run.
[b]Hydrogen and oxygen values are organic only.

Table 3.11 Pertinent pyrolysis oil components (wt.%) in electrostatic precipitator pyrolysis oil produced over catalyst M with in situ regeneration (Mullen et al., 2011).

Catalyst	None	350 g cycles	175 g cycles
Acetic acid	9.5	7.9	4.5
Furfural	0.9	0.8	0.2
Acetol	2.7	1.5	0.4
Levoglucosan	23.5	9.0	3.5
Phenol	0.2	0.7	1.0
Cresols	0.2	1.2	1.5
Dimethyl-phenol + ethyl-phenol	0.3	0.4	0.9
Guaiacol	0.3	0.2	0.1
4-Methyl guaiacol	0.3	0.1	0.2
Syringol	0.3	0.2	0.3
Toluene	Trace	0.1	0.2
Xylenes	Trace	0.2	0.4
Naphthalene	Trace	0.6	1.60
1-Methyl naphthalene	Trace	0.7	2.6

however, guaiacol has been shown to be partially deoxygenated and then alkylated when coprocessed with hydrocarbons over HZSM-5 (Graça et al., 2011). It is, therefore, logical to assume that a similar mechanism where guaiacol is demethoxylated to phenol with transfer of H_2 from another molecule in the feed is responsible for the increases noted in the CFP over β-zeolite catalyst. Furthermore, the fact that the

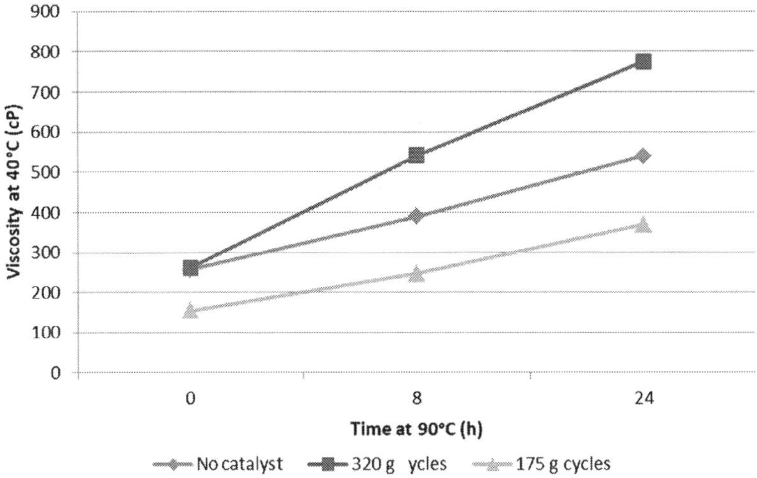

Figure 3.10 Comparison of viscosity changes with accelerated aging of pyrolysis oils produced over catalyst M (β-zeolite) (Mullen et al., 2011).

concentration of guaiacols does not appear to decrease in the upgraded liquid produced over the catalyst indicates the likelihood that the β-zeolite catalyst has the concurrent net effect of further depolymerization of the lignin in the white oak wood than mere thermal pyrolysis alone (Mullen and Boateng, 2010).

As shown in Table 3.8, a large amount of coke, about 25% of the total mass of input biomass, could be deposited on the catalyst because of the loss of hydrogen associated with the dehydration and dehydrogenation occurring during the catalytic pyrolysis over zeolite catalyst (e.g., catalyst M). This accounts for about half of the input carbon. So even though the acid functionalized catalyst is able to deoxygenate the white oak wood pyrolysis vapors, it does so with very low carbon efficiency despite the relatively low production of carbon oxides in the gas. Elemental analysis of the coke deposit reveals an H/C ratio of 0.6/1 suggesting a polyaromatic structure.

Accelerated ageing tests showed that the bio-oil produced over zeolite catalyst such as M in the 350 g cycles had an initial viscosity like that of the pyrolysis oil produced over sand (Fig. 3.10), but the rate of viscosity change was observed to be faster than the pyrolysis oil produced noncatalytically. The pyrolysis oil produced in the shorter cycles, however, was less viscous initially and the rate of viscosity increase was slower than that produced noncatalytically. It is, therefore, enough evidence to conclude that significant partial deoxygenation is required to reduce the rate of aging but does not eliminate bio-oil instability completely.

3.6 Ex situ catalytic pyrolysis

We saw the potential for CFP when the catalyst is placed in situ replacing the sand bed in a fluidized-bed process arrangement. Another way of introducing catalyst is to place a packed bed of heterogeneous catalyst in the gas stream post the reactor bed, that is, ex situ CFP. Because there are several condensers in series, the vapor fractional composition varies at points upstream of the ESP, and it would be important to know the effect of the fractionation on the vapor cracking in order to properly engineer the best insertion point for a fixed-bed of catalyst. An optimization study was carried out to provide the necessary data to inform full-stream upgrading of pyrolysis oils via ex situ vapor cracking (Mihalcik et al., 2011b). The study shows that the extent of deoxygenation in a fractional condensation system like this one is location specific and depends upon temperature and relative concentrations of water, oxygenates, and residual solids, in the approach vapor. There is a concentration gradient of major pyrolysis oil oxygenates, water and solids (trapped biochar) downstream of the condensers which affects catalytic upgrading over zeolite catalyst bed (Fig. 3.11).

Table 3.11 compares the compositional analysis of the bio-oil collected at the ESP and the products of the vapor-phase catalytic upgrading for each location where the catalyst bed was placed using the two downselect catalysts in the screening exercise. As the table shows, almost 100% conversion of the oxygenated compounds is achievable, regardless of the placement of the catalyst-bed column and that only trace amounts of oxygenated compounds were found thereafter in a methanol spray condenser downstream. Most of the hydrocarbon product recovered here was composed of >C6 aromatic compounds, including ethyl benzene, toluene, xylene (o, p), and naphthalene (unsubstituted, 1-methyl). The product recovered from the post-ESP spray condenser showed only trace amounts of the most-prevalent oxygenates (i.e., acetic acid, acetol, and levoglucosan), compared to the approach vapor in the condensers and ESP (untreated) bio-oil. Progressive removal of water vapor, light organics, and solids from the pyrolytic vapor stream allows for a higher concentration of levoglucosan, and other higher order carbon containing pyrolysis components, to undergo a two-step deoxygenation that leads to isomerization, cyclization, and aromatization of the hydrocarbons as per the scheme presented in Fig. 3.12 (Table 3.12).

3.7 Catalytic pyrolysis and metals balance

We saw earlier that catalyst poisoning during in situ CFP of bio-feedstocks is problematic and regeneration is required to ensure consistent product distribution over time on-stream. The efficiency of catalyst regeneration depends, in part, on the reactor design. In FCC systems used for the cracking of heavy oils in the petroleum refinery typically circulating fluidized-bed designs are used that automatically allow for the continuous regeneration of spent catalyst without interrupting the process. For bubbling fluidized-bed reactors, however, like the PDU described throughout this chapter, intermittent regeneration is the only possibility. This has the drawback of process interruption even if it is practical to do since certain heterogeneous

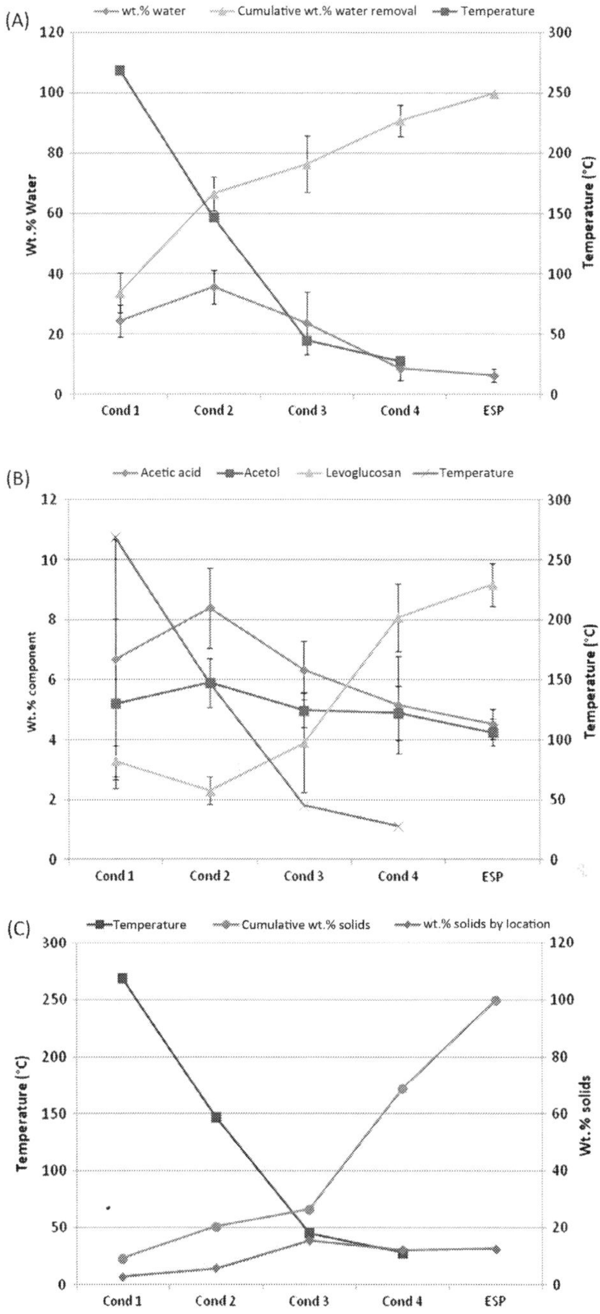

Figure 3.11 (A) Abundance of water by location within condensers 1–4 and ESP (Mihalcik et al., 2011b). (B) Concentration of major oxygenates by location within condensers 1–4 and ESP (Mihalcik et al., 2011b). (C) Abundance of solids by location within condensers 1–4 and ESP (Mihalcik et al., 2011b). *ESP*, Electrostatic precipitator.

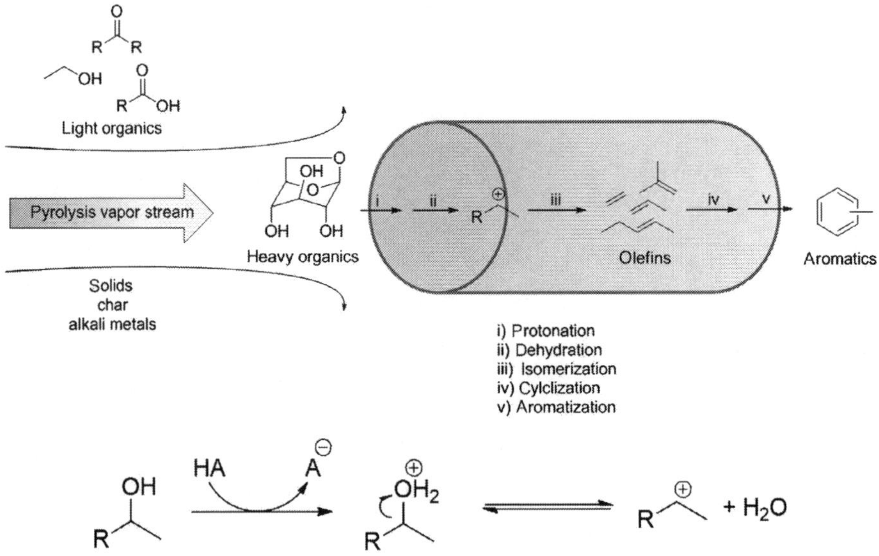

Figure 3.12 Ex situ vapor cracking in a fractional condensation system (Mihalcik et al., 2011b).

catalysts can quickly become inactive in a short period of time (<5 minutes in some instances).

The metal accumulation based on the number of times the same catalyst is exposed to the biomass is an interesting correlation between catalyst poisoning or its longevity. Estimates of the metal loading from the biomass to the catalyst were carried out using repeated runs of the same catalyst with product distribution analyzed. Fig. 3.13 shows the metal accumulation is additive and linear.

Individually, most of the metals found in the biomass correlate positively with metal loading as more biomass is exposed to the same catalyst over time. Because of this accumulation the deoxygenation effect of the catalytic is diminished increasing oxygenates and decreasing the aromatic hydrocarbons in the product distribution (Fig. 3.14).

3.8 Metal-modified ZSM-5 catalysts for biomass pyrolysis

We previously saw that zeolite catalyst such as ZSM-5 is most effective for the CFP process. While the hydrogen exchanged HZSM-5 catalysts have been proven to be the most effective class of zeolites for deoxygenation of pyrolysis vapors and production of aromatic hydrocarbons, further enhancement of the effectiveness of

Table 3.12 Quantitative gas chromatograph/mass spectrometer results from methanol spray condenser and dry ice/acetone (Mihalcik et al., 2011b).

Condenser	–	**1**		**2**		**4**			
Catalyst	**None**	β	**Ca-Y54**	β	**CaY54**	β		**CaY54**	
Run number	**ESP**	**1**	**2**	**3**	**4**	**5**	**5b**	**6**	**6b**
Oxygenates (wt.%)									
Acetic acid	4.46	–[a]	–	–	–	–	–	–	–
Furfural	0.23	–	–	–	0.25	–	–	–	–
Acetol	4.75	–	–	–	–	–	–	–	–
Levoglucosan	9.24	–	–	–	–	–	–	–	–
Phenol	0.13	0.10	–	–	–	–	0.55	0.20	0.40
Cresols (*o*, *m*, *p*)	0.22	0.18	0.12	0.14	0.23	0.11	0.84	0.28	0.46
Guaiacol	0.35	–	–	–	0.47	–	–	–	–
Syringol	1.00	–	–	–	–	–	–	–	–
Aromatics (wt.%)									
Benzene[b]	0.00	0.22	0.31	0.10	0.17	0.10	–	0.40	–
Toluene	0.02	2.72	2.93	2.94	4.82	3.35	9.63	4.55	16.84
Xylenes (*o*, *p*)	0.05	4.05	1.90	3.89	6.93	6.53	8.12	8.63	10.13
Naphthalenes (un-sub, 1-methyl)	0.07	5.46	6.30	4.44	3.74	5.20	9.23	7.14	17.75
Total quantitative hydrocarbons	0.14	12.23	11.13	11.27	15.49	15.08	26.98	20.32	44.72

ESP, Electrostatic precipitator.
[a]Value less than 0.1 wt.%.
[b]Benzene values artificially low because of loss during distillation phase of hydrocarbon recovery.

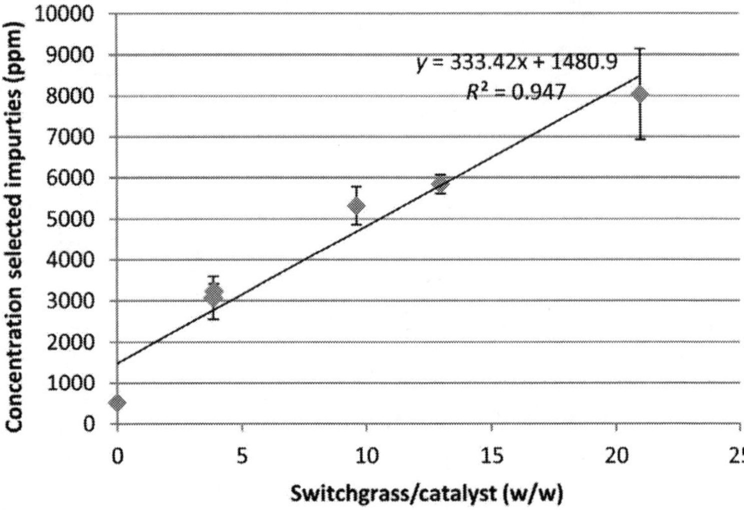

Figure 3.13 Concentration of total Ca, Cu, Fe, K, Mg, Na, and P on HZSM-5 at different levels of catalyst exposure to switchgrass in a fluidized bed at pyrolysis conditions (500°C, N_2). Values are average of three ICP spectrometry measurements. Error bars are one standard deviation (Mullen and Boateng, 2013). *ICP*, Inductively coupled plasma.

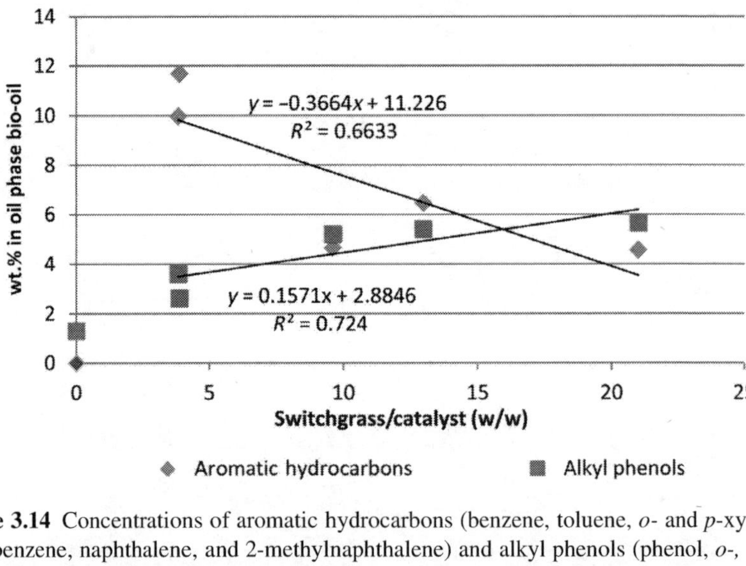

Figure 3.14 Concentrations of aromatic hydrocarbons (benzene, toluene, *o*- and *p*-xylenes, ethyl benzene, naphthalene, and 2-methylnaphthalene) and alkyl phenols (phenol, *o*-, *m*-, and *p*-cresols, 4-ethyl phenol, and 2,4-dimethylphenol) in bio-oil from HZSM-5 catalytic pyrolysis of switchgrass with different exposure rates of the catalyst to switchgrass. Purple points (zero ordinance) are representative of concentrations found in thermal-only pyrolysis oil, provided for comparison (Mullen and Boateng 2013).

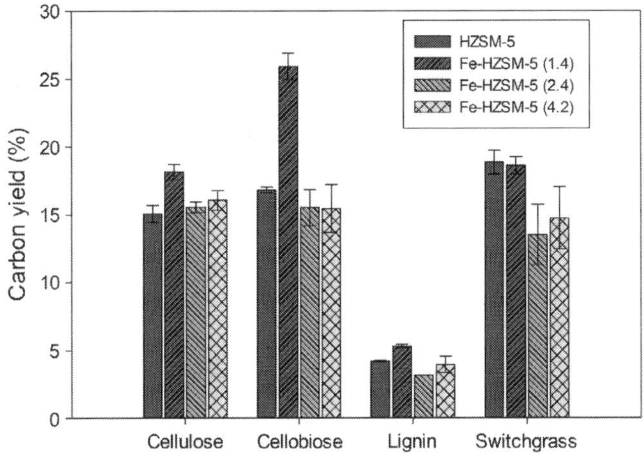

Figure 3.15 Aromatic hydrocarbon production from each biomass feedstock with different catalysts at (catalyst/feedstock = 10 w/w). Average of three replicates; error bars are one standard deviation (Mullen and Boateng, 2015).

the catalysts in terms of activity and yield is possible by metal modification to improve activity or functionality. One of the modifications possible is to exchange a metal cation in place of some of the Bronsted acid (H^+) functionality within the active sites of the catalyst. Ion exchange with Fe(II) cations (Mullen and Boateng, 2015) revealed that natural Fe present in biomass could increase aromatics yield during CFP (Mullen et al., 2014). Iron modification of HZSM-5 at low levels can be effective in increasing the yield of aromatic hydrocarbons from carbohydrates via CFP (Fig. 3.15), but it is feedstock dependent. Of the Fe-loadings studied the smallest loading of 1.4 wt.% proved to be the most effective. Iron modification of the zeolite changes the chemical pathways to favor the formation of benzene and naphthalenes over *p*-xylene and other alkyl benzenes (Table 3.13). Phenol formation was also inhibited by the addition of Fe, which may result in lower coke production. Although the yield increases were not shown to translate from the model carbohydrates to actual biomass (switchgrass), the selectivity changes did. Therefore Fe-modified HZSM-5 catalysts may be effective for applications in which the product mixture rich in those components is desired.

Several other metals have been explored to modify HZSM-5 catalysts and have been studied for their effectiveness in the production of aromatic hydrocarbons (Schultz et al., 2017; Fig. 3.16). The production of aromatic hydrocarbons from catalytic pyrolysis of *Eucalyptus urophylla* increased for gallium modified ZSM-5 at high loading levels (Ga-HZSM-5 B, 6.99 wt.%.Ga), whereas modification with Ni or Zn caused a decrease in the yield of aromatic hydrocarbons. The metal modification of ZSM-5 also affected the selectivity of aromatic hydrocarbons. Opposite of the effect of Fe, shown previously, or Ni or Zn shown below, Ga increased the selectivity to alkylated benzenes (e.g., xylenes) over benzene (Table 3.14, Figs. 3.16 and 3.17).

Table 3.13 Carbon yields and aromatic selectivities for pyrolysis of switchgrass (catalyst/switchgrass = 10 w/w) (Mullen and Boateng, 2015).

Catalyst	None	HZSM-5	Fe-HZSM-5 (1.4)	Fe-HZSM-5 (2.8)	Fe-HZSM-5 (4.2)
Carbon yield (%)					
BTEX[a]	0.07	15.97^{A}	14.30^{AB}	10.84^{C}	11.68^{BC}
Naphthalenes[b]	0.01	2.24^{C}	3.99^{A}	2.54^{BC}	2.89^{B}
Phenols[c]	0.14	0.34^{A}	0.05^{B}	0.03^{B}	0.05^{B}
Methoxylated phenols[d]	1.06	0.08^{A}	0.01^{A}	0.01^{A}	0.01^{A}
Acetic acid	2.12	0.18^{A}	0.00^{A}	0.00^{A}	0.00^{A}
Levoglucosan	2.75	0.03^{A}	0.00^{A}	0.00^{A}	0.03^{A}
HMF[e]	0.58	0.00^{A}	0.00^{A}	0.00^{A}	0.00^{A}
Other oxygenates[f]	1.98	0.16^{A}	0.05^{A}	0.06^{A}	0.05^{A}
CO	1.69	5.88^{A}	5.71^{A}	6.7^{A}	6.57^{A}
CO_2	3.94	5.26^{A}	3.65^{B}	6.27^{A}	5.87^{A}
Olefins[g]	0.25	2.59^{A}	1.72^{A}	2.81^{A}	2.47^{a}
Aromatic hydrocarbon selectivity (%)					
Benzene		8.89^{B}	11.12^{A}	11.45^{A}	11.72^{A}
Toluene		41.95^{B}	39.43^{B}	43.64^{A}	42.64^{A}
p-Xylene		26.32^{A}	20.53^{B}	20.58^{B}	20.03^{B}
o-Xylene		2.22^{B}	4.38^{A}	2.87^{B}	3.25^{B}
Ethylbenzene		5.33^{A}	1.38^{B}	1.70^{B}	1.61^{B}
1,2,4-Trimethylbenzene		3.42^{A}	1.73^{B}	0.97^{C}	1.14^{C}
Naphthalenes[b]		11.88^{D}	21.42^{AB}	18.79^{C}	19.61^{BC}

Values in a row that do not share a superscript capital letter are statistically different based on ANOVA of three replicates.

[a]BTEX = benzene, toluene, ethylbenzene, and xylenes.

[b]Naphthalenes = naphthalene and 2-methylnaphthalene.

[c]Phenols = phenol, *p*-,*o*-,*m*-cresols, 2,4-dimethylphenol, 4-ethylphenol.

[d]Methoxylated phenols = guaiacol, 4-methyl guaiacol, isoeugenol, vanillin, 1-(4-hydroxy-3-methoxyphenyl)acetone, syringol, syringaldehyde and 1-(4-hydroxy-3,5-dimethoxyphenyl) acetone.

[e]HMF = 5-hydroxymethylfurfural.

[f]Other oxygenates = acetol, furfural, furfuryl alcohol, 2,3-butandione.

[g]Olefins = ethylene, propene.

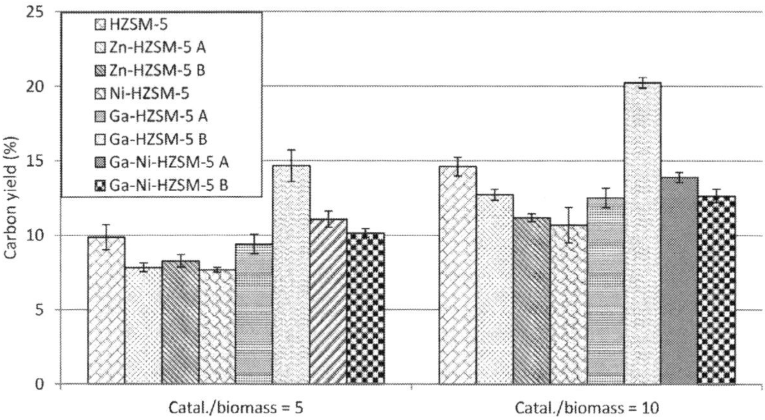

Figure 3.16 Aromatic hydrocarbon production from *Eucalyptus urophylla* with different catalysts based on ZSM-5 (CBV2314) at catalyst/biomass = 5 and 10 w/w. Average of three replicates, error bars are standard deviation (Schultz et al., 2017).

3.9 Start-up challenges of commercial catalytic fast pyrolysis biorefineries

The synergy with petrochemical pathways in producing biofuels from biomass using the CFP processes led to an early rush of commercial CFP plants in the United States that had hoped to fulfill the markets earmarked by the mandated volumes in the RFS2 quotas for advanced biofuels. Perhaps the most prominent are Dynamotive, KIOR, Anellotech, Envergent (UOP-ENSYN joint venture), etc. KIOR aimed to produce renewable gasoline from woody biomass but failed due to their inability to overcome process and yield challenges at the commercial scale required (Perego et al., 2017; Adkins et al., 2012). KIOR could not demonstrate yields above 30 gallons per bone dry ton of biomass due to several challenges that set biomass pyrolysis oil feed streams apart from petroleum. Some of these challenges include those that we have outlined, including oxygenates and associated instabilities, ash, and catalysts poisoning. The propensity to coke formation is higher for biomass processing than petroleum and leads to short catalyst lifetime and shorter times on stream. The added cost of engineering catalyst regeneration systems and the high cost of the catalysts that work effectively for fast pyrolysis like FCC catalysts (\$2−\$3K/metric ton) and ZSM-5 costing \$6−\$8K/metric ton, the path to commercialization became increasingly difficult. After several attempts on fungible fuels, Allenotech is currently attempting to commercialize the conversion of woody biomass to benzene, toluene, ethylbenzene, and xylenes (BTEX) targeting the chemical rather than the biofuels market (Perego et al., 2017). The major drawback faced by these technologies is the inherent low carbon conversion rate for the pyrolysis step upstream of the biorefinery exacerbated by the significant increases in carbon losses to effluent gases and coke formation (Dutta et al., 2015).

Table 3.14 Aromatic selectivities for pyrolysis of *Eucalyptus urophylla* using ZSM-5 catalysts—CBV2314 (catalyst/biomass = 10 w/w) (Schultz et al., 2017).

Aromatic carbon selectivity (%)[a]								
	HZSM-5	**Zn-HZSM-5 A**	**Zn-HZSM-5 B**	**Ni-HZSM-5**	**Ga-HZSM-5 A**	**Ga-HZSM-5 B**	**Ga-Ni-HZSM-5 A**	**Ga-Ni-HZSM-5 B**
Benzene	14.75^{E}	21.71^{B}	21.21^{B}	27.28^{A}	18.24^{CD}	18.81^{C}	16.56^{DE}	16.96^{D}
Toluene	27.89^{D}	32.46^{B}	34.78^{A}	25.87^{E}	26.16^{E}	27.90^{D}	29.75^{C}	30.40^{C}
p-Xylene	23.46^{ABC}	17.89^{E}	19.91^{D}	9.46^{F}	22.51^{C}	24.59^{A}	24.29^{AB}	23.27^{BC}
o-Xylene	3.55^{AB}	2.62^{C}	2.05^{D}	1.03^{E}	3.49^{B}	4.06^{A}	3.29^{B}	3.17^{BC}
Ethylbenzene	1.01^{D}	1.31^{C}	1.70^{B}	1.16^{CD}	1.32^{C}	1.30^{CD}	1.71^{B}	2.03^{A}
TMB[b]	1.66^{C}	1.09^{D}	0.94^{D}	0.41^{E}	2.09^{B}	2.61^{A}	1.87^{BC}	1.66^{C}
NAs[c]	27.68^{B}	22.93^{C}	19.41^{D}	34.79^{A}	26.19^{B}	20.74^{CD}	22.53^{C}	22.50^{C}

Values in a row do not share a superscript capital letter are statistically different ($P < .05$) based on ANOVA of three replicates. *BTEX*, Sum of benzene, toluene, ethylbenzene and xylenes.
[a]Selectivity = (mol C compound(s))/(mol C total aromatics) × 100.
[b]TMB = 1,2,4-trimethylbenzene.
[c]Naphthalenes = sum of naphthalene and 2-methylnaphthalene.

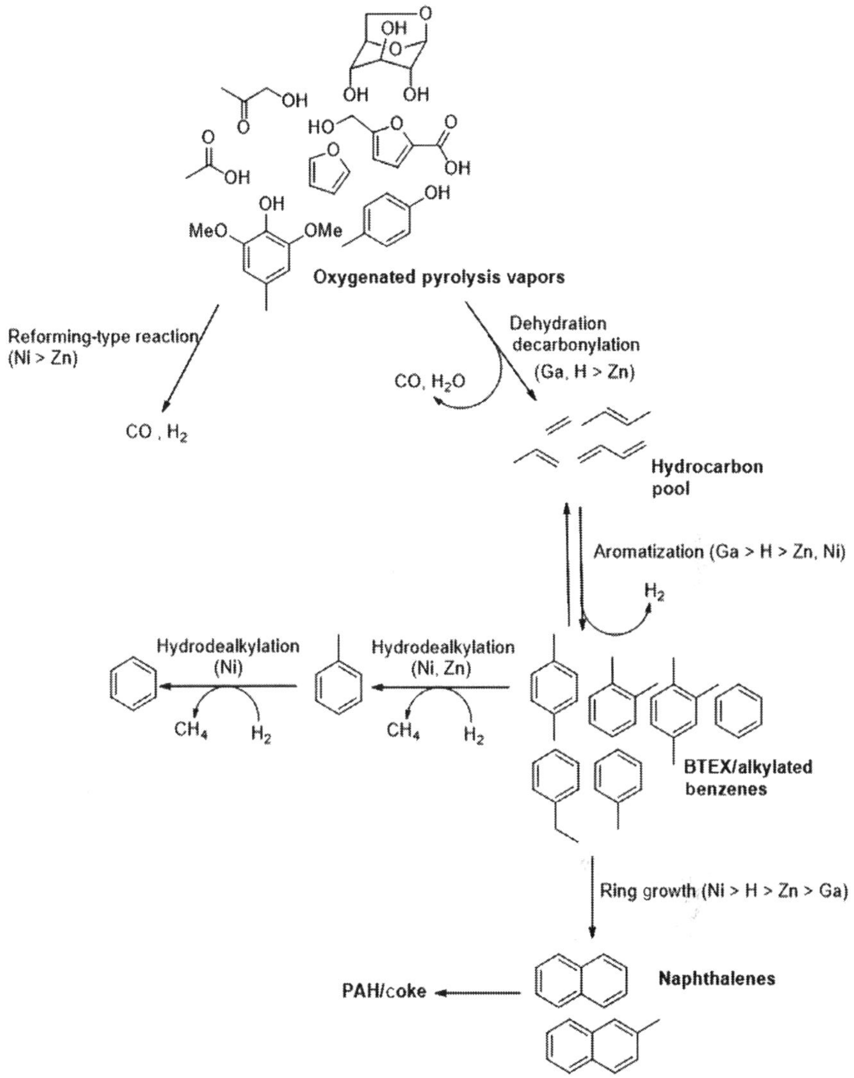

Figure 3.17 A simplified scheme for reaction pathways to produce aromatic hydrocarbons via catalytic pyrolysis of Eucalyptus over metal-modified HZSM-5 catalysts. Potential influences of the various metals and proposed relative acidity levels are noted (Schultz et al., 2017).

Generally, the level of deoxygenation that can be achieved by CFP follows an inverse trend with bio-oil yield (Mukarakate et al., 2014; Venderbosch, 2015), and this reality hinders the potential of these technologies to be economically viable pathways to exclusively produce biofuel products. The US-DOE has estimated that the near-economic price for the thermo-catalytic biorefinery can be placed at \$3.50

per gallon of gasoline equivalent if carbon yield is 44% and oxygen content for the bio-oil intermediate is 6 wt.% (Dutta et al., 2015), but these targets have not been met yet at scale. While incremental improvements in carbon yield of aromatics have been made at the laboratory, for example, at the USDA and elsewhere by modification of zeolite catalysts metal-ion exchanges, they have hardly moved the needle to economic viability for the production of biofuels. Technologies that target carbon efficiency are therefore necessary.

References

Adam, J., Blazso, M., Meszaros, E., Stocker, M., Nilsen, M.H., Bouzga, A., et al., 2005. Pyrolysis of biomass in the presence of Al-MCM-41 type catalysts. Fuel 84, 1494–1502.

Adkins, B., Stamires, D., Bartek, R., Brady, M., Hackskaylo, J., 2012. Improved Catalyst for Thermocatalytic conversion of biomass to liquid fuels and chemicals. WO Patent 2012/142490 A1.

Aho, A., Kumar, N., Eranen, K., Salmi, T., Hupa, M., Murzin, D.Y., 2007. Catalytic pyrolysis of biomass in a fluidized bed reactor: influence of the acidity of H-beta zeolite. Process. Saf. Environ. Prot. 85, 473–480.

Anellotech <https://resource.co/article/facility-advancing-100-cent-bio-based-pet-10764>.

Carlson, R.T., Tompsett, G.A., Conner, W.C., Huber, G.W., 2009. Aromatic production from catalytic fast pyrolysis of biomass-derived feedstocks. Top. Catal. 52, 241–252.

Compton, D.L., Jackson, M.A., Mihalcik, D.J., Mullen, C.A., Boateng, A.A., 2011. Catalytic pyrolysis of oak via pyroprobe and bench scale, packed bed pyrolysis reactors. J. Anal. Appl. Pyrolysis 90, 174–181.

Corma, A., Huber, G.W., Sauvanaud, L., O'Connor, P.O., 2008. Biomass to chemicals: catalytic conversion of glycerol/water mixtures into acrolein, reaction network. J. Catalysis. 247, 163–171.

Demirbas, A., 2000. Mechanisms of liquefaction and pyrolysis reactions of biomass. Energy Convers. Manage. 41, 633–646.

Dutta, A., Sahir, A., Tan, E., Humbird, D., Snowden-Swan L., Meyer, P., et al., 2015. Process design and economics for the conversion of lignocellulosic biomass to hydrocarbon fuels—Thermochemical research pathways with in-situ and ex-situ upgrading of fast pyrolysis vapors. In: NREL/TP-5100-62455.

Fahmi, R., Bridgwater, A.V., Donnison, I., Yates, N., Jones, J.M., 2008. The effect of lignin and inorganic species in biomass on pyrolysis oil yields, quality and stability. Fuel 87, 1230–1240.

French, R., Czernik, S., 2010. Catalytic pyrolysis of biomass for biofuels production. Fuel Process. Technol. 91, 25–32.

Graça, I., Lopes, J.M., Ribeiro, M.F., Ribeiro, F.R., Cerqueira, H.S., de Almeida, M.B.B., 2011. Appl. Catal. B 101, 613–621.

Horne, P.A., Nugranad, N., Williams, P.T., 1995. Catalytic coprocessing of biomass derived pyrolysis vapors and methanol. Journal of Analytical and Applied Pyrolysis 34, 87–108.

Huber, G.W., Corma, A., 2007. Synergies between bio- and oil refineries for the production of fuels from biomass. Angew. Chem. Int. Ed. 46, 7184–7201.

Ingram, L., Mohan, D., Bricka, M., Steele, P., Strobel, D., Crocker, D., et al., 2008. Pyrolysis of wood and bark in an auger reactor: physical properties and chemical analysis of the produced bio-oils. Energy Fuels 22, 614–625.

Jackson, M.A., Compton, D.L., Boateng, A.A., 2009. Screening heterogeneous catalysts for the pyrolysis of lignin. Anal. Appl. Pyrolysis 85, 226–230.

KIOR <http://www.biofuelsdigest.com/bdigest/2016/05/17/kior-the-inside-true-story-of-a-company-gone-wrong/3/>.

Lappas, A.A., Samolada, M.C., Iatridis, D.K., Voutetakis, S.S., Vasaloa, I.A., 2002. Biomass pyrolysis in a circulating fluid bed reactor for the production of fuels and chemicals. Fuel 81, 2087–2095.

Larsen, S.C., 2007. Nanocrystalline zeolites and zeolite structures: synthesis, characterization and applications. J. Phys. Chem. C 111, 18464–18474.

Mihalcik, D.J., Boateng, A.A., Mullen, C.A., Goldberg, N.M., 2011a. Packed-bed catalytic cracking of oak derived pyrolytic vapors downstream of a fluidized-bed pyrolyzer. Ind. Eng. Chem. Res. 50, 13304–13312.

Mihalcik, D.J., Mullen, C.A., Boateng, A.A., 2011b. Screening acid zeolites for catalytic fast pyrolysis of biomass and its components. Anal. Appl. Pyrolysis 92, 224–232.

Mukarakate, C., Zhang, X., Stanton, A.R., Robichaud, D.J., Ciesielski, P.N., Malhotra, K., et al., 2014. Real-time monitoring of the deactivation of HZSM-5 during upgrading of pine pyrolysis vapors. Green Chem. 16, 1444–1461.

Mullen, C.A., Boateng, A.A., 2010. Catalytic pyrolysis-GC/MS of lignin from several sources. Fuel Process. Technol. 91, 1446–1458.

Mullen, C.A., Boateng, A.A., 2013. Accumulation of inorganic impurities on HZSM-5 Zeolites during catalytic fast pyrolysis of switchgrass. Ind. Eng. Chem. Res. 52 (48), 17156–17161.

Mullen, C.A., Boateng, A.A., 2015. Production of aromatic hydrocarbons via catalytic pyrolysis of biomass over Fe-modified HZSM-5 zeolites. ACS Sustain. Chem. Eng. 3, 1623–1631.

Mullen, C.A., Boateng, A.A., Mihalcik, D.J., Goldberg, N.M., 2011. Catalytic fast pyrolysis of white oak wood in a bubbling fluidized bed. Energy Fuels 25, 5444–5451.

Mullen, C.A., Boateng, A.A., Dadson, R.B., Hashem, F.M., 2014. Influence of mineral components of biomass on its conversion to aromatic hydrocarbons by catalytic fast pyrolysis over HZSM-5. Energy Fuels 28 (11), 7014–7024.

Olazar, M., Aguado, R., Bilbao, J. 2000. Pyrolysis of sawdust in a conical spouted-bed reactor with a HZSM-5 catalyst. Reactors, Kinetics, and Catalysis, 5, 1025–1033.

Park, H.J., Dong, J.-I., Jeon, J.-K., Yoo, K.-S., Yim, J.-H., Sohn, J.-M., Park, Y.-K., 2007. Conversion of the pyrolytic vapor of radiate pine over zeolites. Journal of Industrial Chemistry 13, 182–189.

Perego, C., Bosetti, A., Ricci, M., Millini, R., 2017. Zeolite materials for biomass conversion to biofuel. Energy Fuels 31, 7721–7733.

Schultz, E., Mullen, C.A., Boateng, A.A., 2017. Aromatic hydrocarbon production via *Eucalyptus urophylla* pyrolysis over several metal modified ZSM-5 catalysts—an analysis by py-GC/MS. Energy Technol. 5, 196–204.

UOP-Honeywell <https://www.nytimes.com/2016/05/14/business/energy-environment/biofuels-plant-in-hawaii-is-first-to-be-certified-as-sustainable.html https://www.bizjournals.com/pacific/news/2017/03/09/honeywell-to-shutter-its-hawaii-refinery.html>.

Venderbosch, R.H., 2015. A critical review on catalytic pyrolysis of biomass. ChemSusChem 8, 1306–1316.

Weisz, P.B., Haag, W.O., Rodewald, P.G., 1979. Catalytic production of high-grade fuel (gasoline) from biomass compounds by shape-selective catalysis. Science 206, 57–58.

Williams, P.T., Nugranad, N., 2000. Comparison of products from the pyrolysis and catalytic pyrolysis of rice husks. Energy 25, 493–513.

Reactive pyrolysis

4

4.1 Introduction

As seen in Chapter 3, Catalytic Fast Pyrolysis, catalytic fast pyrolysis (CFP) is a facile approach to producing the desired deoxygenated fuel intermediates. This is accomplished by adding oxygen rejecting solid acid zeolite catalysts to the pyrolysis process to promote cracking-type reactions. A large body of work has shown that the general mechanism by which these catalysts work is through the protonation of oxygenates and generation of carbocations through dehydration process steps that produce olefins which aromatize under the catalytic reaction conditions (Mullen et al., 2011; Mihalcik et al., 2011; Carlson et al., 2008; Jae et al., 2011; Cheng et al., 2012, Williams and Nugranad, 2000). We also saw earlier that the removal of hydrogen via these types of reactions from biomass feedstocks that are already deficit of hydrogen results in coke formation. This is important because it the process carbon inefficient as it reduces carbon conversion toward the liquid phase and deposits coke on the catalyst, thereby deactivating it. We showed that in order to mitigate catalyst deactivation in CFP, one must offer a reactor design that provides for continual regeneration of catalysts by burning off the carbon deposits. Such designs often result in a more complex system than that designed for thermal only pyrolysis as they may require additional footprint, process control, and special operational expertise to avoid temperature runaway and related expense to run the operation. Ultimately, these setbacks complicate the opportunity to deploy pyrolysis biorefineries at the biomass source or at farm sites. We also discussed that the early commercial start-up attempts were met with some of these technological challenges amongst them biofouling of catalysts leading to limited lifetime, short times on stream, and low liquid yields. It is therefore desirable if partially deoxygenated stable fuel intermediates could be produced without the use of catalysts. We were made aware that production of aromatics is associated, in part, with hydrogen deficiency. The intentional addition of hydrogen-containing reactive gases to the pyrolysis media may therefore restore hydrogen loss and thence alter the molar H/C ratios that could reverse the coke-forming dehydration and dehydrogenation reactions.

Mante et al. (2012) had previously reported work on the CFP of the woody biomass, hybrid poplar in a fluidized-bed reactor in which the noncondensable gases were recycled back into the reactor. It was claimed that when a pyrolysis gas mixture, composed of CO, H_2, CO_2, CH_4, and other C_1-C_5 hydrocarbons, generated during CFP was mixed with the initial inert N_2 fluidizing gas, a higher quality pyrolysis oil with a marked decrease in O content ($\sim 15\%$) was produced. The

Pyrolysis of Biomass for Fuels and Chemicals. DOI: https://doi.org/10.1016/B978-0-12-818213-0.00004-7

proportional contributions from CFP and the recycled gas were not disclosed; however, we know from the previous chapter that CFP alone has successfully achieved 15%–20% oxygen reduction elsewhere. At the USDA (U.S. Department of Agriculture) researchers developed a fast pyrolysis process in a fluidized-bed reactor using a similar product gas recycling process, but this time without the use of externally added catalyst (Mullen et al., 2013). In this process, called tail gas reactive pyrolysis (TGRP), the recycle gas led to a much larger decrease in O content (~60%) than what had been previously reported for noncatalytic pyrolysis indicating the recycled gas, which was rich in CO and H_2, might have created a reducing environment suitable for deoxygenation reactions. However, there were sweet spots within the operation as the deoxygenation effect diminished from a high of 60% to ~35% when the recycled tail gas concentrations exceeded 90% indicating that the extent of deoxygenation can reach a maximum threshold with product gas recycle. The bio-oil produced by the TRGP process was rich in aromatic hydrocarbons including benzene, toluene, ethylbenzene, and xylenes (BTEX) and naphthalene similar to CFP products. These studies highlight the reducing capability of the pyrolysis gas mixture, and its ability to produce a lower oxygen, more stable pyrolysis oils without the added catalyst. Elsewhere, Zhang et al. (2011) published the first example of biomass fast pyrolysis using a fluidizing gas medium containing individual common gases that TGRP gas is made of (N_2, CO_2, CO, CH_4, and H_2) for the pyrolysis of corn cob. They observed a small decrease in pyrolysis oil yield (~5%) under CO and H_2 atmospheres, but it led to a large improvement in bio-oil fuel quality in terms of energy content [higher heating value (HHV) ~6 MJ/kg], a result which is expectedly due to the reducing nature of the gases. In yet another development, Meesuk et al. (2011) reported the fast pyrolysis of rice husk under N_2 and H_2 atmospheres using a fluidized-bed reactor and observed a decrease in pyrolysis oil yield (~5%) along with a minor decrease in O content (~10%) under a H_2 atmosphere (hydropyrolysis). The literature precedence on the use of individual heteroatom gases suggests that N_2 and CO_2 provide inert atmospheres, while H_2, CO, and light hydrocarbons (i.e., CH_4) provide reactive atmospheres suitable for the deoxygenation of volatiles eluting from biomass pyrolysis. Some of the reactive pyrolysis processes that have found equally deoxygenation applications such as hydropyrolysis, TGRP, and transfer hydrogenation within the incipient pyrolysis step will be presented and discussed in this chapter.

4.2 Tail gas reactive pyrolysis

A USDA discovery discloses that when volatile pyrolysis products are recycled and used as fluidizing gas, and then tuned within certain limits to provide a reducing reaction atmosphere, an autocatalytic effect can occur that partially deoxygenates pyrolysis oils in a manner similar to that obtained by CFP but without the use of externally added catalysts (US Patent 9,434,885 B2). If TGRP is successfully deployed on farm, it can have the potential to reduce capital and operational costs and avoid the expertise required to operate catalytic and catalyst regeneration systems. The developmental study leading toward the TGRP invention is presented herein.

4.2.1 Reaction atmosphere

The USDA fluidized-bed pyrolysis PDU was modified to recycle and utilize the gaseous products, also known as the tail gas, from the biomass fast pyrolysis for use as the fluidization gas. While it is common practice to return the effluent gas to a combustor to generate heat for pyrolysis systems, the tail gas is directed to the reactor space to create a reactive gas atmosphere to replace, in part or in full, the inert nitrogen gas traditionally used in thermal only pyrolysis (Fig. 4.1). Prior to the product gas recycle and to creating the TGRP conditions, fluidization is initiated by looping N_2 through the system. When the feeding of biomass begins the concentration of N_2 in the system is decreased as pyrolysis vapors are formed with product gas eventually replacing N_2 in whole or in part (Fig. 4.2). The advantages gained by employing a reaction environment via TGRP and the quality of the pyrolysis oils produced were tested at the USDA for three agriculturally different biomass feedstocks including white oak, a woody biomass, switchgrass, herbaceous feedstock both low-nitrogen lignocelluloses, and pennycress presscake, a proteinaceous (high nitrogen) biomass. Their elemental compositions are presented in Table 4.1. The gradual replacement of inert atmosphere gas (N_2) with TGRP gas is depicted in Fig. 4.2, and the gas composition of the environment created by the pyrolysis of these feedstocks are presented in Table 4.2.

As shown in Table 4.2, there are differences between these environments and they depend upon the feedstock. For example, the composition of the cumulative product gas for the low-N lignocelluloses feedstock (oak and switchgrass) changes more dramatically with the use of the recycled tail gas as the fluidizing gas than the high-N feedstock. These determine the extent of reducing atmosphere and bio-oil quality.

Compared with the product gas from inert atmosphere, generally the TGRP gas atmosphere results in a higher concentration of H_2, CH_4, and other hydrocarbons with a diminishing concentration of CO_2 in the gas. This, thereby, constitutes an atmosphere that is considered reducing atmosphere, chemically and therefore highly burnable as evidenced by its ability to sustain a flame on its own when flared (Fig. 4.3).

For both oak and switchgrass the HHV of the gas fraction increases from 5.5 to 11.4 MJ/kg and from 7.7 to 12.4 MJ/kg, respectively, when the atmosphere shifted from the N_2 to the optimum recycled tail gas atmosphere, indicating the extent of reactivity the atmosphere possesses and its potential to influence the pyrolysis reactions. It also demonstrates the capacity to use it for process heat to power the endothermic pyrolysis reactions. For the pyrolysis of pennycress presscake, however TGRP had a lesser effect enabling only a minor increase in H_2 and a higher fraction

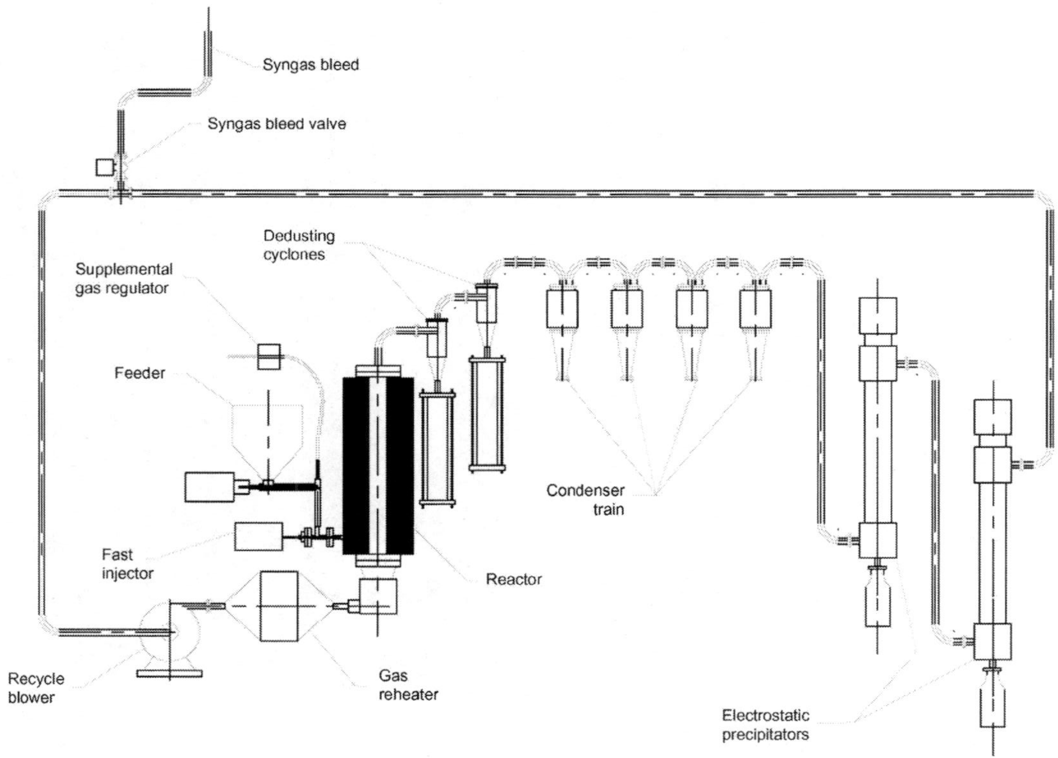

Figure 4.1 Diagram of pyrolysis system designed for recycling of product gases (TGRP) for fluidization and reaction atmosphere. *TGRP*, Tail gas reactive pyrolysis.

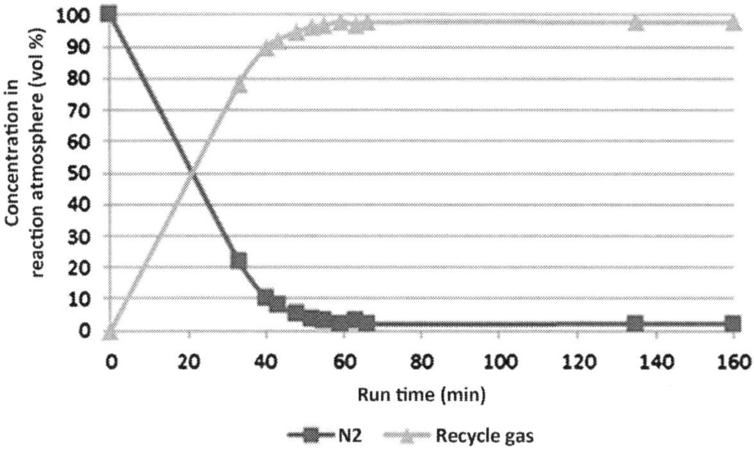

Figure 4.2 Example profile of changing concentration of N_2 and recycle gas in reaction atmosphere (Mullen et al., 2013).

Table 4.1 Feedstock elemental composition (wt.%) (Mullen et al., 2013).

		C	H	N	S	O	H_2O	Ash
Oak	As is	50.12	6.29	0.51	0.00	40.93	1.6	0.59
	Dry ash free	51.23	6.43	0.52	0.00	41.83	–	–
Switchgrass	As is	46.55	5.75	0.48	0.00	42.02	2.60	2.63
	Dry ash free	49.41	6.06	0.51	0.00	44.33	–	–
Pennycress Presscake	As is	44.79	5.32	5.66	1.25	28.504	3.06	12.67
	Dry ash free	52.37	6.22	6.62	1.46	33.33	–	–

Table 4.2 Composition of product gases[a] (mol.%) (Mullen et al., 2013).

Feedstock	Atmosphere	CO	CO_2	H_2	CH_4	C_2H_6	C_3H_8	HHV (MJ/kg)
Oak	N_2	45.6	46.6	1.0	6.6	0	0	5.47
	Recycle	40.3	20.7	22.4	14.7	0.9	1.0	11.36
Switchgrass	N_2	57.6	29.5	5.1	7.8	0	0	7.67
	Recycle	54.6	15.7	12.0	15.6	1.1	0.8	12.41
Pennycress Presscake	N_2	22.6	69.2	6.1	2.0	0.4	Trace	2.33
	Recycle	23.9	48.3	11.1	11.4	2.7	2.6	6.47

[a]Other compounds detected in the recycle stream but not quantified include ethylene, propylene, butenes, acetone, acetic acid, propanal, furan, benzene, and toluene.

Figure 4.3 Flaring of excess tail gas (Mullen et al., 2013).

of CO_2 and consequently a modest increase in HHV compared with the low-N feedstocks but still about three times higher than the control.

In addition to the permanent gases generated from the pyrolysis of the biomass, the recycled tail gas stream contains certain amounts of condensable volatile compounds that would normally escape the bio-oil collection system and which would typically not be detected by an on-line Micro GC. An off-line glass bulb condensation train placed post-ESP and maintained at $-15°C$ shows evidence of the presence of water, butenes, pentenes, acetic acid, acetone, propanal, furan, benzene, and toluene all of which, especially the acids, are not detected but participate in reactive pyrolysis when reintroduced into the atmosphere.

4.2.2 Yield and distribution of tail gas reactive pyrolysis products

Figs. 4.4–4.6 taken from the USDA study show the yield distribution of TGRP products for switchgrass, oak and pennycress presscake, respectively, is notably different for the low-N lignocelluloses compared with the proteinaceous biomass just as their reactor environments differed. With both switchgrass and oak as feedstocks, yields of noncondensable gases and reaction water increased with a concurrent drop in organic liquid yield for the TGRP environment over inert N_2 atmosphere (thermal only pyrolysis). This observation trends similarly with pyrolysis over a solid acid catalyst such as acid zeolite presented in the previous chapter indicating a possible autocatalytic effect of the tail gas recycled process can occur. For both low-N lignocellulosic biomasses studied herein, the autocatalytic behavior continues up until the concentration of TGRP gases in the reaction chamber reached about 70% indicating peak threshold which, perhaps, is feedstock dependent. Specifically, for oak, NCG yield increased from 13 wt.% of input biomass for pyrolysis reactions

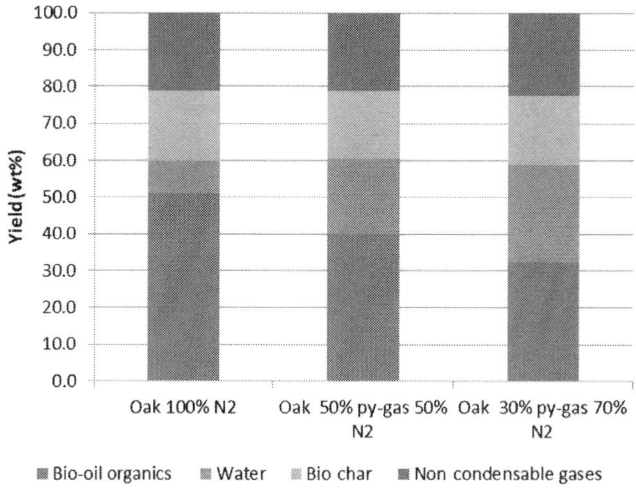

Figure 4.4 Yield distribution of pyrolysis products for oak under varying concentrations of recycled product gas (Mullen et al., 2013).

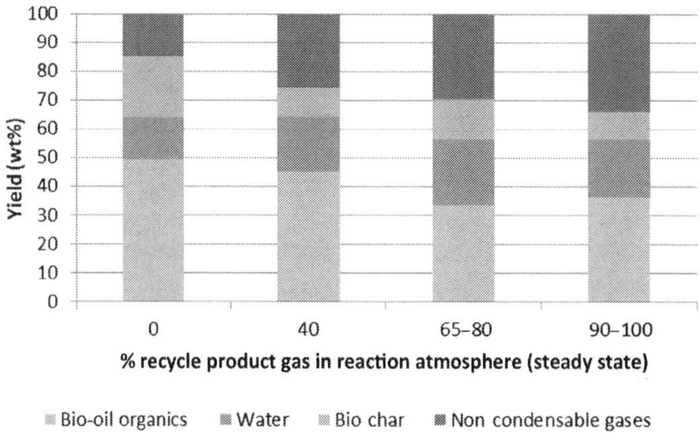

Figure 4.5 Yield distribution of pyrolysis products for switchgrass under varying concentrations of recycled product gas (Mullen et al., 2013).

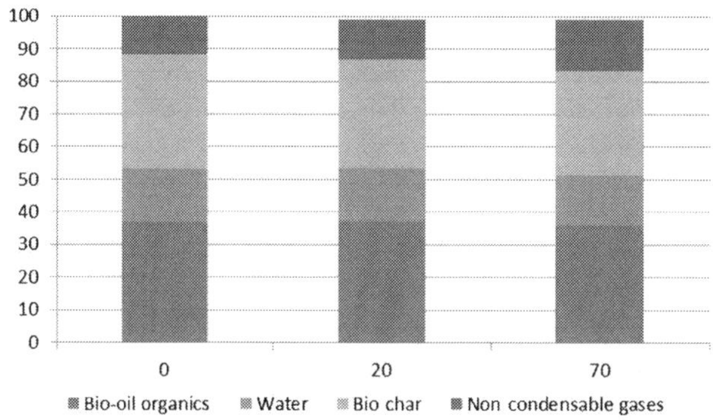

Figure 4.6 Yield distribution of pyrolysis products for pennycress presscake under varying concentrations of recycled product gas (Mullen et al., 2013).

under N_2 to 26 wt.% when TGRP was at its peak concentration of 70%; water yield increased from 7.8% to 26% while the organic yield of the bio-oil decreased from 54% to 32%. Experiments with switchgrass follow similar trends with NCG yield increasing from 22% to 33%, water yield increasing from 14% to 20%, and bio-oil organic fraction yield decreasing from 49% to 33%. However, when the concentration of the recycled tail gas reached between 90% and 99% of the chamber atmosphere, a diminishing effect was realized where the distribution of noncondensable gas produced continued to increase (to 36%), less water was produced (down to 20%) and the yield of the organic fraction of the liquid remaining increasing slightly to 36%, thereby showing a maximum or peak recycle gas concentration where the greatest effect occurs. Such performance is undoubtedly similar to catalytic pyrolysis as we previously discussed in Chapter 3, Catalytic Fast Pyrolysis. For the proteinaceous biomass feedstock the change in atmosphere had little effect on the pyrolysis product distribution, which remained at about 36% organic liquid, 16% water, 30%–35% biochar, and 11%–16% noncondensable gases.

Like CFP, the increases in water yield and that of carbon oxides produced under TGRP atmosphere constitutes a net removal of oxygen from the organic fraction of the bio-oil. As a result of the deoxygenation, bio-oil product quality improves in terms of homogeneity, acidity/stability, and energy content in a way similar to that observed for CFP. The elemental analysis including total acid number (TAN) and energy content of the organic liquids collected by the ESP for the biomass feedstock studied are presented in Tables 4.3–4.5 for TGRP and CFP. There are some parallels that can be drawn between TGRP and zeolite catalyzed fast pyrolysis products. For example, fluidized-bed TGRP of oak at 50:50 tail gas-N_2 atmosphere resulted in an increase in the bio-oil C/O ratio by >60% (from 2.2 to 3.6), thereby increasing its HHV by some 30% (from 23.7 to 31.2 MJ/kg). Introducing more of the recycled tail gas and thereby creating a more reducing atmosphere, up to 70% concentration in the atmosphere resulted, on the one hand, in oxygen content of <10 wt.%, C/O ratio of 11.6, and energy content of 34.0 MJ/kg. On the other

Table 4.3 Elemental analysis, energy content, and total acid number (TAN) of oak pyrolysis oils produced under varying amounts of product gas atmosphere (Mullen et al., 2013).

Percentage recycle gas in reaction atmosphere	N_2	50	70	Catalytic pyrolysis β-type zeolite catalyst, N_2
Water (wt.%)	6.95	6.79	4.79	7.7
Carbon (wt.%, db)	58.10	67.37	80.24	76.5
Hydrogen (wt.%, db)	6.11	5.51	5.88	5.63
Nitrogen (wt.%, db)	0.70	1.90	2.07	0.38
Oxygen (wt.%, db)	35.09	25.22	9.19	17.4
C/O	2.19	3.56	11.64	5.9
H/C	1.26	0.98	0.88	0.90
HHV (MJ/kg, db)	23.7	31.2	34.0	32.3
TAN (mg KOH/g)	138	115	55.8	68

Table 4.4 Elemental analysis, energy content, and total acid number (TAN) of switchgrass pyrolysis oils produced under varying amounts of product gas atmosphere (Mullen et al., 2013).

Percentage recycle gas in atmosphere	N_2	40	65–80	90–99	Catalytic pyrolysis over HZSM-5, N_2
Water (wt.%)	6.7	8.2	3.2	3.6	5.1
Carbon (wt.%, db)	59.82	64.27	80.29	70.71	68.55
Hydrogen (wt.%, db)	6.03	6.00	5.67	5.83	5.74
Nitrogen (wt.%, db)	0.92	1.76	1.50	1.88	0.74
Oxygen (wt.%, db)	40.46	27.88	12.54	21.43	24.97
C/O	1.97	3.07	8.53	4.39	3.67
H/C	1.21	1.12	0.84	0.99	1.00
HHV (MJ/kg, db)	23.4	26.3	33.2	29.0	29.7
TAN (mg KOH/g)	119	96	24	54	51

Table 4.5 Elemental analysis, energy content, and total acid number (TAN) of pennycress presscake pyrolysis oils produced under varying amounts of product gas atmosphere (Mullen et al., 2013).

Percentage recycle gas in atmosphere	N_2	20–30	65–80
Water (wt.%)	7.7	7.2	10.2
Carbon (wt.%)	69.01	66.40	68.37
Hydrogen (wt.%)	8.35	7.78	8.30
Nitrogen (wt.%)	7.14	8.43	8.10
Oxygen (wt.%)	14.75	16.96	14.65
C/O	6.2	5.2	6.2
H/C	1.5	1.4	1.45
HHV (MJ/kg)	31.4	33.1	33.2
TAN (mg KOH/g)	84	85	76

hand, the H/C ratio of the resulting bio-oil trended downward from 1.26 to 0.88 ($>40\%$ reduction) at 70% TGRP condition primarily due to the rejection of oxygen as water. This is analogous to that observed with fast pyrolysis of oak over a β-type acid zeolite in the same system whereby under the best conditions a C/O ratio of 5.9, an H/C ratio of 0.9 and a dry basis HHV of 32.3 MJ/kg were realized. Unlike the ESP product the liquid collected at the condenser points during TGRP process is extremely aqueous representing $>85\%$ water at the optimum TGRP conditions exemplifying oxygen rejection as water.

Similar trends are observed for TGRP of switchgrass (Table 4.4) where C/O, C/H, HHV, and TAN values of the resulting bio-oil increased. Interestingly, comparable values were observed at different tail gas recycle rates, 40% and 65%–80% for switchgrass compared to 50% and 70% concentrations for oak. Beyond the higher threshold, a diminishing return resulted especially when the recycled tail gas

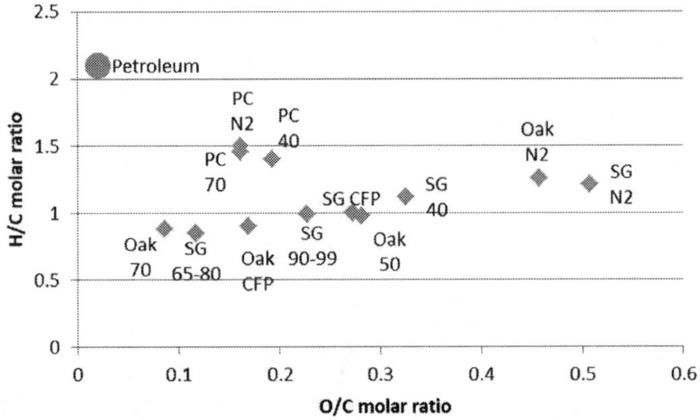

Figure 4.7 Van Krevelen type diagram comparing oxygen and hydrogen contents of pyrolysis oils produced under varying concentrations of recycled product gases. Numbers on data points indicate the percentage of recycle gas in the atmosphere of the reaction at steady state. Data for petroleum and pyrolysis oils from CFP over zeolites provided for comparison (Mullen et al., 2013). *CFP*, Catalytic fast pyrolysis; *PC*, pennycress presscake; *SG*, switchgrass.

atmospheres approached 90%–99%. In these cases the average C/O ratio of the bio-oil dropped to 4.4, the H/C ratio was 1.00, and the average energy content was 29.7 MJ/kg. The Van Krevelen diagram shown in Fig. 4.7 summarizes the C, H, O contents of TGRP deoxygenated bio-oils compared with thermal only pyrolysis liquids, catalytic pyrolysis bio-oils, and petroleum to illustrate the role the various deoxygenation pathways compare with each other in the quest toward reaching petroleum quality product. We note the shift toward lower O/C ratio when we transition from inert atmosphere to TGRP at times surpassing CFP when the TGRP is optimum concentration. However, like CFP little upward mobility toward higher H/C is observed except for the high-N feedstock. This is due to hydrogen deficiency in the low-N lignocellulosic biomass that results in more aromatic compounds than the pennycress presscake that must have some residual hydrogen-containing lipid.

4.2.3 Product analysis

The TGRP bio-oil chemical analyses (Tables 4.6–4.8) again show that the organic oil composition is similar to bio-oil produced over zeolite catalysts. The concentration of highly oxygenated compounds such as acetic acid, acetol, and the dehydrated glucose monomer, levoglucosan decreases as the reducing atmosphere intensifies to <1 and 2 wt.% at 70% TGRP from 17 and 18 wt.% at 100% N_2 atmosphere for oak and switchgrass respectively. Observed concurrently is the formation of aromatic hydrocarbons such as benzene, toluene, xylenes, and naphthalenes as well as increases in the concentration of nonmethoxylated phenols (phenol and cresols). In the case of switchgrass, for example, while only trace amounts of total benzene, toluene, and xylene (BTX) are produced over inert atmosphere as much as

Table 4.6 Concentrations (GC/MS, wt.%) of selected compounds in oak pyrolysis oils produced under varying amounts of product gas atmosphere (Mullen et al., 2013).

Percentage recycle	N_2	50	70	Catalytic pyrolysis β-type zeolite catalyst, N_2
Acetic acid	4.4	3.1	2.3	4.5
Furfural	0.2	Trace	0.1	0.2
Acetol	4.8	0.7	Trace	0.4
Levoglucosan	9.0	Trace	Trace	3.5
Phenol	0.2	1.4	3.0	1.0
Cresols	0.4	2.1	3.7	1.5
Guaiacol	0.2	Trace	0	0.1
Syringol	0.1	0.1	0.1	0.3
Benzene	Trace	0.6	2.1	0.2
Toluene	0.1	0.4	1.0	0.2
Xylenes	0.1	0.2	0.2	0.4
Naphthalene	0.1	0.2	2.4	1.60
1-Methyl napthalene	0.1	0.2	1.0	2.6

Table 4.7 Concentrations (GC/MS, wt.%) of selected compounds in switchgrass pyrolysis oils produced under varying amounts of product gas atmosphere (Mullen et al., 2013).

	0	40	65–80	90–100	Catalytic pyrolysis HZSM-5, N_2
Acetic acid	6.4	4.89	0.27	2.99	0
Furfural	0.18	0.19	0	0.07	0.04
Acetol	5.6	6.47	0	1.19	0.19
Levoglucosan	4.65	4.86	0	0.295	0.16
Phenol	0.47	1.98	3.75	3.22	1.79
Cresols	0.59	2.51	2.51	3.28	1.76
Guaiacol	0.46	0	0	0	0
2,6-Dimethoxyphenol	0.38	0	0	0	Trace
Benzene	0.05	0.89	2.47	2.22	0.30
Toluene	0.04	0	0.97	0.65	0.275
Xylenes	0.06	1.09	0.63	0.45	0.86
Naphthalene	0	0.22	4.36	1.36	3.7
1-Methyl naphthalene	0	0.14	1.0	0.48	2.40

4% BTX and as much as 4% naphthalene could be realized under the optimum TGRP conditions. In the case of the proteinaceous biomass (pennycress presscake in this case) the change in product composition (Table 4.8) appears to be less dramatic, but the trending is similar with a slight increase in the presence of benzene and toluene and a decrease in acetic acid concentration.

The similarities in product composition between TGRP and CFP over zeolite catalysis tend to suggest that the chemical mechanisms for the deoxygenation of bio-oils with the pathway being initiated by the same mechanism described for CFP, that is, protonation of an oxygen functionality followed by dehydration and dehydrogenation reaction to form olefins as we described earlier. The same hydrogen reactions that lead to the formation of aromatics, which is shown in Scheme 4.1.

Whether this pathway could also be initiated by the trace amounts of organic acid in the recycled tail gas is still speculative, at best. But the hypothesis is credible for the fact that the process was highly effective for switchgrass and oak but not for pennycress presscake where the vapors are not as acidic. Regarding CFP, the aromatization of the olefins is supposedly organized within the shape-selective cavity of the zeolite structure which is not in the case for the TGRP process. Also, the diminishing effect observed at high concentrations of recycle tail gas complicates the CFP mechanistic analogy. Given these constraints, another pathway like that suggested by Zhang et al. (2011) could be possible which is perhaps, increased concentration of radicals, for example, from homolytic cleavage of H_2 could be a contributing factor as H_2 is found in higher concentration in the TGRP environment. Other proposed deoxygenation systems involve the role of metal oxides in reduction/oxidation cycles because it is also possible that trace metal oxides present in the biomass could be reduced to their alkaline earth metals by interaction with the

Table 4.8 Concentrations (GC/MS, wt.%) of selected compounds in pennycress presscake pyrolysis oils produced under varying amounts of product gas atmosphere (Mullen et al., 2013).

Percentage recycle	**N_2**	**20–30**	**60–80**
Acetic acid	2.4	1.5	0.5
Furfural	0	0	0
Acetol	0	0	0
Levoglucosan	2.0	0.2	0.3
Phenol	0.3	0.3	0.4
Cresols	0.5	0.5	1.0
Guaiacol	0.1	0.1	0.1
Syringol	0.1	0.1	0.1
Benzene	0.1	0.1	0.7
Toluene	0.1	0.3	0.7
Xylenes	0.1	0.2	0.2
Naphthalene	0	0.1	0.1
1-Methyl naphthalene	Trace	Trace	Trace

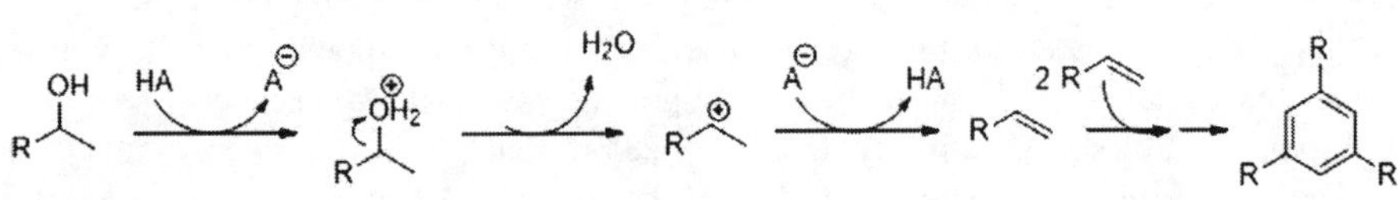

Scheme 4.1

CO and H_2 rich atmosphere and then act as oxygen scavengers on the pyrolysis vapors. However, it is highly unlikely that reducing reactions will proceed under the pyrolysis temperature of 500°C except if there are higher temperature hot spots that might favor metal oxide reduction. The chemical mechanism behind the deoxygenation reactions observed with TGRP is still illusive but there is enough data to support its effectiveness in bio-oil deoxygenation. Ongoing work at the USDA to uncover the scientific underpinnings includes testing gas-phase reactions in down-scaled reactor set-ups including microwave pyrolysis (Tarves et al., 2016; Raymundo et al., 2019).

4.2.4 Static bed and various reactive gas atmospheres

The influence of reactive gases in a static rather than fluidized-bed environment was tested at the USDA using a laboratory-scale microwave reactor (Fig. 4.8). The pyrolysis of switchgrass was carried out in batch, under varying gaseous atmospheres and bio-oils obtained characterized. A batch of 100 g biomass in the form of pellets was pyrolyzed with a 900–1000 W microwave power over the course of 7 minutes in the presence of a microwave absorber, a 10 g of activated charcoal, and an atmosphere varying from N_2 (the control) to various reactive gases (CO, H_2, and CH_4) including a model pyrolysis gas mixture (PyGas) comprising N_2:CO:H_2:CO:CH_4:C_2H_4 in a percentage ratio of 30:22:20:16:10:2 that is characteristically representative of a pyrolysis tail gas.

Like the fluidized-bed TGRP process, the yields of organic liquids decreased (~5%) when the N_2 atmosphere was replaced by H_2 atmosphere. That of the non-condensable gas atmosphere increased (~5%–10%) compared with the N_2 atmosphere. NMR analysis (Table 4.9) shows that the bio-oils obtained from pelletized switchgrass under a N_2 atmosphere contained the lowest percentage of aliphatic (36.9%) and aromatic (36.4%) compounds, but the greatest percentage of alcohols/sugars (18.1%). The atmosphere containing reducing gases composed of H_2 and CO produced bio-oils with the greatest percentages of aliphatic compounds (47.1% and

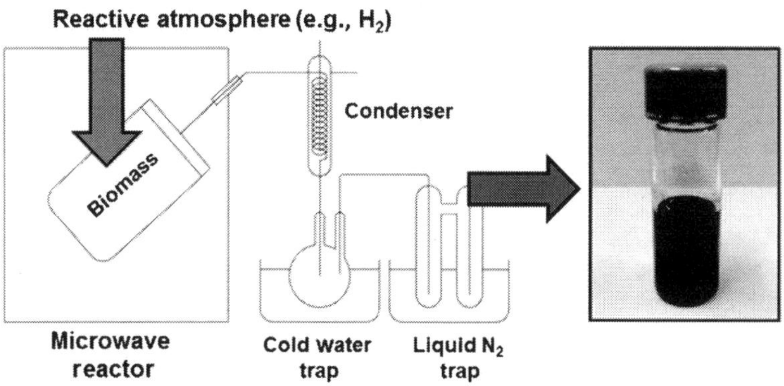

Figure 4.8 Microwave pyrolysis for reactive atmospheric studies (Tarves et al., 2016).

Table 4.9 ^{13}C NMR analysis of microwave pyrolysis of switchgrass under various reactive atmospheres (Tarves et al., 2016).

Gas atmosphere	N_2 (%)	CO (%)	CH_4 (%)	H_2 (%)	PyGas (%)
0–55 ppm (aliphatics)	36.9	43.8	35.2	47.1	36.4
55–95 ppm (alcohols/sugars)	18.1	8.9	9.9	10.1	9.9
95–165 ppm (aromatics)	36.4	41.7	45.0	37.6	47.3
165–180 ppm (acids/esters)	6.1	4.9	6.6	4.9	4.2
180–215 ppm (ketones/aldehydes)	2.6	0.8	3.3	0.3	2.1

43.8%, respectively) and the lowest percentages of ketones/aldehydes (0.3% and 0.8%, respectively). The PyGas and CH_4 atmospheres produced bio-oils with the greatest percentage of aromatic compounds (47.3% and 45.0%, respectively), but comparatively lower aliphatic compounds (36.4% and 35.2%, respectively). Except for the N_2 atmosphere, all the bio-oils contained approximately the same amounts of alcohols and sugars (9.7% $\pm$ 0.5%).

With regard to oxygen removing potential, the H_2 and the PyGas (with 20% H_2) atmospheres yielded the most deoxygenated pyrolysis oils ($\sim$25% decrease in O content) which are expected to give their strong reducing abilities. Likewise, the CH_4 atmosphere produced bio-oils with decreased O content (27.43%) relative to the N_2 control (33.46%), whereas the CO-only atmosphere (32.89%) was not equally effective. Employing a CO-only atmosphere had a negligible effect on the quantity and quality of bio-oils produced, whereas the use of H_2, CH_4, and PyGas atmospheres was effective in producing more deoxygenated products (i.e., BTEX, naphthalenes, etc.) and at a lower oxygen content overall. This follow-up work lends credence to the use of either hydrogen [hydropyrolysis (HYP)] or a mixture of H_2, CO and CH_4 found in the tail gas in the pyrolysis chamber to affect a deoxygenated liquid product compatible with drop-in biofuels.

4.3 Temperature effect on deoxygenation

The microwave atmospheric studies advised the need to revisit the possible effects of temperature spikes in the fluidized-bed TGRP process. A recent systematic study to elucidate the role high temperatures play in favoring vapor deoxygenation via the TGRP was carried out at the USDA using a laboratory-scale pyrolysis system (Raymundo et al., 2019). Two hypotheses were explored including whether temperature excursions above the fast pyrolysis temperature setpoint of 500°C and/or the catalyzing effect of accumulated biochar could be potential key parameters for deoxygenation during TGRP. The study employed a scaled-down version of the PDU (c. Fig. 2.1) to a laboratory-scale fluidized-bed system as depicted in Fig. 4.9. The system comprised additional features such as a filter for biochar separation, an

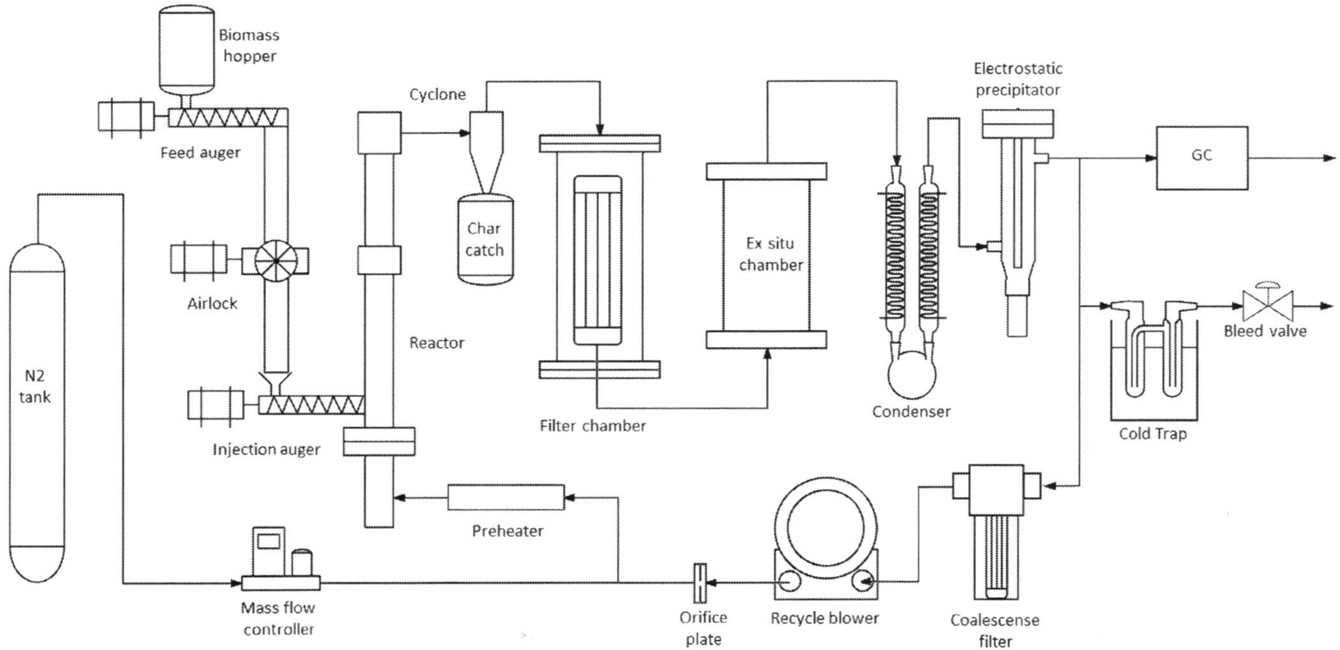

Figure 4.9 Process diagram of laboratory-scale equipment (Raymundo et al., 2019).

ex situ secondary reaction chamber prior to the condensation system. The new features allowed the testing of the effects of a secondary heated zone in the absence or presence of product biochar. It also allowed for variations in the process temperatures in the 500°C–750°C range in both the fluidized-bed pyrolysis reactor (in situ) and in the static, secondary chamber (ex situ) positioned downstream. The results of several experiments (Table 4.10 along with Fig. 4.10) confirm that the oxygen content of the bio-oils produced varied with temperature excursions. For in situ chamber temperatures at 500°C, 600°C, 700°C, a reducing trend of bio-oil oxygen contents were determined as 31, 30, and 19 wt.%, respectively. This trend was

Table 4.10 Experimental identification and operational conditions (Raymundo et al., 2019).

Label/code	5/5/5N_2	6/6/5N_2	7/7/5N_2	5/5/5	6/6/5	5/7/5	7/7/5	5/5/6	5/5/7	5/5/7.5
T_{bed} (°C)	500	600	700	500	600	500	700	500	500	500
$T_{freeboard}$ (°C)	500	600	700	500	600	700	700	500	500	500
$T_{ex\ situ}$ (°C)	500	500	500	500	500	500	500	600	700	750
Carrier gas	N_2	N_2	N_2	Rec.[a]	Rec.	Rec.	Rec.	Rec.	Rec.	Rec.
Est. res. time at highest T (s)	6.27[b]	1.03	0.26	6.27[a]	1.03	0.26	0.93	0.87	0.78	0.74

[a]*Rec.*, Recycled tail gas atmosphere.
[b]Includes reactor, cyclone, hot vapor filter, and ex situ chamber, all components operate at 500°C.

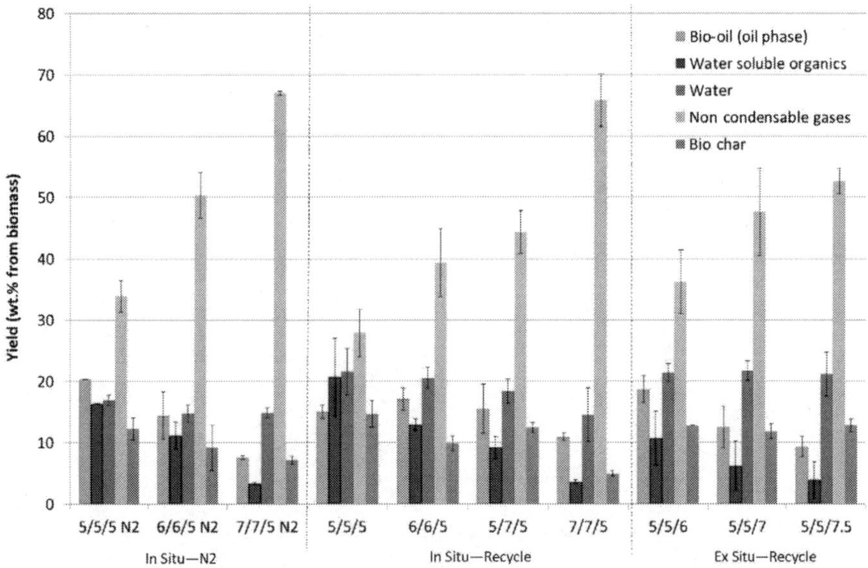

Figure 4.10 Pyrolysis product yields at temperatures higher than perceived optimum of 500°C (Raymundo et al., 2019).

about the same when the ex situ chamber temperature was rather varied. For the same temperature range plus one at 750°C the bio-oil oxygen content was of the order 31 wt.%→27 wt.%→23 wt.%→19 wt.%, respectively. Carbon yields of the bio-oil increased from 20% at the baseline case of 500°C at all points, to 25% and 29% at 600°C for in situ and ex situ conditions respectively, but they decreased to 18% and 20%, respectively, when the temperature jumped to 700°C. As we saw, bio-oil with the lowest oxygen content was obtained with the ex situ chamber set at 750°C but with a low carbon yield of 16%. With regard to the effect of temperature on product distribution the observation is that generally, increasing temperatures, at either the in situ or the ex situ chambers, lead to an increase in hydrocarbon yield such as BTEX and naphthalenes and also an increase in phenols. This is accompanied by a simultaneously decrease in oxygenated compounds such as levoglucosan, acetol, and furfural, which are responsible for bio-oil instability. The discrepancy between the results of the in situ and ex situ cases at 700°C could suggest an effect of the presence of biochar in the fluidized-bed reactor at that temperature. To further analyze the effect of biochar accumulations as TGRP contributor, separate experiments with biochar loaded in the ex situ chamber were performed at 500°C and 600°C. At 600°C the oxygen content of bio-oil produced was as low as 19 wt.%, like those obtained with quartz in the ex situ chamber at 750°C suggesting that temperature spikes and biochar could have a catalytic deoxygenation effect during TGRP in addition to the presence of a reducing atmosphere. This only goes to complicate the mechanistic pathways previously hypothesized. However, the temperature effect of bio-oil quality in TGRP is not unexpected as this was confirmed for thermal only pyrolysis using diffusion order NMR spectroscopy in Chapter 2, Thermal Pyrolysis (Mullen et al., 2016).

4.4 Hydropyrolysis

We are aware of condensed-phase hydrodeoxygenation (HDO) whereby the oxygen content of the formed pyrolysis liquids can be removed by hydrogen as water under high pressure. We will discuss HDO in the coming chapters but observation from reactive pyrolysis products thus far shows evidence that hydrogenation reactions play an important role in the deoxygenation of pyrolysis oil intermediates as they are being formed. In pyrolysis systems like fluidized beds where dilute phase contacting takes place, H_2 can either be applied as the gas atmosphere in an inert reactor bed such as silica sand or over a solid catalyst bed (Fig. 4.11). Hydrogenation reactions in the presence or absence of solid catalyst can lead to bio-oil intermediate product with very low oxygen content with the potential to render downstream processes of the biorefinery more selective and economical. By downstream processing, we mean condensed-phase upgrading to final hydrocarbon fuels such as gasoline and/or chemicals by catalytic HDO, distillation, extraction, and/or other separation forms. The thermal deconstruction of the biomass by pyrolysis under atmospheric pressure hydrogen gas like the TGRP with the purpose to remove

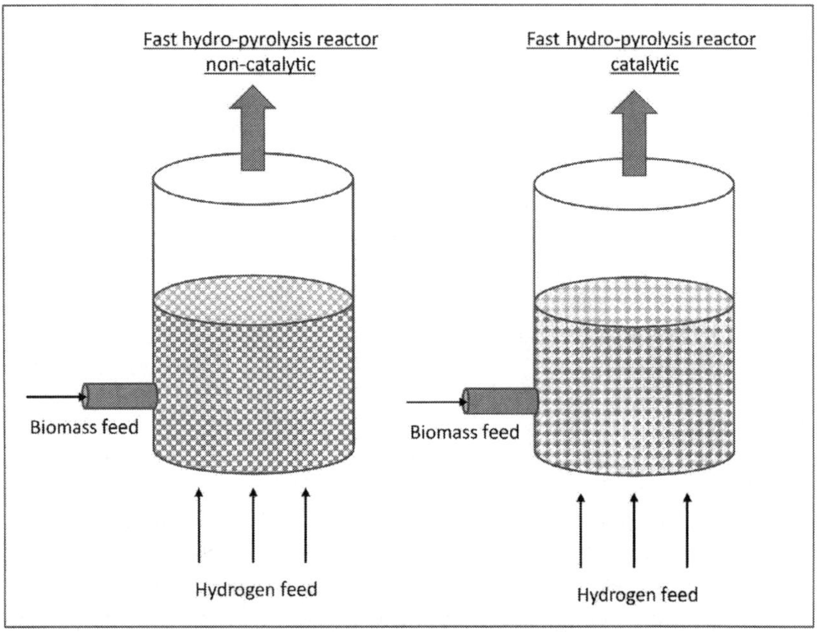

Figure 4.11 Hydropyrolysis configuration with dilute-phase contacting of H_2. Inert bed (left); catalytic bed (right).

oxygen into a slew of liquid and gaseous products is what is termed hydropyrolysis (HYP).

Unlike fast pyrolysis, fast HYP, the rapid decomposition of organic materials under hydrogen atmosphere, as a stand-alone process, has been reported only over the last few years. As we reported earlier, reactive pyrolysis under H_2 atmosphere yields bio-oil rich in aromatic hydrocarbon similar to that by CFP over zeolite catalysts but is still high in oxygen content compared with gasoline. To form a final fuel and/or chemical product such as alkanes and naphthalenes, these liquid fuel intermediates need to be upgraded post-pyrolysis in the condensed-phase via high-pressure HDO.

4.4.1 Noncatalytic hydropyrolysis

The levels of deoxygenation enabled by hydropyrolysis without the assistance of catalysts are what we have reported above; they are equivalent to the levels achieved by CFP. The reactions leading to these products involve H_2 and light oxygenates in the decomposed biomass volatiles where the hydrogen is introduced into the reactor chamber as the fluidizing gas. Because of short contact times and near-atmospheric pressure conditions associated fluidized-bed reactors the kinetics of mild pressure HYP does not favor complete deoxygenation as would be in high-pressure HDO. Increasing the reactor pressure (Güell et al., 1993) has been found to affect product yield and distribution. Like CFP, bio-oil yields from noncatalytic HYP do decrease by relatively small amounts over a pressure range, with significant changes in liquid product structures. Progressively smaller molecular masses with lower polarity have been obtained with increasing pressure. Nevertheless, a preponderance of oxygenates, still exists including syringol and guaiacol derivatives and carboxylic acids, rather than alcohols, ketones, terpenoid, and polynuclear aromatic hydrocarbons.

4.4.2 Catalytic hydropyrolysis

The extent of deoxygenation may be enhanced when HYP is carried out over solid catalysts. One would suspect that hydrogenation catalysts such as nickel or cobalt over molybdenum (NiMo and CoMo, as they are so called) to be appropriate source of solid catalysts given the involvement of hydrogen reactions, but zeolite catalysts used in CFP have also been explored. Herein one can engineer a gas–solid contact reactor system such as fluidized-bed involving hydrogen as a fluidizing gas and solid catalysts as the fluidized-bed medium like the one depicted in Fig. 4.12, reported by RTI for a combination of CFP and mild HYP. Dayton's group at Research Triangle Institute (RTI) (Dayton et al., 2016; Wang et al., 2017) reported work under the US-DOE's National Advanced Biofuel Consortium (NABC) funded under the American Recovery and Reinvestment Act (ARRA) in 2010 to develop biofuel technologies that address the fundamental challenges of converting lignocellulosic biomass feedstocks to drop-in biofuels using a catalytic HYP. In this work entitled reactive CFP (RCFP) of biomass with atmospheric pressure hydrogen was investigated in a lab-scale fluidized-bed reactor with varying reaction conditions including temperature and hydrogen concentration and catalysts. Five solid catalysts were screened including alumina-based solid acid catalysts (SA1), a tungsten-based reducible metal oxide catalyst (RMO1), an iron-based mixed metal oxide catalyst (MMO1), a molybdenum-based reducible metal oxide catalyst (RMO2), and a commercial hydrotreating catalyst (HT1). In all, the presence of atmospheric hydrogen over various catalysts was found to improve the yield and quality of bio-oil or biocrude as they termed it and minimized coke formation than one would encounter in a regular CFP over an inert gas. Expectedly, the group found the molybdenum-based catalyst to be most effective at hydrodeoxygenating biomass pyrolysis vapors to produce a hydrocarbon-rich bio-oil or biocrude intermediate with oxygen content as low as <10 wt.%. As a point of reference, recall about 12 wt.% O content was realized with TGRP at its optimum recycle stream. Dayton's group at RTI found higher hydrogen concentration and moderate reaction temperature (450°C) to favor higher bio-oil yields and quality during catalytic HYP. They report that a yield of 43.2 C% in the C4 + organic fraction with as low as 6.2 wt.% oxygen in the liquid could be obtained under optimized reaction conditions, although this optimum condition was not disclosed. Furthermore, the resulting biocrude contained primarily aliphatic and aromatic hydrocarbons, with only small amounts of simple ketones, furans, and phenols. Up to 41.5% carbon in the biomass feed was successfully converted into gas and condensable hydrocarbons with 29.7% carbon contained in the

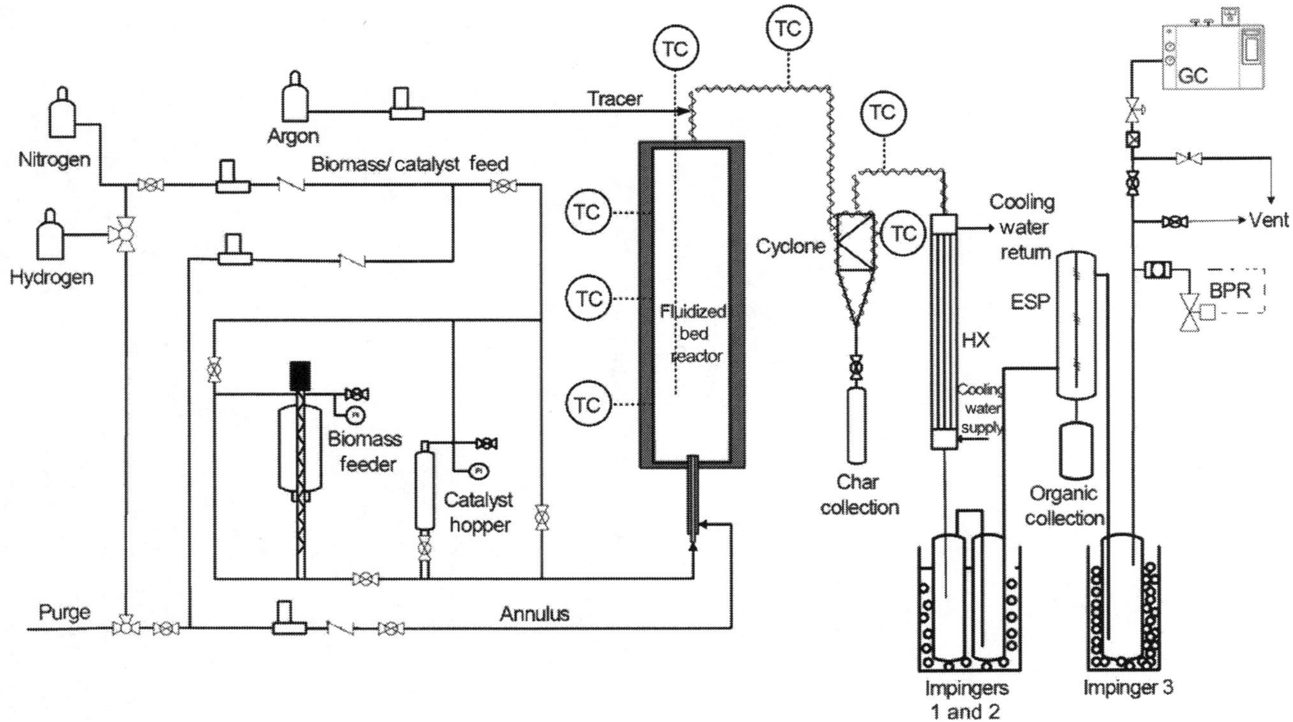

Figure 4.12 Process flow diagram of the bench-scale hydropyrolysis reactor system (Dayton et al., 2016).

condensable C4 + fraction. With this quality, they claimed that such biocrude intermediate would be more compatible with existing petroleum refining infrastructure allowing production of "drop-in" biofuels.

On this point, HYP at elevated pressures has been equally studied by groups other than RTI notably by the Gas Technology Institute, GTI (Marker et al., 2015) and by Perdue University (Venkatakrishnan et al., 2014) all of whom have reported similar results with similar catalysts to further reduce oxygen and improve the overall carbon efficiency. Wang et al. (2017) reviewed and compared data from various groups on the correlation between carbon efficiency and bio-oil oxygen content for CFP, CFP under hydrogen atmosphere (RCFP), and HYP at high-pressure conditions as presented in Fig. 4.13. As the figure suggests, the quality and quantity of desirable organic products from RCFP are higher compared to CFP. Yields of C4 + organics from RCFP are in the range of 30%–43%, and the oxygen content in the biocrude is consistently below 20 wt.% with some data showing as low as 5.7 wt.% under certain reaction conditions. Moreover, the H/C_{eff} of the RCFP biocrude increases, in this case up to 1.07 compared to 0.4–0.8 for typical CFP biocrude. The increased H/C_{eff} indicates improved hydrogen utilization during when CFP is carried out in a mild hydrogen atmosphere. The biocrude yield and quality from HYP are higher compared to RCFP due to high pressure which favors HDO. However, one needs to be mindful that high-pressure operations are severe conditions, and for that, the RCFP would be preferred if the purpose is to produce drop-

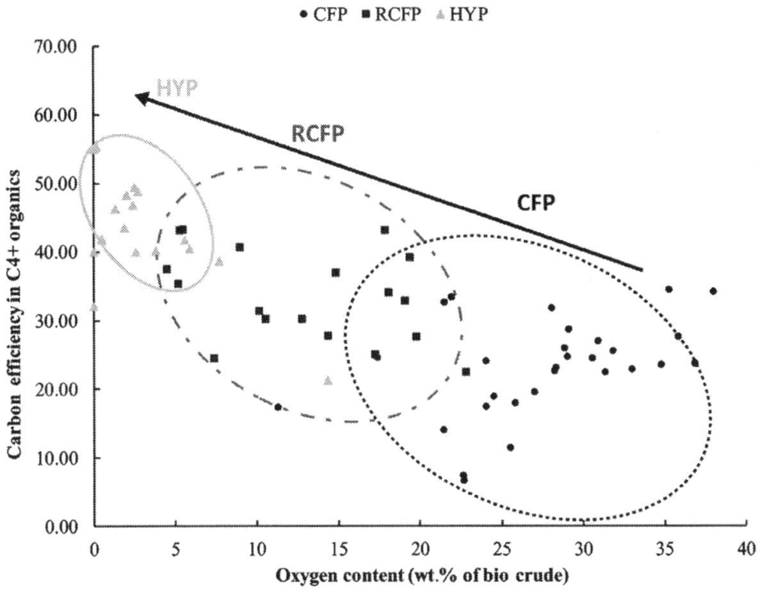

Figure 4.13 Yield of C4 + organics versus oxygen content in the biocrude for CFP, RCFP, and HYP (Wang et al., 2017). *CFP*, Catalytic fast pyrolysis; *HYP*, hydropyrolysis; *RCFP*, reactive catalytic fast pyrolysis.

in fuels, especially when it has been successfully developed to the point where greater than 43% carbon and less than 7 wt.% oxygen can be obtained.

A clear logistical and economic challenge for the RCFP technology is the need for hydrogen in the pyrolysis reactor. Wang et al. (2017) have shown that hydrogen consumption for RCFP can range from 1.0 to 3.2 wt.% of feedstock depending on the reaction conditions and catalysts. This is high and resource challenged but perhaps hydrogen can be resourced from waste streams such as plastics if catalyst stability over longer times on stream can be engineered.

4.5 Copyrolysis with plastics

As seen by now, pyrolysis liquids (biocrude, bio-oil, etc.) produced from the pyrolysis of lignocellulosic biomass can serve as potential intermediates for the production of drop-in renewable advanced biofuels if the appropriate technologies are selected and the right operating conditions are met. However, incompatibility with hydrocarbons and instability resulting from the high concentration of reactive oxygenated components limit its utility. We have discussed CFP as one process to produce deoxygenated pyrolysis liquids with more favorable properties to petroleum. While zeolite catalysts such as HZSM-5 can produce aromatic hydrocarbons from biomass it is plagued by catalyst cost, short catalysts lifetimes, and low carbon efficiencies. But we have also shown the mitigating effect of reactive pyrolysis and HYP with or without catalysts and have seen from various works that CFP under mild pressure hydrogen atmosphere is most effective to produce drop-in biofuel intermediates. However, this process requires a high consumption of hydrogen that poses operational and economic challenges. It is possible that incorporation of carbon and hydrogen-rich coreactants into the biomass CFP process could mitigate these problems. One can only imagine how economical and environment-friendly that process would be if waste plastic materials could be sourced for hydrogen. USDA's interest in plastic disposal in US agriculture prompted investigation into copyrolysis of biomass and agricultural plastics (Dorado et al., 2015a,b). Farmers rely on plastics to increase crop yields, reduce the use of herbicides and pesticides as well as to conserve water. Approximately 2 million tons of agricultural plastics are used annually worldwide, and an estimated 521 million pounds of agricultural plastics are used per year in the United States alone. Aside from being a C–H source, utilization of waste plastics in pyrolysis biorefineries could have the added benefit of alleviating a major waste disposal problem for farmers in a meaningful way. Although there are structural incompatibilities between biomass and plastics as shown in Table 4.11, plastic molecules could serve as donor molecules for biomass to form a new hydrocarbon product with little coking as we saw for HYP earlier.

Table 4.11 Empirical formulas and structures of biomass components and plastics on a C_6 basis (Dorado et al., 2014).

Compound	Empirical formula	Structure
Cellulose	$C_6H_{10}O_5$	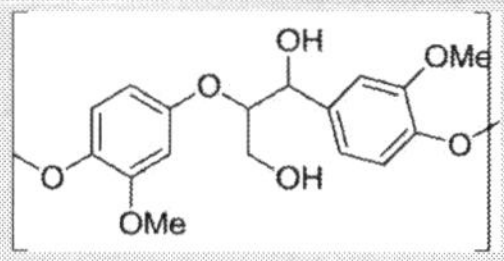
Switchgrass	$C_6H_{10.31}O_{4.03}$	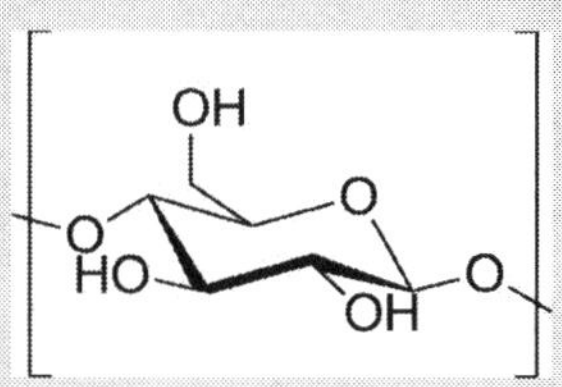
Xylan[a]	$C_6H_{9.6}O_{4.8}$	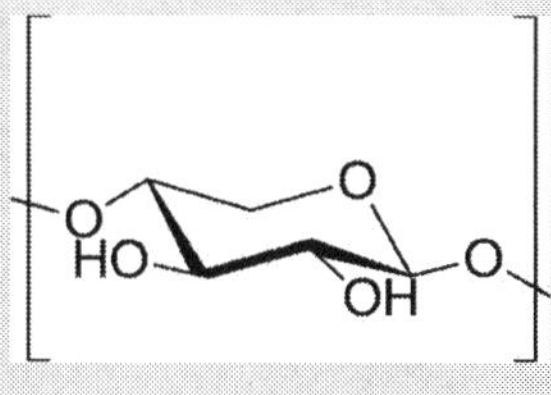
Lignin[a]	$C_6H_{6.55}O_{1.97}$	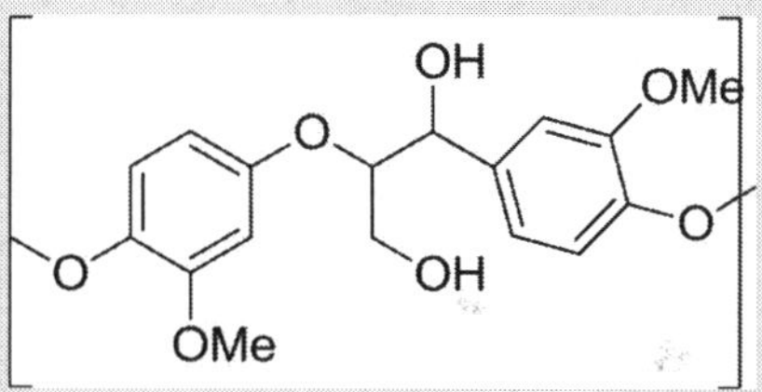
PP	C_6H_{12}	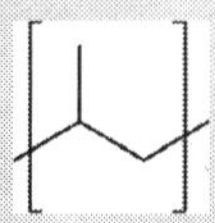
PET	$C_6H_{4.8}O_{2.4}$	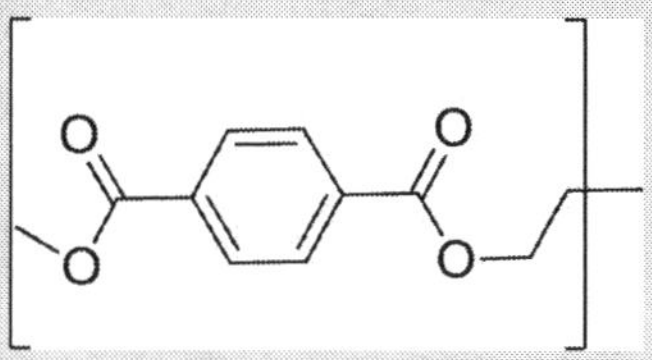
PE	C_6H_{12}	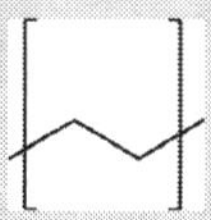

(*Continued*)

Table 4.11 (Continued)

Compound	Empirical formula	Structure
PS	C_6H_6	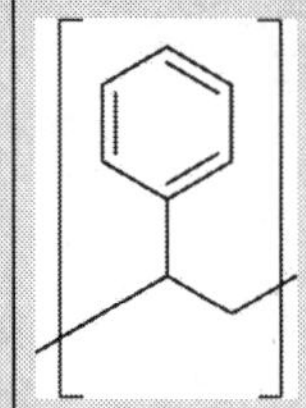

PE, Polyethylene; *PET*, polyethylene terephthalate; *PP*, polypropylene; *PS*, polystyrene.
[a]Xylan and lignin are irregular polymers, sample structure is given.

4.5.1 Biogenic carbon in copyrolysis product pool

Given the hydrogen deficiency of biomass the real concern for biomass and plastic copyrolysis is whether the conversion of biomass to liquid products will be enhanced by the additional hydrogen availability from the presence of plastics. That is, will the products simply be an arithmetic sum of the products from each feedstock or will chemical reactions occur to form a different product. Isotopically labeled reactants were employed in a USDA study to track products, their sources, and the underlying chemistry. In the study the cocatalytic pyrolysis of ^{13}C-labeled cellulose with various plastic forms including polyethylene terephthalate (PET), polypropylene (PP), high-density polyethylene (HDPE), low-density polyethylene (LDPE), and polystyrene (PS) was carried out. With the help of mass spectral data gathered from PY-GC/MS experiments the fractions of plastic and cellulose-derived carbon that ended up in the condensable product pool of aromatic hydrocarbons including benzene, toluene, ethylbenzene, *p*-xylene, *o*-xylene, naphthalene, and 2-methylnaphthalene during their catalytic fast copyrolysis (CFP) over HZSM-5 were determined. In addition, fractional contribution of the reactants toward the formation of noncondensable gas products of the copyrolysis such as carbon monoxide, carbon dioxide, methane, ethylene, ethane, propylene, propane, and butane must provide the efficacy of the use of plastics as C–H source and hydrogen donor.

These fundamental studies show that the CFP of biomass and plastic over HZSM-5 leads to the production of aromatic hydrocarbons, with a yield enhancement or selectivity change over the sum of their individual CFP products. When mixtures of ^{13}C cellulose and plastic were subjected to CFP, they produced aromatic hydrocarbons which contained molecules of mixed origins. Overall, when the mixtures of cellulose and plastic were pyrolyzed together, the amount of biogenic carbon that ended up in the six aforementioned aromatic hydrocarbon products depended on the plastic source (Figs. 4.14–4.18). As these figures indicate, biogenic carbon in the aromatic product pool could be as much as 49% for HDPE, 52% LDPE, 59% for PP, 55% for PS, and 65% for PET. Considering that approximately 56% of the carbon in each prepyrolyzed mixture was ^{13}C from cellulose, it was readily found that the polyethylene (PE) and PS mixtures contribute the highest

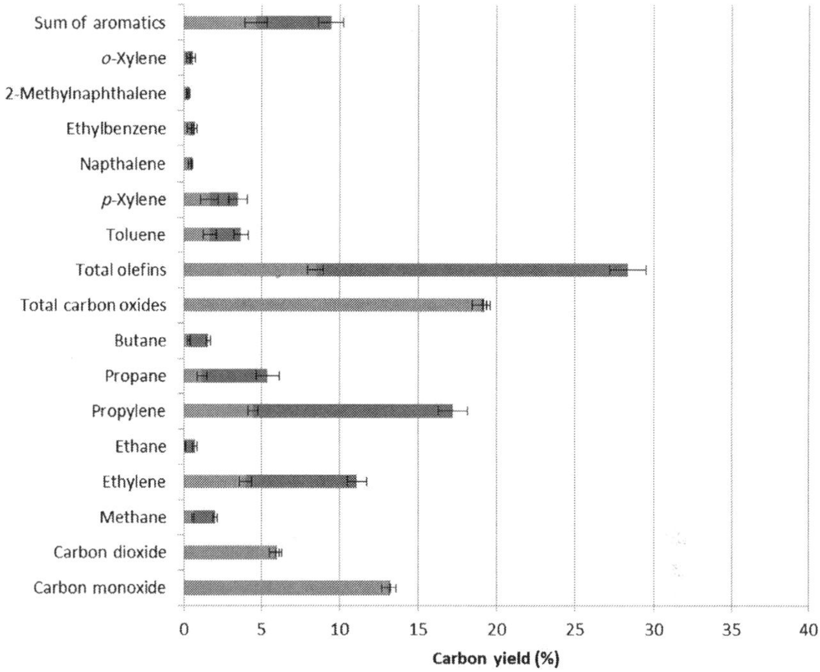

Figure 4.14 The distribution of carbon from cellulose (■) and carbon from HDPE (■) in the condensable and noncondensable gas products from the CFP of both feedstocks in the presence of HZSM-5 (Dorado et al., 2015a). *CFP*, Catalytic fast pyrolysis; *HDPE*, high-density polyethylene.

yield of biogenic carbon to the aromatic products (8.2% for HDPE, 8.8% for LDPE and 8.6% for PS), while the PET mixtures contributing the least fraction of 3.8%. The effective use of PE or PP as a coreactant is attributed to the reaction scheme depicted in Fig. 4.19 whereby the addition of olefins from the cracking of the plastic to the hydrocarbon pool that effectively reacts with the oxygenated biomass primary pyrolysis vapors in a scheme that avoids coke formation. Production of these olefins leads to yields of the largest amounts of toluene and *p*-xylene among all the plastics that can participate in the formation of aromatics via HZSM-5 catalysis (Fig. 4.19). The pyrolysis of nonpolyolefin polymers such as PS and PET, however, produces a lower concentration of olefins and thereby contribute lesser amounts to that pool, lessening their potential to interact in this manner. For PE and PP the carbon monoxide and carbon dioxide yields are composed of mostly carbon from cellulose, with the total carbon yield of oxide gases being less than 20%, a slight decrease compared to PS and PET. This can be attributed to the combination of the small oxygenated species from the pyrolysis of cellulose combined with the olefins produced from the pyrolysis of these plastics in a Diels–Alder type of reaction to form aromatic products (Fig. 4.20), subjecting fewer of the oxygenates to conversion to carbon oxides via cracking over HZSM-5.

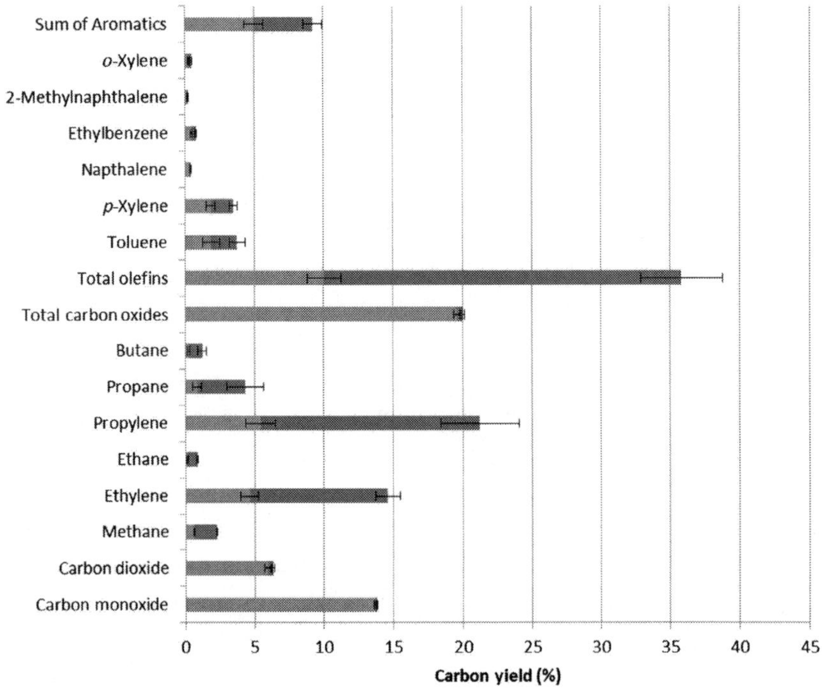

Figure 4.15 The distribution of carbon from cellulose (■) and carbon from LDPE (■) in the condensable and noncondensable gas products from the CFP of both feedstocks in the presence of HZSM-5 (Dorado et al., 2015a). *CFP*, Catalytic fast pyrolysis; *LDPE*, low-density polyethylene.

In recent literature, Diels–Alder reactions have been suggested as the mechanism that leads to an increase in the selectivity of aromatics produced under CFP. The reaction of furans, produced from the pyrolysis of biomass, with olefins, such as those produced from the thermal degradation of polyethylene, reacts in a Diels–Alder-type reaction to produce the corresponding aromatic compounds with increased selectivity. In the case where yield enhancement by the plastics occurs, the reason for some of the enhanced aromatic product formation, particularly toluene, xylenes, and ethylbenzene, may be explained by the Diels–Alder type reaction mechanism where oxygenated products, such as furans derived from biomass, can react with olefins derived from plastics to produce aromatic hydrocarbons (Dorado et al., 2014).

4.5.2 Biomass–plastic copyrolysis via tail gas reactive pyrolysis

The analytical studies described in Section 4.2.1 are effective in informing the fundamental chemistry for the catalytic copyrolysis of biomass and plastics of all types but the real challenge is whether engineering a scaled system design to produce

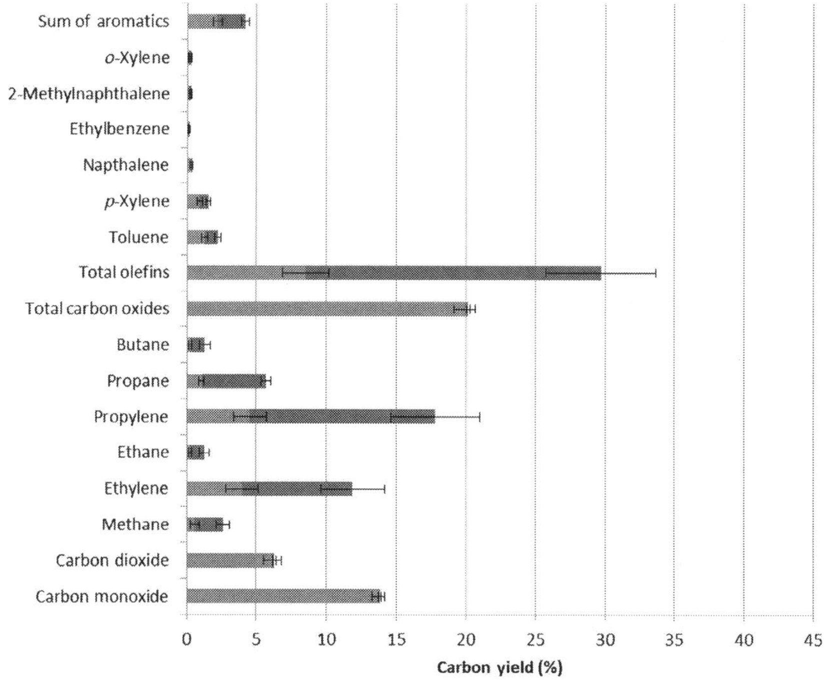

Figure 4.16 The distribution of carbon from cellulose (■) and carbon from PP (■) in the condensable and noncondensable gas products from the CFP of both feedstocks in the presence of HZSM-5 (Dorado et al., 2015a). *CFP*, Catalytic fast pyrolysis; *PP*, polypropylene.

copyrolysis products is practical. For that, coprocessing of biomass and plastic was tested at the Pilot Scale using the USDA PDU shown earlier in Fig. 2.1 (Dorado et al., 2015b). Mixtures of agricultural plastic waste in the form of polyethylene hay bale covers (PE) (4%–37%) and switchgrass were investigated using fast pyrolysis and TGRP at varying wt.% plastic (0–37 wt.%), recycled gas percentage and various reactor temperatures (400°C–570°C) as presented in Table 4.12. Under inert (N_2) fast pyrolysis atmosphere the polyethylene was unable to be completely depolymerized at temperatures near 550°C, interrupting the process with waxy solid deposits formed throughout the system as shown in Fig. 4.21 top. However, using the TGRP process atmosphere, not only was the fouling of the vessel alleviated (Fig. 4.21 bottom) but high yields of deoxygenated pyrolysis oils were also produced.

The yield distribution of primary pyrolysis products in comparison with nonplastic is presented in Fig. 4.22. It shows high bio-oil yields comparable to regular TGRP of biomass could be achieved with plastic copyrolysis TGRP up to 16 wt.% beyond which fouling was severe and compromising the reactor. While carbon conversion was equally comparable, the oxygen content of the TGRP, plastic copyrolysis liquid product was very low. Table 4.13 shows that the best TGRP PE/biomass

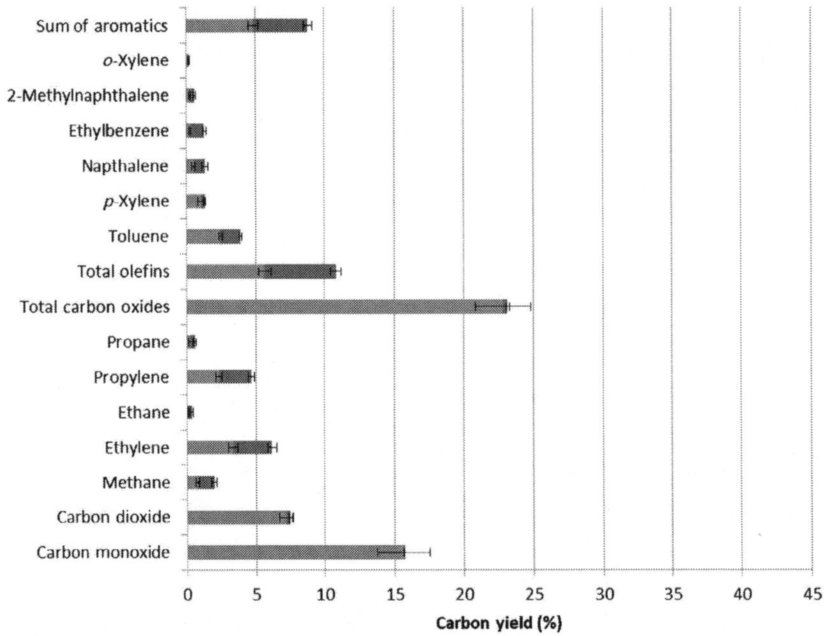

Figure 4.17 The distribution of carbon from cellulose (■) and carbon from PS (■) in the condensable and noncondensable gas products from the CFP of both feedstocks in the presence of HZSM-5 (Dorado et al., 2015a). *CFP*, Catalytic fast pyrolysis; *PS*, polystyrene.

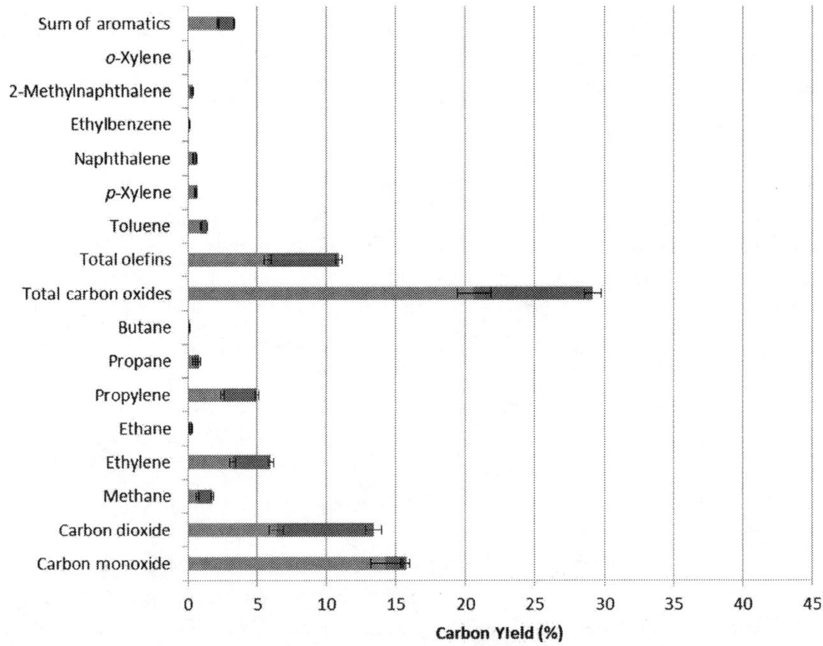

Figure 4.18 The distribution of carbon from cellulose (■) and carbon from PET (■) in the condensable and noncondensable gas products from the CFP of both feedstocks in the presence of HZSM-5 (Dorado et al., 2015a). *CFP*, Catalytic fast pyrolysis; *PET*, polyethylene terephthalate.

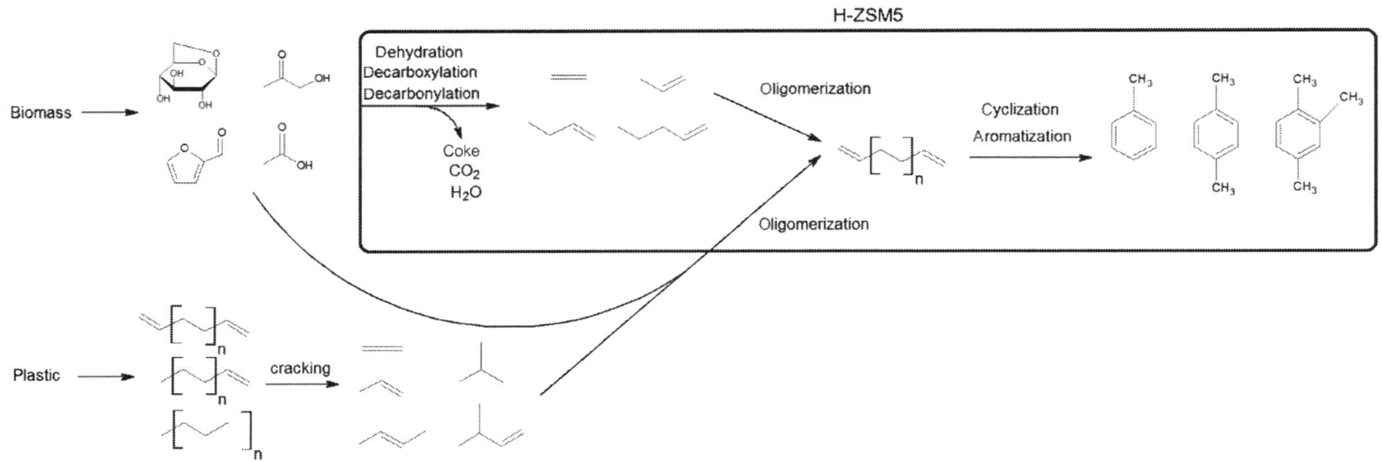

Figure 4.19 Reaction pathways for the breakdown of biomass and plastic and the formation of aromatic compounds via HZSM-5 (Dorado et al., 2014).

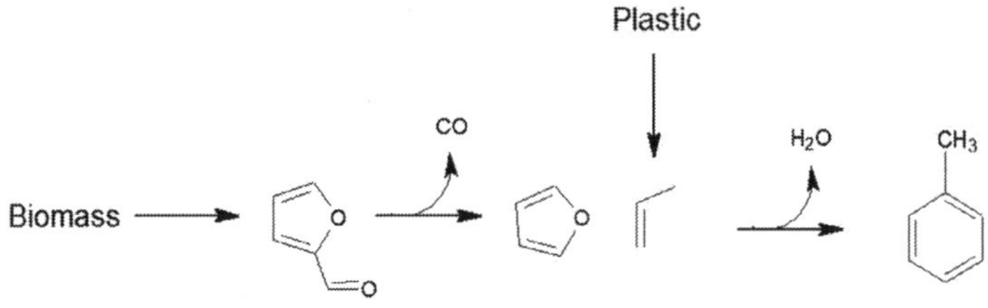

Figure 4.20 Reaction scheme of biomass pyrolysis product and plastic pyrolysis product participating in a Diels–Alder type reaction to form an aromatic product (Cheng et al., 2012).

blend was 16 wt.%, 500°C–570°C, and 54%–76% recycle gas, the experimental condition 2 of Table 4.13. This results in a high-quality drop-in fuel with energy content equivalent to gasoline. The produced pyrolysis oil constituents are shown in Table 4.14. At the optimum condition described above (no. 2), only trace amounts of oxygenated compounds were found. A characteristic TGRP atmosphere for this condition depicts a fuel-rich atmosphere containing primarily H_2, CO, ethylene, and other light hydrocarbons resulting in the production of highly deoxygenated and aromatic bio-oil, but in low yields. While higher bio-oil yields could be obtained at lower tail gas recycle rates, it comes at the expense of bio-oil quality (more oxygenated) although this quality is still significantly better compared with traditional biomass pyrolysis oils. The data suggest that hydrogen-releasing aromatization is the driving force for deoxygenation of copyrolysis of biomass and plastic via the TGRP process enhancing the conversion of biomass to deoxygenated liquids. However, there is no conclusive evidence that hydrogen transfer from polyethylene is the major pathway of this process. This suggests that more work is required to optimize the coprocessing of biomass and polyolefin plastics, a successful application of which can have a huge impact on our environmental footprint given the menace of plastic waste disposal causes.

4.6 Other reactive environments—solvents

Aside from the use of solids as C–H sources copyrolysis with liquid solvents, the liquefaction thermolysis that was discussed earlier, has found equally facile application of the decomposition of biomass. One interesting study is the depolymerization of lignin via copyrolysis with 1,4-butanediol that is intended for the same purpose (Tarves et al., 2017). The production of valuable compounds from low cost but abundant residual lignin has proven to be challenging and the lack of effective biochemical lignin depolymerization processes has led to an increased focus on thermochemical conversion methods to address it. When a bench-scale pyrolysis of lignin was carried out in a microwave the liquid products obtained were composed of smaller polymeric components and moderate yields of monomeric phenols. However, upon the addition of 1,4-butanediol, repolymerization reactions that limit

Table 4.12 Feedstock composition and experimental conditions for the regular and tail gas reactive pyrolysis (TGRP) of agricultural plastic [polyethylene (PE)] and switchgrass (Dorado et al., 2015b).

Experiment	**1**	**2**	**3**	**4**	**5**	**6**	**7[a]**	**8[a]**
wt.% PE of mixture	37	16	16	16	8	4	0	0
Total feedstock (kg)	2.57	2.73	2.74	2.70	2.78	2.85	~3.0	~3.0
Feed rate (kg/h)	0.69	1.05	1.06	0.80	1.05	1.19	1.5	1.5
Temperature (°C)	~540	500–570	400–450	420–540	480–550	520–540	450–500	450–500
Pyrolysis	Regular	TGRP	TGRP	TGRP	TGRP	TGRP	Regular	TGRP
Recycle gas (%)	N_2	54–76	42–76	42–69	44–70	42–58	N_2	65–80

[a]Data taken from Mullen et al. (2013) for comparison.

Figure 4.21 Images of soft brown waxy solid material collected in the first ESP (A) and the funnel of the first ESP (B) after the regular pyrolysis of agricultural plastic (PE) and switchgrass. Images of a thin film of oil collected in the first ESP (C) and the relatively clean funnel of the first ESP after the TGRP of PE and switchgrass (Dorado et al., 2015b). *ESP*, electrostatic precipitator; *PE*, polyethylene; *TGRP*, tail gas reactive pyrolysis.

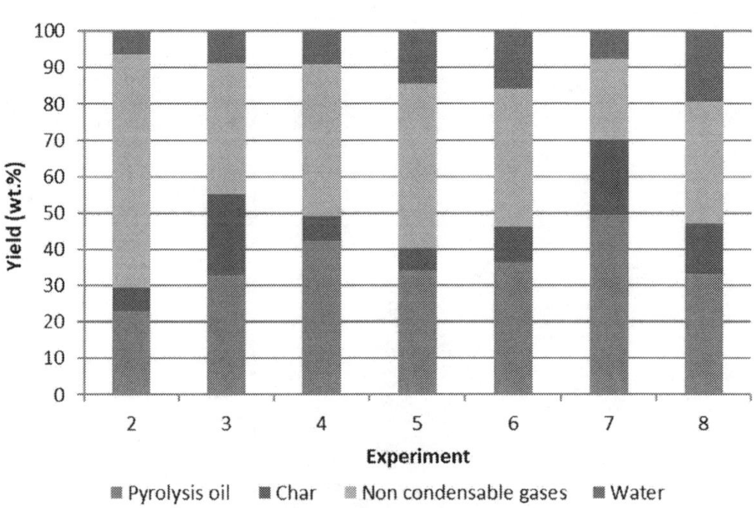

Figure 4.22 Yield distribution of products from the regular and TGRP of agricultural plastic (PE) and switchgrass with experimental condition described in Table 4.12 (Dorado et al., 2015b). *PE*, Polyethylene; *TGRP*, tail gas reactive pyrolysis.

Table 4.13 Water content, elemental analysis, total acid number (TAN), and energy content [higher heating value (HHV)] of oil produced from the regular and tail gas reactive pyrolysis of agricultural plastic (polyethylene) and switchgrass (Dorado et al., 2015b).

Experiment	1	2	3	4	5	6	7[a]	8[a]
Water (wt.%)	4.3	0.5	5.3	1.6	6.4	4.4	6.7	3.2
C (wt.%, db)	72.2	87.7	63.0	78.1	78.4	68.2	59.8	80.3
H(wt.%, db)	9.4	4.9	7.2	5.5	5.6	5.4	6.0	5.7
N (wt.%, db)	0.5	0.4	0.6	0.8	1.5	1.4	0.9	1.5
O (wt.%, db)[b]	17.9	7.1	29.2	15.6	19.4	25.0	33.3	12.5
C/O (mol)	5.4	16.6	3.2	6.7	5.4	3.6	2.0	8.5
H/C (mol)	1.6	0.7	1.8	0.8	0.9	1.0	1.2	0.8
TAN (mg KOH/g)[c]	68.2	16.4	54.0	33.6	52.3	79.3	119.0	24.0
HHV (MJ/kg, db)	26.3	39.4	28.7	36.4	34.0	31.6	23.4	33.2

[a]Values taken from Mullen et al. (2013) for comparative analysis.
[b]Oxygen determined by difference.
[c]Titration end point average.

Table 4.14 Concentration of selected compounds (GC/MS, wt.%) from the oil and aqueous fractions from the regular pyrolysis and tail gas reactive pyrolysis of switchgrass and agricultural plastic (polyethylene) (Dorado et al., 2015b).

Experiment	1	2	3	4	5	6	7[a]	8[a]
Oil phase								
Acetic acid	6.2	Trace	3.0	0.2	0.6	0.4	6.4	0.3
Acetol	3.5	Trace	1.1	0.1	0.1	0.2	5.6	nd
Levoglucosan	2.1	Trace	5.1	Trace	0.4	1.2	4.5	nd
Phenol	1.1	0.2	1.3	0.9	1.8	1.4	0.5	3.8
o, *m*-, *p*-Cresols	1.0	0.1	1.5	0.8	1.8	3.1	0.6	2.5
BTX[b]	Trace	0.3	Trace	0.5	0.3	0.3	0.1	4.1
Styrene	Trace	0.8	0.3	0.7	0.5	0.9	NR	NR
Selected naphthalenes[c]	0.1	7.7	1.4	6.3	3.1	1.9	nd	5.4
Selected PAHs[d]	0.1	4.8	0.8	4.3	2.0	4.7	NR	0.8
Aqueous phase								
Acetic acid	6.1	0.0	4.3	5.5	5.1	3.1	7.8	1.1
Acetol	3.2	0.0	1.7	3.8	0.9	0.9	10.4	0.7
Levoglucosan	0.3	0.0	5.1	4.0	0.3	0.1	1.7	nd
Phenol	1.2	0.3	0.1	1.3	0.6	0.4	nd	nd
Cresols[b]	1.1	0.0	0.2	0.3	0.2	0.1	nd	nd
Water	4.9	97.6	22.9	34.0	76.8	80.2	27	84.9

BTX, Benzene, toluene, and xylene; *nd*, not detected; *PAH*, polyaromatic hydrocarbon.
[a]Values taken from Mullen et al. (2013) for comparative analysis.
[b]Benzene, toluene, *o*- and *p*-xylene.
[c]Naphthalene, 1-methyl and 2-methylnaphthalene.
[d]Indene, biphenyl, fluorene, and anthracene.

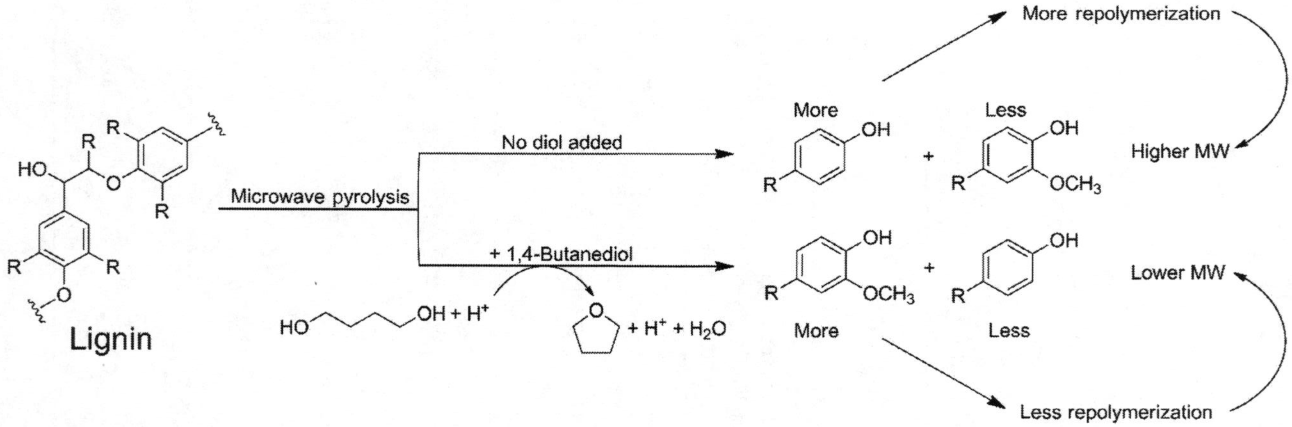

Figure 4.23 Effects of butanediol on lignin pyrolysis pathways.

the yield of monomeric and other reduced molecular weight products are inhibited. A great reduction in the average molecular weight (~85%–90% decrease) of the liquid products was produced in addition to an overall increase in liquid yield. At the optimized ratio of 2:1 lignin to 1,4-butanediol (w/w) the yield of selected monomeric phenols increased threefold on feedstock basis, while the yield of monoaromatic hydrocarbons decreased by approximately 90%. The addition of the diol coreactant also led to a significant shift in selectivity toward the production of methoxy-phenols (guaiacols, syringols) over nonmethoxylated alkyl-phenols (phenol, cresols, etc.) following the scheme proposed in Fig. 4.23. These results show the promise of producing valuable chemicals via copyrolysis of biomass with solvents that are capable of C–H donors.

References

Carlson, T.R., Vispute, T.P., Huber, G.W., 2008. Green gasoline by catalytic fast pyrolysis of solid biomass derived compounds. ChemSusChem 1, 397–400.

Cheng, Y.T., Jae, J., Shi, J., Fan, W., Huber, G.W., 2012. Production of renewable aromatic compounds by catalytic fast pyrolysis of lignocellulosic biomass with bifunctional Ga/ZSM-5 catalysts. Angew. Chem. Int. Ed. 51, 1387–1390.

Dayton, D.C., Hlebak, J., Carpenter, J.R., Wang, K., Mante, O.D., Peters, J.E., 2016. Biomass hydropyrolysis in a fluidized bed reactor. Energy Fuels 30, 4879–4887.

Dorado, C., Mullen, C.A., Boateng, A.A., 2014. HZSM-5 catalyzed co-pyrolysis of biomass and plastics. ACS Sustain. Chem. Eng. 2 (2), 301–311.

Dorado, C., Mullen, C.A., Boateng, A.A., 2015a. Investigation of HZSM5 catalyzed fast pyrolysis of cellulose and plastic blends using isotopic labeling. Appl. Catal. B: Environ 162, 338–345.

Dorado, C., Mullen, C.A., Boateng, A.A., 2015b. Co-processing of agricultural plastic waste and switchgrass via tail gas reactive pyrolysis. Ind. Eng. Chem. Res. 54, 9887–9893.

Güell, A.J., Li, C.Z., Herod, A.A., Stokes, B.J., Hancock, P., Kandiyot, R., 1993. Effect of H_2-pressure on the structures of bio-oils from the mild hydropyrolysis of biomass. Biomass Bioenergy 5, 155–171.

Jae, J., Tompsett, G.A., Foster, A.J., Hammond, K.D., Auerbach, S.M., Lobo, R.F., et al., 2011. Investigation into the shape selectivity of zeolite catalysts for biomass conversion. J. Catal. 279, 257–268.

Mante, O.D., Agblevor, F.A., Oyama, S.T., McClung, R., 2012. The influence of recycling non-condensable gases in the fractional catalytic pyrolysis of biomass. Bioresour. Technol. 111, 482–490.

Marker, et al., 2015. Refinery upgrading of hydropyrolysis oil from biomass. Technical Report, for U.S. DOE Award DE-EE-0005992. The Gas Technology Institute (GTI).

Meesuk, S., Cao, J.-P., Sato, K., Ogawa, Y., Takarada, T., 2011. Fast pyrolysis of rice husk in a fluidized bed: effects of the gas atmosphere and catalyst on bio-oil with a relatively low content of oxygen. Energy Fuels 25, 4113–4121.

Mihalcik, D.A., Boateng, A.A., Mullen, C.A., Goldberg, N.M., 2011. Packed-bed catalytic cracking of oak derived pyrolytic vapors. Ind. Eng. Chem. Res. 50, 13304–13312.

Mullen, C.A., Boateng, A.A., Mihalcik, D.L., Goldberg, N.M., 2011. Catalytic fast pyrolysis of white oak wood in a bubbling fluidized bed. Energy Fuels 25, 5444–5451.

Mullen, C.A., Boateng, A.A., Goldberg, N.M., 2013. Production of deoxygenated biomass fast pyrolysis oils via product gas recycling. Energy Fuels. 27 (7), 3867–3874.

Mullen, C.A., Boateng, A.A., Goldberg, N.M., 2016. Methods for Producing Bio-Oil. US Patent US 9434885 B2.

Raymundo, L.M., Mullen, C.A., Strahan, G.D., Boateng, A.A., Trierweiler, J.O., 2019. Deoxygenation of biomass pyrolysis vapors via TGRP and in-situ and ex-situ thermal and bio-char promoted upgrading. J. ACS Energy Fuels 33, 2197–2207.

Tarves, P.C., Mullen, C.A., Boateng, A.A., 2016. Effects of various reactive gas atmospheres on the properties of bio-oils produced using microwave pyrolysis. ACS Sustain. Chem. Eng. 2016 (4), 930–936.

Tarves, P.C., Mullen, C.A., Strahan, G.D., Boateng, A.A., 2017. De-polymerization of lignin via co-pyrolysis with 1,4-butanediol in a microwave reactor. ACS Sustain. Chem. Eng. 5, 988–994.

Venkatakrishnan, V.K., Degenstein, J.C., Smeltz, A.D., Delgass, W.N., Agrawal, R., Ribeiro, F.H., 2014. High-pressure fast-pyrolysis, fast-hydropyrolysis and catalytic hydrodeoxygenation of cellulose: production of liquid fuel from biomass. Green. Chem. 16, 792–802.

Wang, K., Dayton, D.C., Peters, J.E., Mante, O.D., 2017. Reactive catalytic fast pyrolysis of biomass to produce high-quality bio-crude. Green. Chem. 19, 3243–3251.

Williams, P.T., Nugranad, N., 2000. Comparison of products from the pyrolysis and catalytic pyrolysis of rice husks. Energy 25, 493–513.

Zhang, H., Xiao, R., Wang, D., He, G., Shao, S., Zhang, J., et al., 2011. Biomass fast pyrolysis in a fluidized bed reactor under N_2, CO_2, CO, CH_4, and H_2 atmospheres. Bioresour. Technol. 102, 4258–4264.

Condensed-phase pyrolysis oil upgrading

5

5.1 Introduction

Instability of biomass-derived pyrolysis oil (py-oil, bio-oil) is problematic and has been the main barrier to its large-scale utility as fuel or chemicals. The strong tendency to repolymerize has been attributed to the many oxygenated compounds of the bio-oil with these compounds also contributing to the oil's acidity and rate of viscosity increases over time. Thus far, we have discussed various methods of reducing bio-oil oxygen content right at the incipient pyrolysis steps, including in situ and ex situ catalytic pyrolysis over solid acid catalysts and reactive pyrolysis with a reducing gas atmosphere or both. Although all these processes can be classified as some form of upgrading techniques capable of deoxygenation, they do not completely get rid of all oxygen contained in the product bio-oil as found in petroleum. We saw, for example, that catalytic fast pyrolysis (CFP) can reduce the oxygen content from 35%-50% encountered in thermal-only pyrolysis to 10%−20% range; tail gas reactive pyrolysis (TGRP) can do the same up to about 10%−12% and hydropyrolysis can reduce the oxygen to even lower levels when done over certain molybdenum-base catalysts. Meanwhile, the oxygen content of diesel fuels is about 0.01%, which means that these upgrading methods that are associated with influencing the incipient pyrolysis process can only improve the quality of the resulting bio-oil in terms of its stability but not entirely eliminate it. These are therefore not the final product but rather a biocrude oil intermediate that needs to undergo liquid- or condensed-phase upgrading to remove all the oxygen in the liquid to be compatible with petroleum. An important goal is to reduce the oxygen content in the first pyrolysis step to low levels enough to allow coprocessing of the biocrude with petroleum feeds in a colocated integration. The general method of upgrading the liquid phase includes physical, chemical, and catalytic processes. Physical upgrading includes filtration, solvent addition, emulsion, and distillation, some of these have been used to reasonably achieve meaningful results depending on the starting oxygen content. Chemical upgrading involves processes such as esterification, cracking, and even gasification. Some physical/chemical upgrading technologies such as emulsification, solvent, and/or surfactant addition are simple but go only to improve stability and increase bio-oil's shelf life. Gasification routes such as steam reforming, for example, are pertinent upgrading technologies if gaseous products such as hydrogen are the targeted goal. But to convert pyrolysis oil into fungible fuels that are seamless with existing petroleum high-pressure catalytic upgrading of the liquid phase is necessary. Catalytic upgrading encompasses a variety of pathways, and several methods have been proposed and explored for converting pyrolysis oil into transportable fuels. If based on its elemental composition pyrolysis oil is represented by

Pyrolysis of Biomass for Fuels and Chemicals. DOI: https://doi.org/10.1016/B978-0-12-818213-0.00005-9

a compositional formula, for example, $C_6H_8O_4$, then catalytic upgrading can involve either cracking, where the oxygen is rejected as oxides of carbon, or hydrogenation, also known as hydroprocessing or hydrodeoxygenation (HDO), whereby the oxygen is rejected as water leaving a hydrocarbon (HC) only liquid product.

$$\text{Catalytic cracking } C_6H_8O_4 = 4.6CH_{1.2} + 1.4CO_2 \quad (5.1)$$

$$\text{Hydrogenation } C_6H_8O_4 + H_2 = 6CH_{1.2} + 4H_2O \quad (5.2)$$

As it can be recognized, Eq. (5.1) is similar to CFP over zeolite as we saw earlier but herein involves cracking pyrolysis liquids akin to fluid catalytic cracking (FCC) of petroleum fractions where zeolite acid catalysts have been used to change the decomposition pathways to partially or fully deoxygenate the product via simultaneous dehydration–decarboxylation reactions. Current approaches include coprocessing of the bio-oil by cracking with FCC feedstocks such as vacuum gas oils. This process does not require external hydrogen; it operates at atmospheric pressure and it rejects oxygen through the loss of carbon via gases such as CO_2, but it has the disadvantage of high coke formation. High-pressure hydrogenation (Eq. 5.2), however, requires the supply of H_2 at high pressure; it rejects oxygen as H_2O, and coking of the catalyst may be prevalent during the hydroprocessing. However, unlike cracking, it preserves the carbon. Since biomass is hydrogen deficient, hydrogenation provides needed hydrogen to enrich the HC content. The desired products of catalytic hydrotreating are ideally aliphatic and aromatic HCs, but there can be overconsumption of expensive hydrogen due to undesired hydrogenation of aromatic rings. HDO is best suited for the upgrading of condensed-phase pyrolysis liquids because it is similar to the practice of nitrogen and sulfur removal (hydrodenitrogenation or hydrodesulphurization, respectively), also known as HDN/HDS in existing petroleum refineries and is therefore infrastructure ready. However, the challenge is that petroleum-derived feeds are typically less than 2 wt.% oxygen, whereas that of biomass-derived pyrolysis liquid feeds can be as high as 50 wt.%. The efficacy of the pyrolysis HDO therefore depends on the starting oxygen content.

5.2 Screening hydrodeoxygenation catalysts with model compounds

HDO has been used in the petrochemical refineries for many years, but its application to biomass pyrolysis oil is relatively new and is still evolving. Because pyrolysis oil is a complex mixture of more than 400 compounds, catalysis experts are at the crossroads of understanding the underlying chemical pathways involved in condensed-phase hydrogenation. It is therefore not surprising that the literature is overwhelmed with studies on model compounds that are purported to be representative parts of real pyrolysis oil. While these efforts may not be adequate to design an HDO process, they are useful in screening what HDO catalysts to target for specific compounds found in pyrolysis oils and down-select what might be useful for producing target products as well as their yield and/or selectivity. Si et al. (2017) reviewed the various works on catalytic hydrogenation of pyrolysis oil model compounds and compiled a table of results related to catalyst and HDO products. HDO catalysts by metal sulfides and HDO catalysts by noble metal catalysts are shown in Tables 5.1 and 5.2, respectively.

As shown in the tables, a large portion of the body of work addresses the chemistry of catalytic hydrotreating using model compounds containing oxygen, many of which are relevant to bio-oil. These include compounds such as phenolics and aromatic ethers. The work provides some understanding of the mechanisms one would expect of the HDO of the various oxygenates. For example, it helps to hypothesize that the HDO of furfural, a prevalent chemical in bio-oil, would yield both furfural alcohol and furan via hydrogenation and that a C–C bond cleavage will occur followed by a sequential hydrogenation to tetrahydrofurfuryl alcohol and tetrahydrofuran in addition to which *n*-butanol, *n*-butanal, ethanol, and HC can be derived (Fig. 5.1). One model compound study suggests that because there are two kinds of C–O bond [hydroxyl group (–OH) and methoxy group ($-OCH_3$)], the suggested and logical scheme for the HDO of an oxygenated compound such as guaiacol is that the HDO reactions will proceed sequentially via demethylation, demethoxylation, hydrogenation, and methyl transfer. However, many other studies have indicated that the catalyst composition will play a vital role in the product distribution. Since bio-oils are complex mixtures efforts like these can only inform how the particular model compound behaves. The focus should be on processing of actual bio-oil as the necessary path to commercial success as these compounds coexist in a pyrolysis oil pool. However, the model compound results should provide catalyst selection guidance.

5.3 Hydrodeoxygenation of pyrolysis oils

Hydroprocessing of biomass-derived pyrolysis oils differs from the hydroprocessing of petroleum or coal-derived liquids due to the emphasis on deoxygenation, as opposed to denitrogenation or desulfurization. Hydrogenation of traditional pyrolysis oils has several issues because of the high oxygen content of the starting material. Several technologies have been developed at the US-DOE's Pacific Northwest National Laboratory (PNNL) on their own and in collaboration with UOP, including the use of a two-stage process over CoMo catalysts. However, earlier procedures developed within UOP were only effective in reducing bio-oil formed from thermal pyrolysis containing 40% oxygen to 10% on dry weight basis, with a product yield of 30% (Meier et al., 2013). The historical developments in the hydroprocessing of biomass-derived bio-oils at PNNL were published by Elliott (2007). He surmises that a wide range of heterogeneous catalyst materials tested over a 25-year period, including conventional sulfided catalysts developed for petroleum hydroprocessing as well as precious metal catalysts led to the identification of important processing differences and adjustments required for bio-oil hydrotreating when compared to petroleum feedstocks. These tests ultimately allowed PNNL to conclude that the application of biomass-derived hydroprocessing can be

Table 5.1 Hydrodeoxygenation catalysts by metal sulfides.

Catalyst	Reactor	Reaction conditions			Reactant	Concentration (%)	Major product
		Pressure (bar)	Temperature (°C)	WHSV (h^{-1})			
MoS	Batch	28	350	–	Phenol	71	Benzene, cyclohexane, cyclohexene
CoMoS	Batch	28	350	–	Phenol	98	Benzene, cyclohexane, cyclohexene
NiMoS	Batch	28	350	–	Phenol	96	Cyclohexane, benzene, cyclohexene
NiS	Batch	28	350	–	Phenol	35	Cyclohexane, cyclohexene, benzene
MoS_2	Batch	28	350	–	4-Methylphenol	52	Toluene, 2,4-dimethylphenol
MoS_2	Fixed-bed	40	300	–	Guaiacol	100	Phenol, cyclohexane, benzene, methylcyclopentane
CoMoS	Fixed-bed	40	300	–	Guaiacol	95	Phenol, benzene
NiMoS/Al_2O_3	Fixed-bed	21	280	–	2,3-Dihydrobenzofuran	50	2-Ethylphenol, ethylcyclohexane
CoMoS/MgO	Batch	50	350	–	Phenol	17	2-Cyclohexylphenol, Cyclohexylbenzene, cyclohexanol
MoS_2/AC	Batch	50	300	–	Guaiacol	50	Phenol, catechol, cyclohexene
CoMoWS/SBA-15	Fixed-bed	30	310	24.5	Anisole	38	Phenol, cresol, xylenol
ReS_2/AC	Batch	50	300	–	Guaiacol	40	Phenol, catechol

LHSV, liquid hourly space velocity;
WHSV, Weight hourly space velocity [h^{-1}].
Source: Modified from Si, Z., Zhang, X., Wang, C., Ma, L., Dong, R., 2017. An overview on catalytic hydrodeoxygenation of pyrolysis oil and its model compounds. Catalysts 7(6):169–190.

Table 5.2 Hydrodeoxygenation catalyzed by noble metal catalysts.

Catalyst	Reactor	Reaction conditions			Reactant	Concentration (%)	Major product
		Pressure (bar)	Temperature (°C)	WHSV (h^{-1})			
Ti_xPd/SiO_2	Fixed-bed	20	300	25	Guaiacol	100	Methoxycyclohexanol, cyclohexanol, cyclohexane
Pt/TiO_2	Fixed-bed	20	350	200	Cresol	82	Cyclohexanol, 3-methyl, cyclohexanone, 3-methyl, cyclohexane, methyl
$Pd-FeO_x/SiO_2$	Fixed-bed	60	450	≈1.3	Furans		2-Methyl-decane
Ru/ZrO_2-$La(OH)_3$	Batch	10	170	–	Guaiacol	100	Cyclohexanol, 2-methoxycyclohexanol, methane
$PtMo/Al_2O_3$	Fixed-bed	5	250	–	Cresol	≈95	Methylcyclohexane, 3-methylcyclohexanol
Pd/HZSM-5	Batch	20	200	–	Cresol	100	Methylcyclohexane
Pt/Hβ	Fixed-bed	1	350	2	Cresol	100	Toluene
Ru/CARF[a]	Batch	40	250	–	Guaiacol	97	Cyclohexanol, 2-methoxycyclohexanol
Ru/MWCNT[b]	Batch	40	270	–	Vanillin	100	4-Methylcyclohexanol, 4-propylcyclohexanol, methylcyclohexane
Ru/ZrO_2	Fixed-bed	64	200	1	Propanoic acid	94	Ethane, propane, methane
Ru/Al_2O_3	Fixed-bed	64	200	1	Propanoic acid	58	Propanol, propane, ethane
Ru/C	Fixed-bed	64	190	1	Propanoic acid	94	Ethane, methane, propane
Pt/MgO	Fixed-bed	1	300	11	Guaiacol	6	Phenol, catechol, cyclopentanone

(Continued)

Table 5.2 (Continued)

Catalyst	Reactor	Reaction conditions			Reactant	Concentration (%)	Major product
		Pressure (bar)	Temperature (°C)	WHSV (h^{-1})			
Pt/MZ-5[c]	Fixed-bed	40	200	6 (LHSV)	Dibenzofuran	98	Bicyclohexyl, cyclopentylmethyl, cyclohexane
Pt–Sn/CNF/Inconel[d]	Fixed-bed	1	400	0.3	Guaiacol	100	Phenol, benzene
Pt/γ-Al_2O_3	Fixed-bed	1	300	20	Guaiacol	6	Catechol, phenol, 3-methylcatechol
Zn/Pd/C	Batch	21	150	–	Vanillyl alcohol	>99	2-Methoxy-4-methylphenol
Pt/Al_2O_3	Fixed-bed	29	225	–	Glycerol	90	Ethanol, 1,2-propanediol, carbon dioxide
Rh/SiO_2–Al_2O_3	Batch	40	250	–	Guaiacol	100	Cyclohexane
Ru/SiO_2–Al_2O_3	Batch	40	250	–	Guaiacol	100	Cyclohexane, cyclohexanol
RhPt/ZrO_2	Batch	80	100	–	Guaiacol	100	1-Methyl-1,2-cyclohexanediol, cyclohexanol

LHSV, liquid hourly space velocities [h^{-1}]; *WHSV*, Weight hourly space velocity [h^{-1}].
[a]CARF = Carbon aerogel.
[b]MWCNT = Multiwalled carbon nanotubes.
[c]MZ-5 = Mesoporous ZSM-5.
[d]CNF = Carbon nanofiber.
Source: Modified from Si, Z., Zhang, X., Wang, C., Ma, L., Dong, R., 2017. An overview on catalytic hydrodeoxygenation of pyrolysis oil and its model compounds. Catalysts 7(6):169–190.

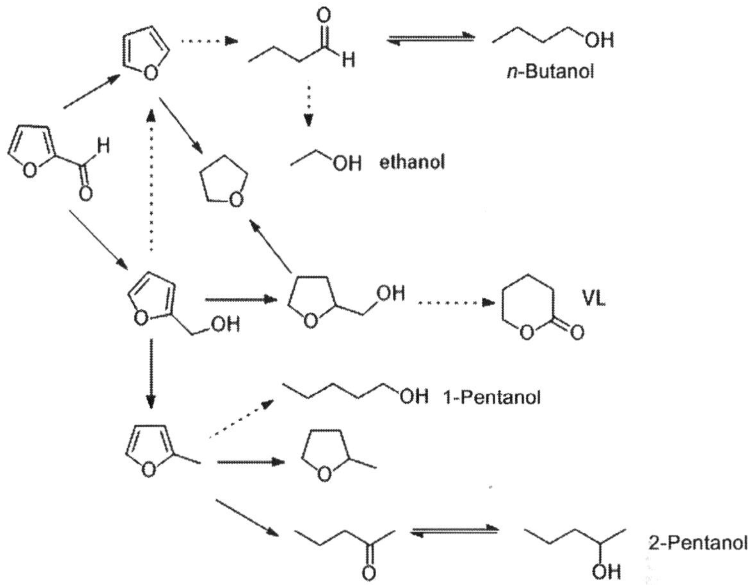

Figure 5.1 Reaction pathway proposed for the hydrogenation of furfural as pyrolysis model compound.
Source: Adapted from Si, Z., Zhang, X., Wang, C., Ma, L., Dong, R., 2017. An overview on catalytic hydrodeoxygenation of pyrolysis oil and its model compounds. Catalysts 7 (6):169–190.

seen as an extension of petroleum processing establishing that system requirements are not far outside the range of conventional hydroprocessing that one would find in the petroleum infrastructure. This finding appears reassuring but certainly not without challenges.

Thus far, among the current gold standards for the HDO of biomass-derived pyrolysis oil in its pure form at 35–45 wt.% oxygen content is the aforementioned two-stage process first developed by PNNL (Elliott and Baker, 1989; Elliott, 2007) (Fig. 5.2). In this development the first reactor would typically comprise a fixed-bed of a noble metal (e.g., Ru) on carbon catalyst operated at 150°C–200°C, the purpose of which is to first stabilize the reactive acids and aldehyde groups of the pyrolysis liquids. The second stage uses high temperature and high-pressure hydrogen (370°C, 70–150 atm) to deoxygenate the stabilized bio-oil over a fixed-bed of sulfided catalyst (e.g., Co/Mo). The two-stage process has demonstrated deoxygenation of bio-oil to <1% oxygen with a time-on-stream of 600–700 hours (including regeneration cycles) before significant plugging occurs. This meets the petroleum industry's operational and oxygen level standards. Over the years, PNNL has developed even a larger and a continuous flow reactor (up to 7 L) for pilot-scale hydrotreatment demonstration of the same two-stage process. Their hydrotreating data shown in Fig. 5.2 indicate that the two-stage hydroprocessing is capable of reducing bio-oil with starting oxygen content of 34.5% to HDO biofuel product with <1% oxygen content in the

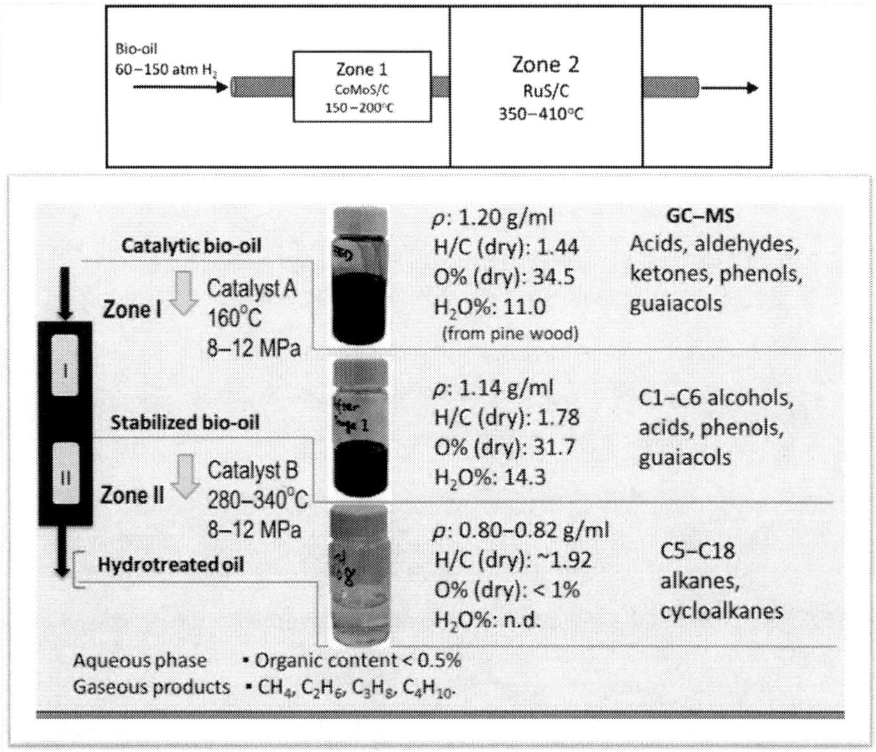

Figure 5.2 Two-stage pyrolysis oil upgrading developed by PNNL. *PNNL*, Pacific Northwest National Laboratory.
Source: DOE Bioenergy Technologies Office (BETO), Project peer review. Available from: https://www.energy.gov/sites/prod/files/2015/04/f21/thermochemical_conversion_abdullah_231401.pdf.

stage-2 reactor after being prestabilized in stage 1 to 31.7%. Although this is carried out with efficient hydrogen consumption, environmental and regulatory constraints imposed on sulfur discharge make the use of sulfide catalysts unfavorable. Nonsulfided catalysts are therefore hotly pursued, but these have not been fully tested in terms of their catalyst lifetimes and their techno-economic potential (Furimsky, 2000, 2013; Bu et al., 2012) leaving room for more research. Instead of a two-stage process, single-step hydroprocessing technologies over noble metal catalysts are vigorously being pursued to fill this void. It means that integrated pyrolysis biorefineries that begin with some of the noncondensed-phase upgrading processes such as CFP and TGRP feeding a low- or mid-level oxygen (MLO) bio-oil into the single-step HDO process (Venkatakrishnan et al., 2014; Boateng et al., 2015, 2016) could be the next-generation technologies to pursue. Other techno-economic opportunities could also involve the use of biomass that inherently produces moderate levels of oxygen content, that is, processes whereby the starting bio-oil feeding into the hydroprocessor is already stabilized. Such includes bio-oil from feedstocks such as algae and guayule.

HDO of fully and partially stabilized fast pyrolysis bio-oils of varying oxygen content produced from various feedstocks over carbon-supported catalysts was studied at the U.S. Department of Agriculture (USDA). This study sought to elucidate the relationships between biomass feedstock type or the suitability of the bio-oils and HDO upgrading (Elkasabi et al., 2014). Pyrolysis oils derived from various agricultural biomass feedstocks, including switchgrass, *Eucalyptus benthamii*, and horse manure (equine waste) created using thermal-only pyrolysis, catalytic (HZSM-5) fast pyrolysis as well as reactive pyrolysis (TGRP), were batch-treated using HDO reactions at 320°C under ~2100 psi H_2 atmosphere for 4 hours, over Pt, Ru, or Pd on carbon supports in a benchtop Parr Series 4598 100 mL reactor. Table 5.3 informs the analyses of the starting bio-oil and that of the HDO product as well as the extent of deoxygenation that a single-step HDO offers based on the feedstock source and commercially available noble metal catalysts.

As shown in the table, the starting bio-oil (HDO feedstock) varies with biomass type in terms of starting oxygen content and so are their related products after hydroprocessing. While the starting oxygen content for the herbaceous (switchgrass) and woody (*Eucalyptus*) biomasses are about the same, at around 33–34 wt.%, that of bio-oil from manure was mid-level at 23 wt.% about 25% less. Importantly, O/C ratio is about 50% less than that of the lignocelluloses. Upon HDO over these commercial noble metal catalysts oxygen content of the resulting oil products could be reduced to 15–20 and 21–23 wt.% for the switchgrass and *Eucalyptus* products, respectively, that is, a reduction of 40%–50% but the oxygen reduction was only about 30% for the manure-derived oil. The starting oxygen contents of the CFP and TGRP bio-oils were equivalent to the HDO product of the switchgrass bio-oil created by thermal pyrolysis. The HDO of these bio-oils over noble metal catalysts failed to provide further oxygen reduction in the batch processing although there were clear advantages over product stability indicating a possible two competing HDO reactions underway, that is, deoxygenation and hydrogenation that need to be analyzed separately for the purpose of carbon efficiency evaluation as they affect the HDO product yield and quality. Elkasabi et al. (2014) resolved this by quantifying and comparing the relative and absolute deoxygenation extent as follows:

$$\%\mathrm{DO}_{\mathrm{rel}} = \left(1 - \frac{x_{\mathrm{O}}}{x_{\mathrm{O},i}}\right) \times 100 \tag{5.3}$$

$$\%\mathrm{DO}_{\mathrm{abs}} = \left(1 - \frac{x_{\mathrm{O}} m_{\mathrm{oil}}}{x_{\mathrm{O},i} m_{\mathrm{oil},i}}\right) \times 100 \tag{5.4}$$

where the variables x_{O} and $x_{\mathrm{O},i}$ represent final and initial oxygen content in wt.%, whereas, m_{oil} and $m_{\mathrm{oil},i}$ represent the final and initial mass of the oil phase, respectively. While $\%\mathrm{DO}_{\mathrm{rel}}$ indicates the percent change in oxygen relative to other

Table 5.3 Product analysis and inspection of bio-oils from different feedstocks before and after hydrodeoxygenation (HDO) over commercial noble metal catalysts (Elkasabi et al., 2014).

	Thermal pyrolysis bio-oil										CFP bio-oil		TGRP bio-oil	
	Switchgrass				*Eucalyptus benthamii*			Horse manure			Switchgrass			
	Raw	**HDO**			**Raw**	**HDO**		**Raw**	**HDO**		**Raw**	**HDO**	**Raw**	**HDO**
	Bio-oil	Pt/C	Ru/C	Pd/C	Bio-oil	Pt/C	Ru/C	Bio-oil	Pt/C	Ru/C	Bio-oil	Pt/C	Bio-oil	Pt/C
Label		SwG-Pt	SwG-Ru	SwG-Pd		Eucal-Pt	Eucal-Pt		Manu-Pt	Manu-Pt		CatSwG-Pt		SwG-rec-Pt
Feed and product elemental composition (wt.% dry basis)														
N	0.52	1.08	0.87	0.84	0.13	0.36	0.29	2.31	2.09	1.96	0.73	0.99	1.96	1.36
C	53.81	75.4	71.64	72.04	59.45	70.87	69.24	67.35	72.14	73.99	76.89	75.04	73.67	72.86
H	5.36	8.48	7.99	7.83	5.99	7.7	7.3	6.82	7.91	7.91	6.1	7.36	5.9	7.13
O	33.21	15.04	19.5	19.29	34.43	21.07	23.17	23.52	17.87	16.14	16.27	16.61	18.46	18.65
Feed and product analysis														
H/C[a]	1.196	1.349	1.339	1.306	1.209	1.303	1.266	1.215	1.315	1.283	0.953	1.177	0.961	1.174
O/C[a]	0.463	0.15	0.204	0.201	0.434	0.223	0.251	0.262	0.186	0.164	0.159	0.168	0.188	0.192
K-F	7.98	2.1	2.245	2.35	6.9	4.31	6.4	7.31	2.95	2.18	4.17	1.72	3.08	1.38
TAN	73	35	44	40	70	37	33	97	48	32	12	16	35	21

CFP, Catalytic fast pyrolysis; *TGRP*, tail gas reactive pyrolysis.
[a]Mol basis; K-F = Karl-Fischer moisture (wt.%); TAN = total acid number (mg KOH/g).

atoms, $\%DO_{abs}$ is an absolute mass balance that represents percent change in oxygen atoms. Fig. 5.3 illustrates that hydroprocessing of bio-oil from switchgrass over Pt/C (SwG-Pt) gives the best deoxygenation and carbon retention in the batch system even if the starting oxygen is high. Bio-oil from manure hydroprocessed over Pt/C catalyst (Manu-Pt) exhibits a greater loss in carbon compared to HDO over R/C (Manu-Ru) (Fig. 5.3A), even though the absolute oxygen losses are similar (Fig. 5.3B). In both catalytic pyrolysis using switchgrass, TGRP bio-oils hydroprocessed over Pt/C catalyst (catSwG-Pt and SwG-rec-Pt, respectively) exhibited negative $\%DO_{rel}$ values, indicating that the extent of deoxygenation is small compared to other atomic losses, primarily carbon.

Aside from carbon and oxygen balance, hydrogen consumption during hydroprocessing not only informs the mechanism of HDO but also the viability and potential

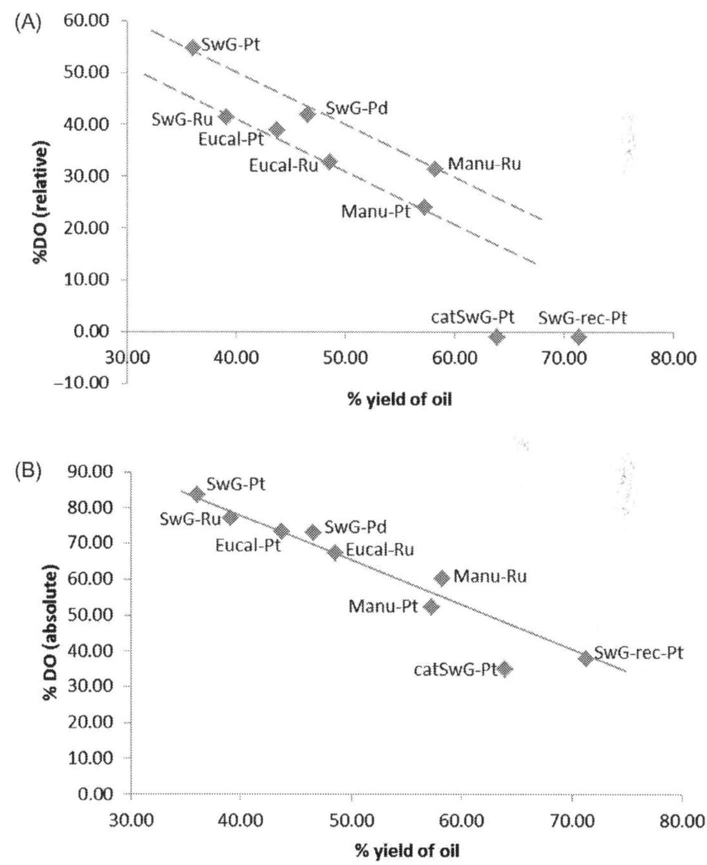

Figure 5.3 (A) Relative deoxygenation and (B) absolute deoxygenation percentages are plotted with respect to the associated product yield of each HDO reaction. Switchgrass with platinum performs the best, in terms of maximizing deoxygenation and minimizing carbon loss (Elkasabi et al., 2014). *HDO*, Hydrodeoxygenation.

success of a continuous process and the economics associated with hydrogen use. Since the total moles of hydrogen consumed in a batch process comes from either the reaction atmosphere or directly from that supplied by a reservoir tank typically used to maintain isobaric conditions, the relationship between available hydrogen and that consumed may be established from the ideal gas law as:

$$n_{\text{consumed}} = \Delta n_{\text{reservoir}} + \sum_{i=1}^{n} x_i n_{\text{gas}} \tag{5.5}$$

where x_i is the mole fraction of gas component i. This allows one to relate hydrogen consumption to the extent of deoxygenation (Fig. 5.4). The dotted line should represent the stoichiometric equivalent of the moles of hydrogen needed for every oxygen lost as water. The fact that the points are above the line indicates oxygen loss will include physical transfer to the aqueous phase, and since hydrogen consumption includes direct hydrogenation, the relative efficiency of hydrogen consumption does not purely reflect HDO reactions. The most efficient processes should be the points that lie within the lower right section of the plot. The two processes in the set that are most efficient are therefore the hydroprocessing of manure bio-oil over ruthenium on carbon catalyst (Manu-Ru) and switchgrass bio-oil hydroprocessed over platinum on carbon support (SwG-Pt). Of the two, however, the location on the plot shows that the latter, SwG-Pt, gives the best relative deoxygenation for the associated H_2 consumed. The plot also shows that *Eucalyptus* bio-oil (hardwood biomass-derived bio-oil) was most inefficient with regards to hydrogen consumption probably due to their high S:G ratio typically known to producing much more permanent, noncondensable gases (NCG).

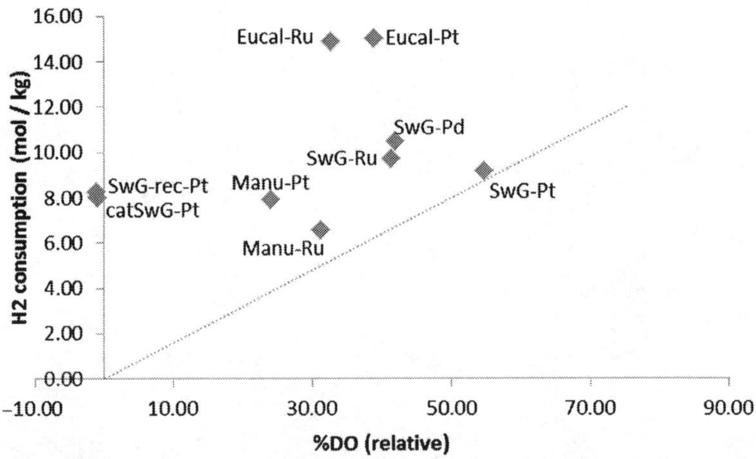

Figure 5.4 Plot of hydrogen consumption for each experiment, as function of associated deoxygenation percentage (relative). The dotted line represents the theoretical stoichiometric consumption of one hydrogen molecule for every oxygen atom removed.

Consequentially, high-syringol feedstocks such as *Eucalyptus* will require different catalysts such as Ru/TiO_2 that can transalkylate methyls from the methoxy groups to the ring or employ a continuous rather than batch processing. The overall performance of the feedstock and HDO over noble metal on carbon-supported catalysts can be evaluated by defining total extent of condensed-phase upgrading as a function of the degree of absolute deoxygenation and degree of hydrogenation (DH) (Elkasabi et al., 2014):

$$\text{TEU} = \frac{\%\text{DO}_{\text{abs}} + \%\text{DH}}{2} \tag{5.6}$$

where the DH is calculated as:

$$\%\text{DH} = \left(\frac{(\text{H}:\text{C}) - (\text{H}:\text{C})_i}{(\text{H}:\text{C})_{\max} - (\text{H}:\text{C})_i}\right) \times 100 \tag{5.7}$$

With this definition, one can draw some conclusions on the effect of the bio-oil source and the HDO catalyst for the batch system by plotting the relative extent of condensed-phase upgrading for the various starting materials in terms of feedstock and their relative starting oxygen contents (Fig. 5.5). Regardless of the yield, catalyst, or feedstock, all the samples tested do fall on a single linear trend line. Just as we saw for hydrogen consumption, SwG-Pt gives the highest TEU, but its associated yield is the lowest. An optimal set of conditions will be a compromise between high yields and high deoxygenation rates.

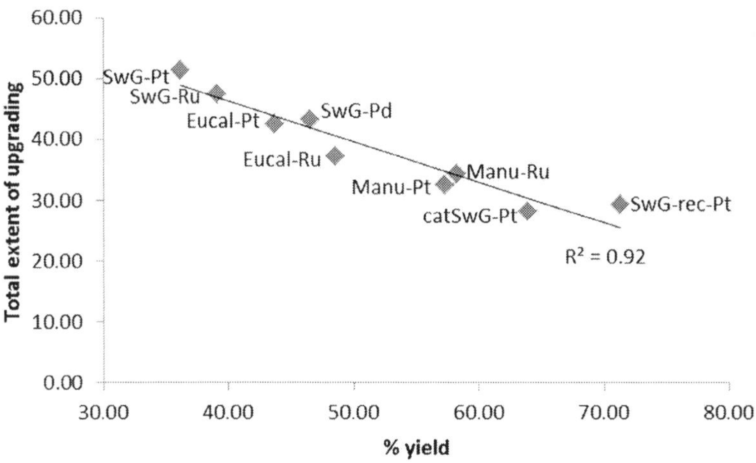

Figure 5.5 Total extent of reaction plotted with respect to the associated yield of upgraded bio-oil. The trend line is nearly independent of the feedstock type and/or catalyst (Elkasabi et al., 2014).

Overall the mass balances reveal little or no information about coke formation for the HDO process. We also observe that switchgrass bio-oil hydroprocessed over Pt/C performs best in terms of hydrogen consumption efficiency, deoxygenation efficiency, and types of upgraded bio-oil compounds. Woody biomass, that is, *Eucalyptus* feedstocks consistently consume more than twice the normal amount of hydrogen, perhaps due to the elevated syringol content characteristic of hardwoods. CFP bio-oils with MLO content deoxygenate poorly over Pt/C in batch but hydrogenate more extensively than the others. Although the relative deoxygenation (% DO_{rel}) may vary based on feedstock and catalyst used, the absolute deoxygenation (%DO_{abs}) depends only on the overall product yield with the total extent of upgrading (hydrogenation + deoxygenation) remaining independent of feedstock and catalyst used.

We will visit more efficient, continuous HDO upgrading of low-oxygen content bio-oils later, but for now let us stay with batch hydroprocessing of pyrolysis oils with the use of some of the upcoming and novel catalysts other than the commercial ones employed for the abovementioned experiments. Some of these novel HDO catalysts have been explored under model compound studies presented earlier. For this purpose a hydrothermally stable metal catalysts with controlled particle size and distribution were synthesized, with the goal of determining which of these catalysts can selectively catalyze the production of aromatics from fast pyrolysis bio-oil (Elkasabi et al., 2017). For this purpose, both precious and base transition metal catalysts (Ru, Pt, Ni, Cu, 2/1-Pt:Ru, NiCu) were deposited on mesoporous alumina and on carbon, respectively, using strong electrostatic adsorption (SEA) as characterized in Table 5.4. The underlying hypothesis stems from the thesis that controlled bimetallic combinations, precious and/or base metal, have the potential to enhance the HDO behavior compared with the commercial noble metal catalysts. With a successful deposition of the metal particles in the nanoparticle size range, HDO of pyrolysis bio-oil was carried out for 3 hours at 300°C in the same batch Parr reactor in an aqueous environment. Tables 5.5 provides a summary of the key results in using these in-house SEA catalysts for the HDO of traditional pyrolysis oils. The product distribution shows aromatic HCs produced at a greater extent than those of Ni/C, or Cu/C, or commercial Ru/alumina with the SEA catalysts generating significantly less NCG than commercial Ru/alumina, Ni/C, or Cu/C, except for Pt/(mesoporous alumina). However, the bimetallic 2/1-Pt:Ru/(mesoporous alumina) did not exhibit significantly greater activity than Pt or Ru, whereas NiCu demonstrated improved oil quality and yields over single-metal Ni and Cu. Post-HDO analysis found that while partial conversion of mesoporous alumina into boehmite phase occurred, the catalysts' particles remained between 2 and 3 nm postreaction, an indication of a high degree of anchoring confirming catalyst stability. Also, the base metal particles tended to sinter more over a carbon support.

The foregoing shows that heterogeneous catalysts produced by the SEA method can deoxygenate biomass-derived pyrolysis oils toward products that are different from that of traditional commercial catalysts. This is evidenced by the significantly greater concentrations of aromatic compounds and marked absence of saturated compounds (whether intermediate or final products). The result also teaches that

Table 5.4 Hydrodeoxygenation catalysts tested, with composition and properties noted (Elkasabi et al., 2017).

Metal support	Ru/mA	Pt mA	PtRu/mA	Ni/mA	Cu/mA	NiCu/ mA	Ni/C	Cu/ C	NiCu/C
Loading ratio (wt.%)	3.0	4.2	2.0/1.0	2.0	2.0	1.0/1.0	2.0	2.0	1.0/1.0
Surface area (m^2/g)	245	245	237	226	226	226	280	280	280
Particle size(s) (nm)	< 1.5	< 1.5	< 1.5	4.2 (NiO)5.6 (NiAl)	3.7	4.4	3.7	3.8	3.3
Metal	Ru		Pt		PtRu		NiCu		NiCu
Support	Al_2O_3 (commercial)		Al_2O_3 (commercial)		mA (DI)		mA (DI)		C (DI)
Loading ratio (wt.%)	5.0		5.0		2.0/1.0		1.0/1.0		1.0/1.0
Surface area (m^2/g)	245		245		226		226		226
Particle size(s) (nm)	5.6		4.7		7.5 (Pt)4.5 (Ru)		2.4–4.6		5.9 (Ni)8.6 (Cu)

mA, Mesoporous alumina; *DI*, dry impregnation (synthesized by dry impregnation).

Table 5.5 GC–MS concentrations of some dominant compounds detected in bio-oil hydrodeoxygenation reactions (Elkasabi et al., 2017).

Cat:oil (g/g)	Pt/mA 0.373	Ru/mA 0.267	2Pt1Ru/mA 0.267	Commercial Ru/A 0.267	Commercial Pt/A 0.291	RuPt/mA (DI) 0.286	2Ru/mA 0.268	Bio-oil
Acetic acid (wt.%)	2.07	1.67	1.92	0.68	1.14	0.76	1.50	5.54
Acetol	0.03	0.01	0.02	–	–		–	2.06
Furfural	0.07	0.02	0.17	0.02	–		–	0.7
Levoglucosan	0.02	0.02	0.04	0.02	–		–	2.66
Cyclopentanone	0.28	0.12	0.25	0.17	0.25	0.22	0.27	0.03
2-Methyl-2-cyclopenten-1-one	0.46	0.01	0.58	0.01	0.28	0.56	–	0
2-Methyl-cyclopentenone	0.43	0.34	0.50	0.61	0.53	0.27	0.58	0.02
Phenols/cresols	4.37	2.22	6.79	3.42	3.68	3.48	3.44	1.5
Methoxyphenols	0.55	0.30	0.81	0.41	0.85	0.79	0.70	0.12
Aromatic hydrocarbons	0.14	0.11	0.21	0.09	0.56	0.38	0.43	0.05
BTEX	0.08	0.07	0.09	0.07	0.45	0.30	0.33	0.05

BTEX, Benzene, toluene, ethylbenzene and xylenes.

particle sizes for precious metals exhibit minimal sintering under hydrothermal conditions, while base metal catalysts exhibit significant sintering and dealloying. Precious metal SEA catalysts on mesoporous alumina remain active similar to bimetallic catalysts. In addition, base metal SEA catalysts on carbon support exhibit enhanced reactivity and production of desired aromatic compounds than their single-metal counterparts; however, the single metal–base metal commercial control catalysts such as Ru/C show some activity, albeit at poor product yields. We learned that the utilization of bimetallic base metal SEA catalysts for HDO will require further stabilization steps. With respect to deoxygenation, however, generally, the SEA catalysts did not fare differently from the commercial catalysts. Bimetallic nanoparticle catalysts did not perform differently indicating that single-stage hydroprocessing does not match the two-stage HDO process with commercial noble metal catalysts nor does the bimetallic nanocatalysts when it comes to real pyrolysis oil HDO even if the starting oxygen content is low. To avoid sulfide catalysts, other two-step approaches combining distillation followed by HDO or vice versa are doable and perhaps economic if the oxygen content of starting bio-oils are lower than that found in traditional pyrolysis oils.

5.4 Distillation

Irrespective of the similarities in system requirements between petroleum and biomass-derived upgrading as pointed out by Elliott (2007), little, if any, continuous separations or fractionation have been incorporated into the biorefinery process flow stream for efficient improvement of downstream processes such as HDO, as would normally be practiced in a petrochemical refinery. A significant component of petrochemical refining relies on distillation as the first fractionation step. But this has been elusive for the pyrolysis biorefinery because of the thermal instability of the feeds. Depending on the quality of oil processed, petroleum is fractionated into several different fractions of varying volatility, which substantially streamlines downstream process variables and combinations thereof. Fractionation into streams of narrow product distribution also makes it possible to improve catalytic HDO yields. However, the reason why distillation has not been applied to pyrolysis biorefinery is because of thermal instability created by the high concentration of reactive oxygenates. At 30–45 wt.% oxygen the heating of pyrolysis oils results in oligomerization of the oxygenated molecules thereby increasing the molecular weight. Laboratory studies on traditional bio-oil distillation indicate a dominance of unwanted reactions and loss of product due to the elevated temperatures. Some research groups, including USDA, have successfully used MLO bio-oils produced from CFP and other reactive pyrolysis such as TGRP as distillation feeds with the distillates feeding into downstream HDO (Elkasabi et al., 2019). The general observation is that prior simple distillation of the bio-oil can increase HDO yields by 30%. One major cause is attributed to the fact that the heavy-molecular-weight compounds are separated a priori. It has been demonstrated that a continuous distillation process based on short-path flash distillation is possible when

MLO bio-oil intermediates are employed as feeds and that a priori fractionation would reduce the eventual costs of HDO catalyst consumption. This means integrated processes whereby the HDO step is preceded by some of the preupgrading technologies such as CFP and TGRP feeding a low-oxygen and stabilized starting bio-oil can undergo a complete deoxygenation in a single step using on-the-shelf noble metal catalysts. This can result in a more carbon-efficient, economically viable, and environmental friendly thermo-catalytic biorefinery. The environmental benefit stemming from the use of noble metal catalysts rather than sulfide-based catalysts used for high-level oxygen bio-oil intermediates in the two-stage process. The following studies reported by the USDA group present us with some interesting results.

A one-stage vertical flash drum was successful in distilling pyrolysis oils with varying oxygen contents as would be produced by some of the pyrolysis technologies mentioned earlier. Bio-oils produced by CFP performed the best with an overall organic yield of ~80 wt.%, due primarily to the low contents of oxygenated residues such as levoglucosan. Bio-oils with intermediate oxygen levels, typically, ~20 wt.% oxygen could be continuously distilled toward high yields instead of distilling in batch. This allows for the possibility of continuous production of useful residual solid coproducts such as biocoke due to process intensification that comes with continuous distillation.

Table 5.6 presents some of the results showing the overall mass distribution and mass balance closure of the flash distillation products of bio-oils from various biomasses with varying oxygen content, based on averages of values at steady state and/or total collected post-run with a default drum temperature at 380°C (Elkasabi et al., 2014).

Based on the recorded total yields shown in the table, a total mass balance closure could be established at as high as 84%–90% when MLO bio-oil feeds are used as compared with the high-oxygen traditional bio-oil from thermal-only pyrolysis. In this case the 10%–16% yield losses are attributed to issues such as production of NCG

Table 5.6 Mass balance closures for hydrodeoxygenation (HDO) experiments (Elkasabi et al., 2014).

Bio-oil source	Process	Feed (g)		Products (g)			% Closure
		Bio-oil	H_2	Gas	Organic	Aqueous	
Switchgrass	HDO-Ru/C	9.08	0.17	1.45	3.56	2.11	76.9
Switchgrass	HDO-Pd/C	8.37	0.17	2.53	3.9	1.73	95.5
Switchgrass	HDO-Pt/C	9.53	0.17	1.99	3.45	1.96	76.4
Manure	HDO-Pt/C	10.25	0.16	1.00	6.42	1.19	82.7
Manure	HDO-Ru/C	8.85	0.12	0.85	5.17	1.46	83.4
Eucalyptus	HDO-Pt/C	8.82	0.27	3.76	3.86	0.84	93.2
Eucalyptus	HDO-Ru/C	8.67	0.25	2.76	4.19	0.42	82.6
Switchgrass	CFP/HZSM-5	9.25	0.15	1.35	5.93	1.43	92.6
Switchgrass	TGRP	9.03	0.15	1.27	5.46	1.09	85.3

CFP, Catalytic fast pyrolysis; *TGRP*, tail gas reactive pyrolysis.

due to cracking, adhesion of oil tar/coke to the walls, and/or losses of coke during post-run collection. Generally, however, continuous distillation is likely to produce a relatively small yield of aqueous phase, as the original starting oils have little water when more stabilized bio-oils such as CFP and TGRP are used. As the oxygen content of bio-oil increases, the yield of aqueous phase tended to increase, which indicates the proportionality of bio-oil reactivity. Simultaneously, a decrease in recovered organic distillates may occur likely because the organic components in high-oxygen bio-oil undergo condensation polymerization to produce the excess water. When comparing total organic yields with steady-state organics yields, the biggest differences occurred for bio-oil with moderate-level oxygen such as that produced from hardwood using the TGRP process; such bio-oil did not reach an entirely distinct steady state. Nonetheless, the steady-state yield recorded is 20% greater than the total overall yield, which starkly contrasts with traditional bio-oil with greater than 30 wt.% oxygen.

5.4.1 Distillation pre- and posthydrodeoxygenation

Medium to highly deoxygenated bio-oil intermediates afford the opportunity to carry out process intensification in the biorefinery whereby distillation or other separation technologies can be carried out prior to HDO or vice versa and/or even both to maximize product targeting for both fuels. Fig. 5.6 depicts the downstream unit operations in a typical pyrolysis biorefinery flow sheet. It includes physical separation of solids, fractionation (by distillation) followed by HDO, and product distribution by distillation cuts made possible due to high-level stability of the bio-oil intermediates produced from guayule bagasse using the TGRP process.

Table 5.7 presents the characterization of the centrifuged, distilled, and continuous HDO products from the products of on-the-shelf noble metal catalysts on carbon support while Table 5.8 shows the distillation products of the Ru/C-catalyzed HDO of bio-oil from guayule bagasse, residue of a rubber plant.

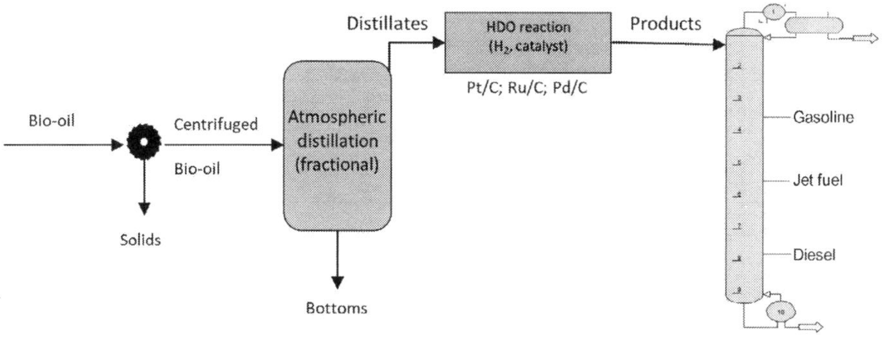

Figure 5.6 Process flow for upgrading of the organic phase of guayule bagasse bio-oil derived from TGRP. The yields are typical over Pt/C, Ru/C, and Pd/C. *TGRP*, Tail gas reactive pyrolysis.

Table 5.7 Characterization of the centrifuged, distilled, and continuous hydrodeoxygenation products from the on-the-shelf noble metal catalysts on carbon support (Boateng et al., 2015).

	Centrifuged oil	Distilled oil	Pt/C	Pt/C	Ru/C	Ru/C
Voil (mL/min)			0.5	0.75	0.45	0.64
H_2 flow (sccm)			3000	3000	3000	3000
P (psi)			1830	1830	1880	1880
LHSV (1/h)			0.46	0.69	0.46	0.66
% Yield			84	79	63	60
Density (g/mL)	1.04	0.94	0.849	0.858	0.81	0.81
K-F (wt.%)	11.27	1.12	0.35	0.44	0.23	0.22
TAN (mg KOH/g)*	32	31.6	5.90*	7.5	20*	14.2*
HHV (MJ/kg)	31.7	38.2	41.3	41.4	43.4	42.9
wt.% O (dry)	12.56	10.42	4.78	5.79	0.66	0.21
H/C (mol)	1.487	1.37	1.7	1.681	1.73	1.71
O/C (mol)	0.124	0.098	0.043	0.053	0.004	0.001
GC–MS (wt.%)						
Acetic acid	0	0.66	0	0	0	0
Acetol	0	0.25	0	0	0	0
Phenols	6.12	7.88	7.85	9.84	0.67	1.77
Naphthenes	0.11	0.13	11.39	8.08	7.93	6.9
Paraffins	0.06	0.05	6.74	5.485	3.06	2.75
Aromatic hydrocarbons	6.72	12.09	22.08	23.57	17.64	17.07

*Reported titration end point value for TAN. *HHV*, High heating values; *K-F*, Karl-Fischer; *LHSV*, liquid hourly space velocities; *TAN*, total acid number.

Table 5.8 Distillation products of the Ru/C-catalyzed hydrodeoxygenation of the said guayule bio-oil (Boateng et al., 2015).

Fraction	1	2	3	4	5	6
Density	0.73	0.78	0.84	0.87	0.88	0.86
HHV	43.2	44.4	44	41.1	44.1	45.0
K-F (wt.%)	0	0	0.24	0.13	0.07	0.15
% Distilled	13.9	16.5	20.1	17	18.4	5.6
T_{ov} (°C)	30–80	80–105	105–130	130–155	155–160	160–260 (170 + vac)
wt.% (dry) N	0.14	0.11	0.11	0.1	0.07	0.07
C	83.95	85.55	84.56	85.27	85.89	85.71
H	13.42	12.48	11.47	10.59	11.49	12.11
O	2.49	1.86	3.86	4.03	2.55	2.11
H/C (mol)	1.918	1.75	1.627	1.487	1.605	1.695
O/C (mol)	0.022	0.016	0.034	0.036	0.022	0.019
GC–MS wt.%						
Acetic acid	0	0	0	0	0	0
Phenols	0.05	0.53	1.71	2.77	0.05	0.06
Naphthenes	17.85	19.39	13.71	5.44	0.14	0.01
Paraffins	14.18	5.15	2.5	3.39	> 5.25	> 2.75
Aromatic hydrocarbons	13.29	28.49	24.62	5.58	2	0.28
Octane number (est)	83.5	92.48	90.09			
Cetane number (est)			63.5	84.3	95.33	106.67

HHV, High heating values; *K-F*, Karl-Fischer.

For all the hydrotreating catalysts tested, reasonable yields and good fuel properties are attained when liquid hourly space velocities are between 0.46 and 0.61 $hour^{-1}$, which is significantly larger than the standard 0.1 and 0.2 h^{-1} required for effective hydrotreatment of traditional bio-oil (Elliott, 2007). Since both processes use similar catalysts, this large improvement is suggested to be a reflection of the starting bio-oil quality and not of the catalyst. From the catalysts tested on the guayule bagasse pyrolyzed bio-oil, the ruthenium-catalyzed HDO produced a product with the lowest moisture, oxygen, and acidity, as well as having the highest organic phase liquid yield. As indicated by GC–MS results in Table 5.8, all catalysts produced significant amounts of fuel-quality compounds, including naphthenes, paraffins, and aromatics, with Ru/C HDO yielding the largest concentrations.

Although all bio-oil HDO products were distillable, the results from the distillation of the optimal Ru/C-catalyzed HDO products are interesting (Table 5.8). For all products emanating from the hydroprocessing of the centrifuged raw bio-oil, distillation resulted in greater than 95% total yield for which most could be quantified by GC–MS. Octane and cetane numbers reported for each fraction are averages of the dominant components, based on calculated numbers and/or reported literature values for the said compounds (Albahri et al., 2002; Do et al., 2007). Based on boiling points, only the relevant parameters such as octane and/or cetane numbers accompany the data for each fraction. Fractions 4–6 account for heavier straight-chain compounds in the jet/diesel fuel range (Fig. 5.7).

The chromatograms for fractions 5 and 6 indicate that they are nearly entirely straight-chain paraffins (indicated by C_x values), which is very atypical of pyrolysis oil derived from regular lignocelluloses. The composition of the guayule bagasse,

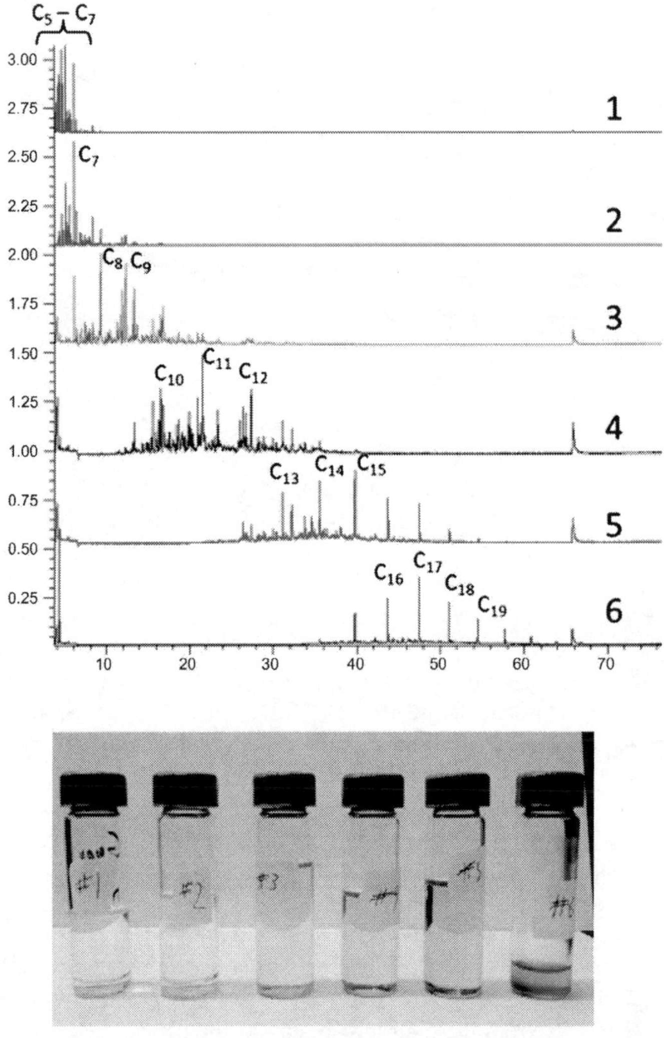

Figure 5.7 Distillate fractions and carbon number distribution of organic phase of guayule bio-oil hydrogenated over Ru/C HDO catalyst. *Approximate cuts*: 1, 2, 3—gasoline; 4, 5—jet/diesel; 6—diesel (Boateng et al., 2015). *HDO*, Hydrodeoxygenation.

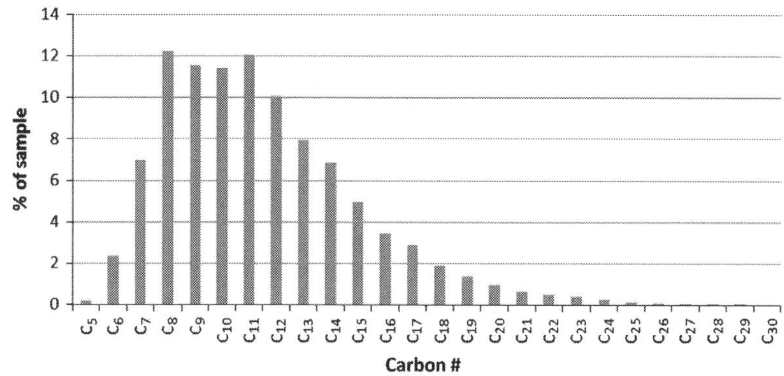

Figure 5.8 Molecular weight distribution by carbon number of the upgraded guayule TGRP bio-oil. *TGRP*, Tail gas reactive pyrolysis.

particularly its resin content, contributes to unusually beneficial outcome making guayule particularly valuable for specialty (e.g., aviation) fuel applications, which require straight-chain compounds. Fig. 5.8 illustrates how the final HDO distillates approach elemental properties required for fuel-grade HCs. Extra oxygen in fraction 4 is likely due to residual unconverted phenols, which are effectively separated from the remaining fractions.

When fractional distillation precedes HDO, the product properties get significantly closer to petroleum quality fuel in terms of their higher high heating values, lower density, and much lower moisture. In addition, distillation directs the very high-molecular-weight compounds into the residual bottoms, which normally would end up in coking, plugging, and deactivating the catalyst bed. Distillation followed by HDO produces higher HDO yields, more deoxygenated products, and product with greater volatility as measured by a higher percentage of compounds that are GC detectable.

5.5 Infrastructure compatibility of hydrodeoxygenation products

The infrastructure readiness of the products of pyrolysis biorefinery that starts with MLO or highly deoxygenated bio-oil intermediates such as TGRP of guayule bagasse is assessed using ASTM Standards developed for traditional fuel-quality control. Table 5.9 is the analysis of the mixture consisted of the HDO products catalyzed by Ru/C and Pt/C. The ASTM results indicate a fuel moisture content of 2216 ppm, a TAN of 1.08 mg/g KOH, 33.8°API gravity, <1 specific gravity, and copper strip corrosion of 2C gives a general compatibility with refinery standard. However, the presence of some sulfur at 335 ppm is reflective of the relatively high

Table 5.9 Summary of ASTM tests performed on a mixture of upgraded guayule tail gas reactive pyrolysis bio-oil.

ASTM method	Specification	Value
D-6304	Karl-Fischer moisture (ppm)	2216
D-664	TAN (mg/g KOH)	1.08
D-4052	API gravity (°API)	33.8
D-4052	Density (g/mL)	0.8553
D-130	Copper strip corrosion	2C
D-2622	Sulfur (ppm)	335
D-4737	Cetane number	21.6

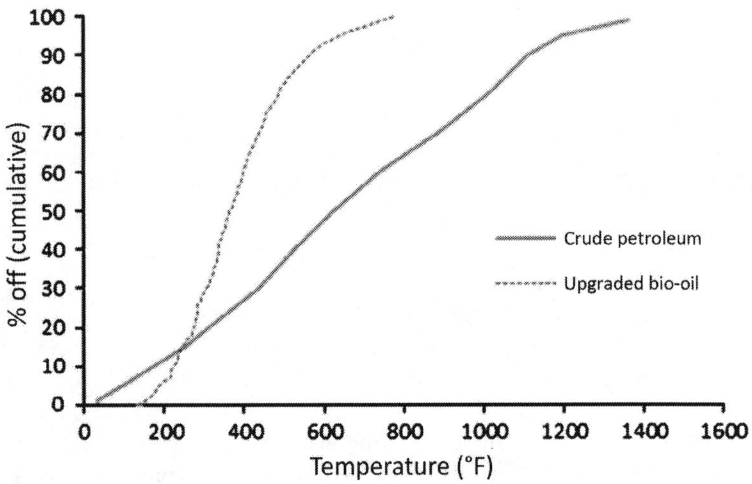

Figure 5.9 Distillation curve of the upgraded guayule TGRP bio-oil, in comparison with typical crude petroleum. *TGRP*, Tail gas reactive pyrolysis.

sulfur content of the guayule bagasse feedstock (0.12% by ASTM D-4239). Sodium sulfite is normally used as an antioxidant in the latex extraction process, contributing to high sulfur in the bagasse (Cornish, 1996). The low cetane number of 21.6 established with ASTM D-4737 is only an indication that the majority of the sample (66%) falls at C_{12} and below, which falls within the gasoline (naphtha) range that we verified by simulated distillation (Fig. 5.9). Beyond C_{12} the molecular weights increase through the diesel range, with C_{37} being the highest observable molecular weight. The greatest fraction of naphtha falls within the C_8-C_{10} range. Based on this figure, it can be conclusively deduced that all of the high-molecular-weight matter from the bio-oil has been eliminated, in contrast to typical crude petroleum. The carbon distribution of the heavy fraction (fraction 6) is presented in Fig. 5.10.

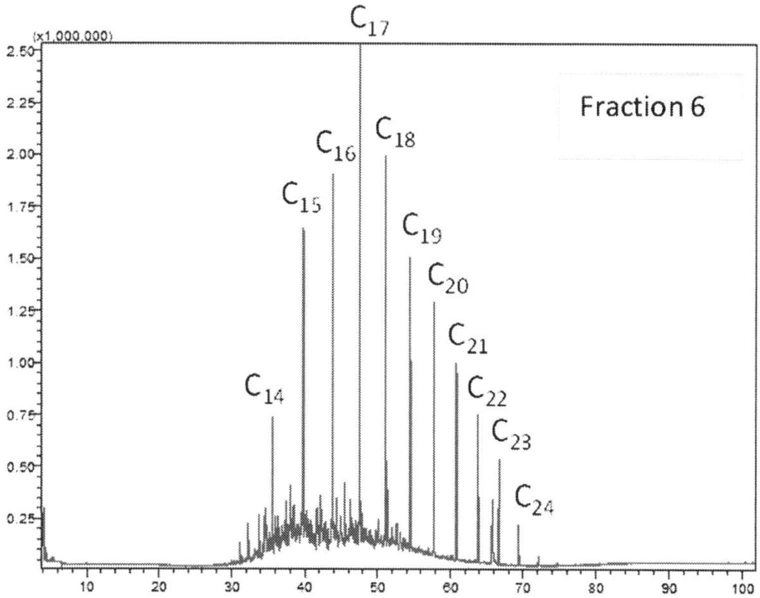

Figure 5.10 Carbon number distribution of heavy distillate.

5.6 Extraction

Owing to the narrow product distribution found within MLO bio-oil intermediates, it is possible to use aqueous-phase NaOH extraction to separate their distillates into phenolic salts and HCs. This allows for the production of HCs that consist primarily of mono- and bicyclic aromatics, which are mostly oxygen-free (<1.0 wt.%), and possess low moisture (<1.0 wt.%) and low acidity (TAN < 5.0 mg KOH/g). The phenolic salts can be reacidified to produce phenols with low moisture (~2.5 wt.%) and with narrow product distribution. To produce naphtha compounds the HC fraction can simply undergo a mild hydrogenation step with sponge nickel base metal catalyst in water. The resulting product can be appropriate for direct use as drop-in fuel and/or refinery blend-stock thereby increasing carbon utilization, efficiency, and product diversity. A potential path to extraction of MLO intermediate extraction is depicted in Fig. 5.11 (Elkasabi et al., 2015).

5.7 Transfer hydrogenation

Transfer hydrogenation is an alternative to the conventional high-pressure hydrogenation, which often involves extreme pressures, temperatures, and hydrogen sources. Transfer hydrogenation is a catalytic addition of hydrogen from an organic

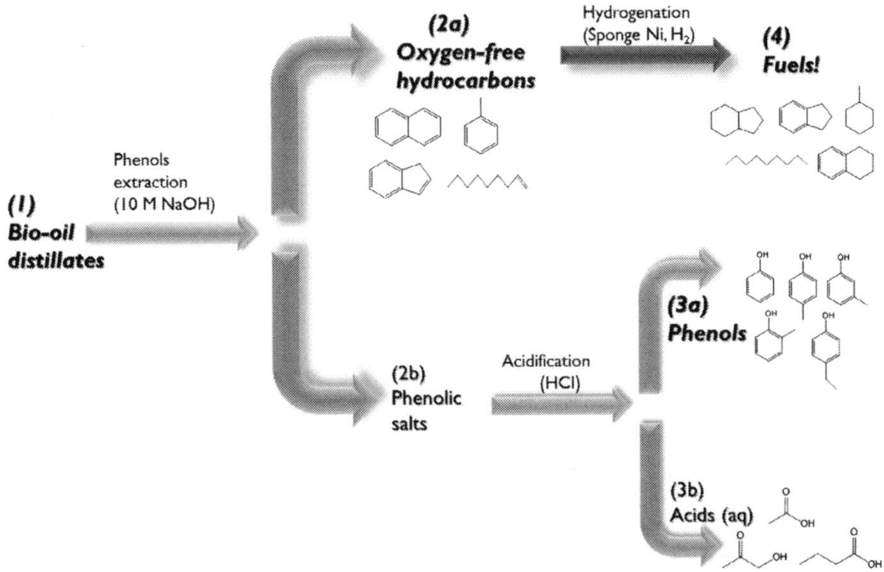

Figure 5.11 Processing steps for extraction of MLO bio-oil into hydrocarbons and phenols.

molecule rather than H_2, typically called reducing agent or hydrogen donor. An example application of transfer hydrogenation is coal liquefaction that uses tetralin as a donor solvent. Commonly used metals for heterogeneous and homogeneous catalytic transfer hydrogen are palladium, ruthenium, rhodium, iridium, and nickel. Recognized reducing agents include the use of formic acid to form CO_2 and alcohols to form ketones (isopropanol to acetone) (Brieger and Nestrick, 1974).

Transfer hydrogenation may take place at atmospheric pressure, and in many different solvents at temperatures as low as 20°C. Instead of using hydrogen gas for the HDO step of the biorefinery, hydrogen donors such as alkali formates and formic acid, some alcohols, and a plethora of organic molecules may be used to supply hydrogen. Preferred donors are typically stable, easily storable and transportable solids or liquids. The original appeal envisioned for pyrolysis biorefinery is that many of these potential hydrogen donors are already present in pyrolysis oil so little, if any, additional hydrogen donor compounds may need to be added to the bio-oil to assist in its hydroprocessing/hydrotreating. Furthermore, it is anticipated that operation at mild temperature conditions would facilitate stabilization of bio-oil with reduced competition from thermally induced reactions that lead to instability and increased viscosity. But the use of hydrogen transfer for bio-oil stabilization has not shown much promise (Elliott and Baker, 1987). According to Elliott, although triethylsilane and palladium/carbon can lower the viscosity of bio-oil, the method is not considered to be economical due to the lack of an economical process to recycle or regenerate the hydrogen donor.

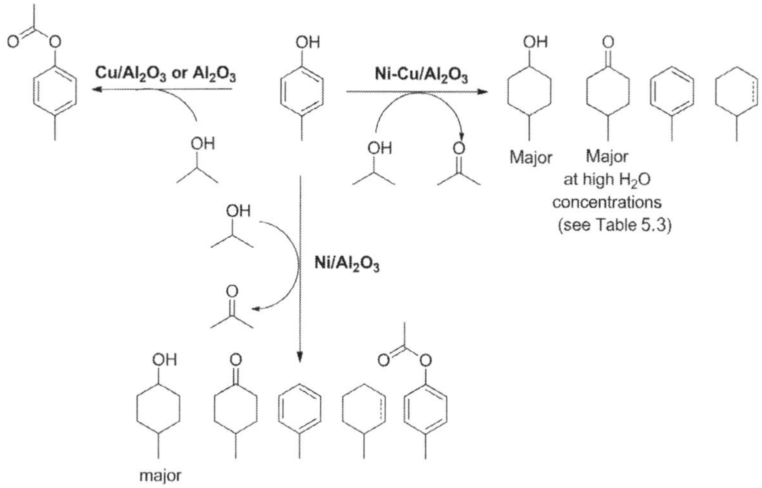

Figure 5.12 Transfer hydrogenation of *p*-cresol using isopropanol over Al_2O_3, Ni, Cu, and Ni–Cu/Al_2O_3.

In a USDA study, catalytic transfer hydrogenation for stabilization of bio-oil oxygenates was tested on *p*-cresol and furfural, model bio-oil compounds, as well as real bio-oil over bimetallic Ni–Cu catalysts using isopropanol (Kannapu et al., 2015). For the study, γ-alumina and carbon-supported mono- and bimetallic Ni and Cu catalysts were synthesized and applied to the reduction of *p*-cresol and furfural via transfer hydrogenation. The developed reaction system was applied to alkylphenol-rich pyrolysis oils produced from the TGRP with switchgrass and oak wood as feedstock. The catalysts used were characterized pre- and postreactions using XRD, TPR, TEM, and TGA. When isopropanol was used as the hydrogen donor solvent, yields of 95% of a mixture of products from the reduction of *p*-cresol were achieved using the Ni–Cu/Al_2O_3 catalyst. This product mixture included ring hydrogenation products (4-methylcyclohexanol and 4-methylcyclohexanone) as well as deoxygenated products (methylcyclohexane and toluene), with 4-methylcyclohexanol being the major product (Fig. 5.12). The activity remained high in the presence of water with very high levels of water concentration resulting in a higher selectivity toward the ketone product. The system was also effective for the reduction of furfural to furfuryl alcohol, although lower temperatures were required to prevent polymerization of the furfural. When applied to real bio-oil, although increases in H/C and C/O ratios and energy content were realized, effective reduction of the alkylphenols in the bio-oils was below expectation. Rather, solid formation, a result of the polymerization of bio-oil compounds, was observed. Effort to extend the successes observed with the model compounds to improve the application of transfer hydrogenation to real bio-oils was equally not entirely promising as a commercially viable pathway for pyrolysis biorefinery.

References

Albahri, T.A., Riazi, M.R., Alqattan, A.A., 2002. Octane number and aniline point of petroleum fuels. Fuel Chem. Div. Prepr. 47, 710–711.

Boateng, A.A., Mullen, C.A., Elkasabi, Y., McMahan, C., 2015. Guayule (*Parthenium argentatum)* pyrolysis biorefining: production of hydrocarbon compatible bio-oils from guayule bagasse via tail-gas reactive pyrolysis. Fuel. 158, 948–956.

Boateng, A.A., Elkasabi, Y., Mullen, C.A., 2016. Guayule (*Parthenium argentatum*) pyrolysis biorefining: fuels and chemicals contributed from guayule leaves via tail gas reactive pyrolysis. Fuel. 163, 240–247.

Brieger, G., Nestrick, T., 1974. Catalytic transfer hydrogenation. Chem. Rev. 74 (5), 567–580.

Bu, Q., Lei, H., Zacher, A.H., Wang, L., Ren, S., Liang, J., et al., 2012. A review of catalytic hydrodeoxygenation of lignin-derived phenols from biomass pyrolysis. Bioresour. Technol. 124, 470–477.

Cornish, K. 1996. Hypoallergenic natural rubber products from Parthenium argentatum (Gray) and other non-Hevea brasiliensis species. U.S. Patent 5,580,942.

Do, P.T.M., Crossley, S., Santikunaporn, M., Resasco, D.E., 2007. Catalytic strategies for improving specific fuel properties. Catalysis 20, 33–64.

Elkasabi, Y.E., Mullen, C.A., Pighinelli, A.L.M.T., Boateng, A.A., 2014. Hydrodeoxygenation of fast pyrolysis bio-oils from various feedstocks using carbon supported catalysts. Fuel Process. Technol. 123, 11–18.

Elkasabi, Y., Mullen, C.A., Boateng, A.A., 2015. Aqueous extractive upgrading of bio-oils created by tail-gas reactive pyrolysis to produce pure hydrocarbons and phenols. ACS Sustain. Chem. Eng. 3 (11), 2809–2816.

Elkasabi, Y., Liu, Q., Choi, Y.S., Strahan, G., Boateng, A.A., Regalbuto, J.R., 2017. Bio-oil hydrodeoxygenation catalysts produced using strong electrostatic adsorption. Fuel 207, 510–521.

Elkasabi, Y., Mullen, C.A., Boateng, A.A., Brown, A., Timko, M.T., 2019. Flash distillation of bio-oils for simultaneous production of hydrocarbons and green coke. Ind. Eng. Chem. Res. 58 (5), 1794–1802.

Elliott, D.C., 2007. Historical developments in hydroprocessing bio-oils. Energy Fuels 21, 1792–1815.

Elliott, D.C., Baker, E.G., 1987. Chapter 42: Hydrotreating biomass liquids to produce hydrocarbon fuels. Energy From Biomass and Waste X. IGT Chicago, pp. 765–784, Chapter 42.

Elliott, D.C., Baker, E.G., 1989. Process for Upgrading Biomass Pyrolyzates. U.S. Patent Number 4,795,841.

Furimsky, E., 2000. Catalytic hydrodeoxygenation. Appl. Catal. A: Gen. 199, 147–190.

Furimsky, E., 2013. Hydroprocessing challenges in biofuels production. Catal. Today 217, 13–56.

Kannapu, H.P.R., Mullen, C.A., Elkasabi, Y., Boateng, A.A., 2015. Catalytic transfer hydrogenation for stabilization of bio-oil oxygenates: reduction of p-cresol and furfural over bimetallic Ni-Cu catalysts using isopropanol. Fuel Process. Technol. 137, 220–228.

Meier, D., van Beld, B., Bridgwater, A.V., Elliott, D.C., Oasmaa, A., Preto, F., 2013. State-of-the-art of fast pyrolysis in IEA bioenergy member countries. Renew. Sustain. Energy Rev. 20, 619–641.

Si, Z., Zhang, X., Wang, C., Ma, L., Dong, R., 2017. An overview on catalytic hydrodeoxygenation of pyrolysis oil and its model compounds. Catalysts 7 (6), 169–190.

Venkatakrishnan, V.K., Degenstein, J.C., Smeltz, A.D., Delgass, W.N., Agrawal, R., Ribeiro, F.H., 2014. High-pressure fast-pyrolysis, fast-hydropyrolysis and catalytic hydrodeoxygenation of cellulose: production of liquid fuel from biomass. Green. Chem. 16, 792–802.

Combustion applications of pyrolysis liquids

6

6.1 Introduction

As we have seen by now, instability of pyrolysis oils has been the major factor limiting their refining in existing petroleum refineries and commercializing pyrolysis oil as a transportation fuel (Schwietzke et al., 2008). While a successful development of upgrading technologies could overcome this limitation, immediate uses of "neat" pyrolysis oil as combustion fuel could provide a near-term application that would allow for immediate revenues and reduction of carbon footprint, easing the pressure on the research leading toward transportation fuels.

Although it is apparent that due to its heterogeneity, neat pyrolysis oil is unsuitable for direct use in diesel engines, several combustion trials in direct injection stationary engines have been conducted (Solantausta et al., 1993; Shihadeh and Hochgreb, 2000). However, stabilized or partially stabilized pyrolysis oil as well as pyrolysis oil blends with solvents, and surfactants may address many of the challenges encountered in using neat pyrolysis oil in combustion systems, including compression-ignition (CI) engines, gas turbines, boilers, and kilns, employed in the power- and material-processing industries. The major issues for combustion of pyrolysis oil "as is" or as fuel oil substitute are related to the bio-oil's acidity, high water content, high oxygen content, wide volatility distribution, and the presence of particulates such as char particles all of which cause spray atomization problems. The difficulty in breaking down the liquid into small droplets, needed for spray combustion, causes ignition delay, propensity to coking, and particulate emissions. Droplet combustion rates for pyrolysis oils are about 2 or 3 factors slower than that for light diesel fuel (Shaddix and Hardesty, 1999; Shihadeh and Hochgreb, 2000, 2002). Although this results in high mass density and latent heat of vaporization, the wide range of volatility of the numerous chemical components results in droplet fragmentation and micro-explosions, which, in turn, reduces the burning rate. Droplet evaporation and burning rates for some select pyrolysis oils and their admixtures were estimated in the late 1990s at Sandia National Laboratories, in Albuquerque, New Mexico and compared with number 2 diesel fuel as well as water as shown in Table 6.1.

Low combustion rates typically characterize flames that are long, lazy, and of low intensity such as that depicted in Fig. 6.1. Such flames produce more particulate matter or soot than diesel fuel fired under similar conditions.

Irrespective of the problems associated with using biomass-derived pyrolysis oil for commercial power generation, field trials involving engine manufacturers, stationary gas turbine operators, and boiler operators interested in cofiring pyrolysis

Pyrolysis of Biomass for Fuels and Chemicals. DOI: https://doi.org/10.1016/B978-0-12-818213-0.00006-0

Table 6.1 Combustion characteristics of select pyrolysis oils and admixtures (Shaddix and Hardesty, 1999).

Liquid fuel	Density (g/mL)	Heat of vaporization q_v (J/g)	Evaporation rate K_v (mm^2/s)	A/F mass ratio	Specification combustion enthalpy, q_c (J/g)	Burning rate K_c, (mm^2/s)
Diesel number 2	0.86	267	0.56	12.6	41	0.99
Water	1.0	2257	0.10	N/A[a]	N/A	N/A
NREL 154 (Oak)	1.2	613	0.25	5.6	17.6	0.52
NREL 175 (poplar)	1.2	711	0.23	6.3	16	0.45
NREL 157 (switchgrass)	1.2	887	0.19	8.2	19.3	0.4
NREL 175 + water[a]	1.18	842	0.20	5.7	14.6	0.42
NREL 175 + methanol[a]	1.16	738	0.23	6.3	16.3	0.45
NREL 175 + ethanol[a]	1.16	720	0.23	6.4	16.8	0.46

[a]Assumes 10% addition (volume basis).

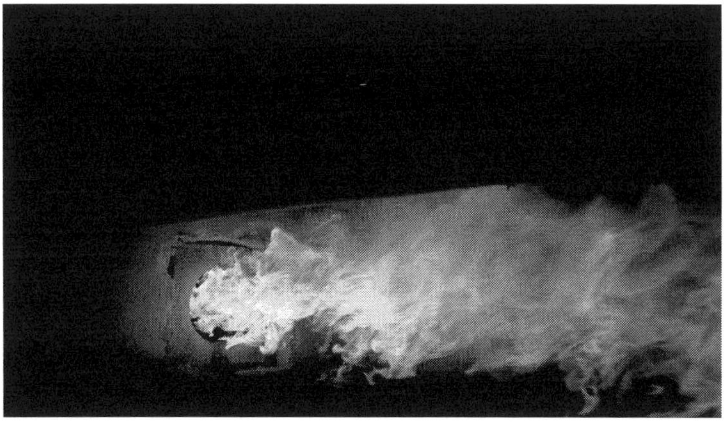

Figure 6.1 Pyrolysis oil flame created in the laboratory (Courtesy of Mississippi State University).

oils with coal, and/or natural gas have been conducted (Czernik and Bridgwater, 2004; Chiaramonti et al., 2007). For example, Wartsila, a large engine manufacturer (http://www.wartsila.com/), previously conducted extensive testing of pyrolysis oils as fuel substitutes in their stationary engine series called VASA. However, their plans to package an energy production supply chain comprising wood-waste to pyrolysis plant to stationary diesel power was aborted due to insurmountable technical problems. These include acid attacks on storage tanks, gaskets, and seals in pumps; preheating problems associated with oligomerization and subsequent lacquering at pistons and nozzles, among others. Similar tests at Massachusetts Institute of Technology (MIT), the Canada Centre for Mineral and Energy Technology (CANMET), and at a company called PyTEC, the developer of containerized pyrolysis waste disposal systems, and others using various internal combustion (IC) engine types have all yielded similar unfavorable results. The conclusion of these earlier tests taught us that in order to successfully operate IC engines solely on pyrolysis oils or admixtures thereof, a standard engine will have to be substantially modified to include changing the materials of construction; a prospect that the engine manufactures found uneconomically viable. Cofiring pyrolysis oil in gas turbines and boilers has not performed well either, largely due to similar materials of construction issues (Lopez Juste and Salva Monfort, 2000; Chiaramonti et al., 2007).

Despite the setbacks encountered in the earlier years, greater success has been achieved with the use of pyrolysis oil in some industrial furnaces such as kilns and boilers (Li et al., 2004). Renewed interests in the use of pyrolysis oil as diesel fuel replacement in furnaces led to the American Standard Test Method specification (D7544-09) that was discussed earlier in Chapter 2, Thermal Pyrolysis, released to guide the design and operation of industrial burners equipped to handle pyrolysis oils. Although the specification does not currently include pyrolysis oil use in residential heaters, small commercial boilers, stationary and/or marine engines, these applications, especially residential heaters, could be most impactful in reducing our carbon footprint given that the cold regions of the Northeast are normally where biomass is also most abundant. The chapter presents some efforts in understanding the use of neat pyrolysis oil and their blends thereof in spray combustion and turbulent diffusion flame applications.

6.2 Characteristics of pyrolysis oil—fuel blends for combustion

We previously learned that biomass-derived pyrolysis oils in their neat form are not miscible with conventional fossil fuels such as gasoline and diesel due to their oxygenated compound complex. Such inhomogeneities are the sources of spray atomization issues in combustion. But pyrolysis oils are miscible with alcohols, and blending can reduce their viscosity allowing for droplet combustion. Another way is to employ the use of certain surfactants to aid in blending pyrolysis oil with hydrocarbon fuels such as diesel oils to create emulsions that are burnable in industrial furnaces. Some characteristics of fuel blends are presented to inform our knowledge base.

Fig. 6.2 presents a ternary diagram of pyrolysis oil, ethanol, and number 2 fuel oil or biodiesel (fatty acid methyl esters) made from rapeseed in the blend along with visual/pictorial observations of the blend's homogeneity (Martin and Boateng, 2014). Visual observation reveals that all the pyrolysis oil from the selected pool form a homogeneous blend with number 2 fuel oil only when the mixture contains 70% or less of pyrolysis oil on mass basis. The ternary diagram indicates that when the concentration of pyrolysis oils is below 70%, a homogeneous blend could be formed with at least 10% fuel oil mix with the remainder consisting of ethanol. If the fuel oil is increased to around 20%, the mixture separates into two phases. If the pyrolysis oil is maintained at 10% a homogeneous blend containing up to 20% number 2 fuel oil is possible; this is also true at 5% pyrolysis oil and 30% number 2 fuel oil. Biodiesel exhibits the same behavior as number 2 fuel oil does when pyrolysis oil concentration is 50% and above. However, at pyrolysis oil concentrations within 20%–40% range, much more biodiesel could be blended homogeneously than number 2 fuel oil does before phase separation occurs. The highest concentration of biodiesel found in a homogeneous blend may be capped at 50%, with 20% pyrolysis oil and 30% ethanol blend. At concentrations of 10% and 5% pyrolysis oil, there appears to be no clear phase separation regardless of the amount of biodiesel blended, but the pyrolysis oil would not mix completely with the biodiesel at biodiesel concentrations above 50%. With these blends a portion of the pyrolysis oil would remain suspended in the mixture at the highest concentrations of biodiesel, and the pyrolysis oil would adhere to the sides of the container.

6.2.1 *Viscosity changes*

Given the homogeneity of the binary mixture of pyrolysis oil and ethanol, the combustion characteristics such as spray atomization followed by actual combustion of

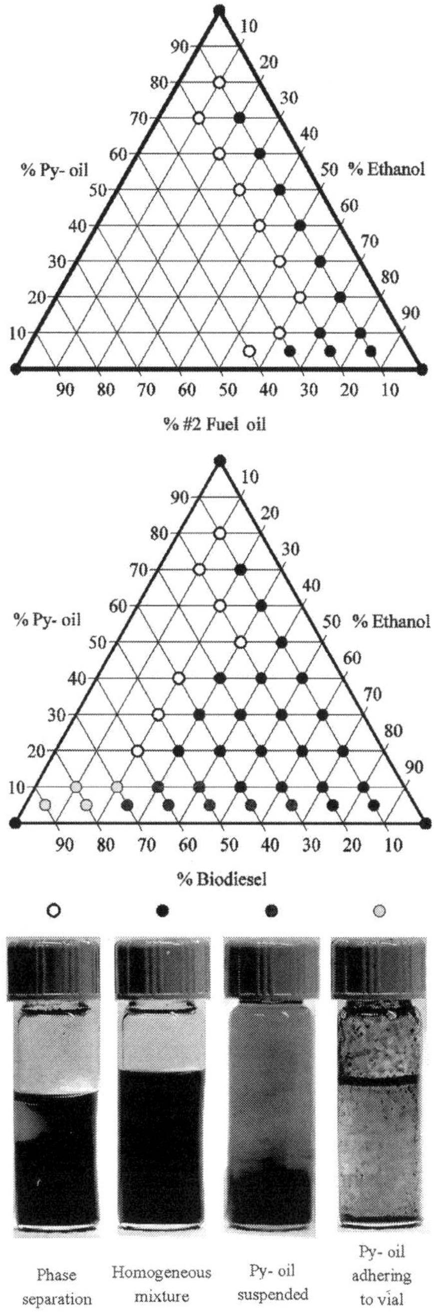

Figure 6.2 Ternary blends of pyrolysis oil, ethanol, and number 2 fuel oil (top)/biodiesel (bottom). Below the plots is a key with examples of blend behavior (Martin and Boateng, 2014).

such blends are worth exploring. With the empirical equation developed by Naidu and Krishnan (1996) (Eq. 6.1), one can establish the bio-oil/ethanol mixture amount that meets a specific viscosity for combustion in a specific engine (e.g., SAE 40 or 50). The equation relates mixture viscosity with volume fractions of the blend and temperature as follows:

$$\ln(\mu_{\text{blend}})V_{\text{blend}} = x_{\text{eth}}\ln(\mu_{\text{eth}})V_{\text{eth}} + x_{\text{py}}\ln(\mu_{\text{py}})V_{\text{py}} + x_{\text{eth}}x_{\text{py}}A/RT \tag{6.1}$$

in which, V represents the molar volume of the liquid, R is the ideal gas constant, T represents temperature in Kelvin, and A is an empirical coefficient accounting for the interaction energy between the two components. This allows the values of the variables A, V_{py}, and V_{blend} which are unknown for pyrolysis oils due to their complexity to be assessed by fitting experimental data in a regression analysis.

In this light, regression curves shown in Fig. 6.3 represent temperature-dependent viscosities of pyrolysis oil/ethanol blends most suitable for use as combustion fuel replacement for a residential boiler approved to burn motor oils with viscosities up to and including SAE grade 50. The plots map the viscosities corresponding to a blend containing 80%–90% pyrolysis oil and 10%–20% ethanol; they show that for pyrolysis oil concentrations higher than 20% the blend-stock will be too viscous for the said boiler to use. The plot also shows the severity of the dependence of viscosity on temperature for pyrolysis oil blends. Increases in temperature, with increasing percentage of pyrolysis oil in the blend, indicates that preheating of the fuel will result in greater reductions in fuel viscosity, which will impact negatively on spray atomization quality.

The spray atomization quality of the mixture is characterized by the Sauter mean diameter (SMD) and is estimated from the surface tension (σ), density (ρ), and mass flow rate (W) of the fuel and the air, the diameter of the nozzle orifice (D_o), and the velocity of the air relative to the fuel (U_R) at the point where air and fuel first mix. With the desired viscosities known the quality of atomization that could be obtained with each fuel blend may be established for plain-jet airblast atomizers as (Lefebvre, 1980):

$$\text{SMD} = 0.95\left[\frac{(\sigma_l W_l)^{0.33}}{\rho_l^{0.37}\rho_A^{0.30}U_R}\right](1+W_l/W_A)^{1.70} + 0.13\left(\frac{\mu_l^2 d_O}{\sigma_l \rho_l}\right)^{0.5}(1+W_l/W_A)^{1.70} \tag{6.2}$$

The properties of sprays also depend on the atomizing nozzle geometry. Sprays emanating from pyrolysis oil (bio-oil) blends with ethanol using an air-assisted atomization nozzle were characterized by Lujaji et al. (2016a,b) to explore the potential for pyrolysis oil combustion in industrial and residential furnaces. Twin-fluid externally mixed nozzles labeled SU2, SU4, and SU5 with liquid orifice areas of 0.40, 1.82, and 5.07 mm^2, respectively (Fig. 6.4), were used to investigate bio-oil/ethanol blends with concentrations of 20:80 and 40:60 vol.% and compared with neat ethanol and number 2 diesel. The liquid and atomizing air flow rates as well as

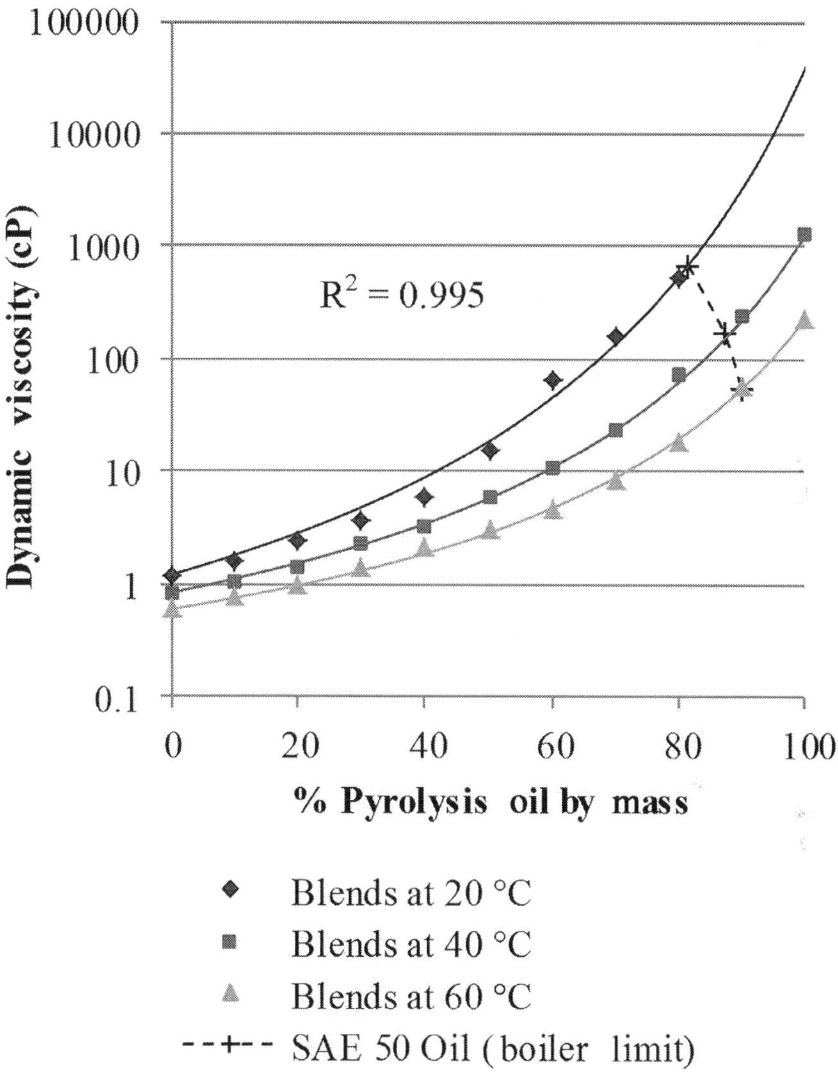

Figure 6.3 Calculated viscosity of py-oil/ethanol blends, fit to Eq. (6.1) (Martin and Boateng, 2014). *Py-oil*, Pyrolysis oil.

temperature were controlled to maintain constant liquid flow rates (cc/s) equivalent to 30 and 50 kWth energy input. The impetus is to characterize the spray atomization patterns underlying diffusion flames.

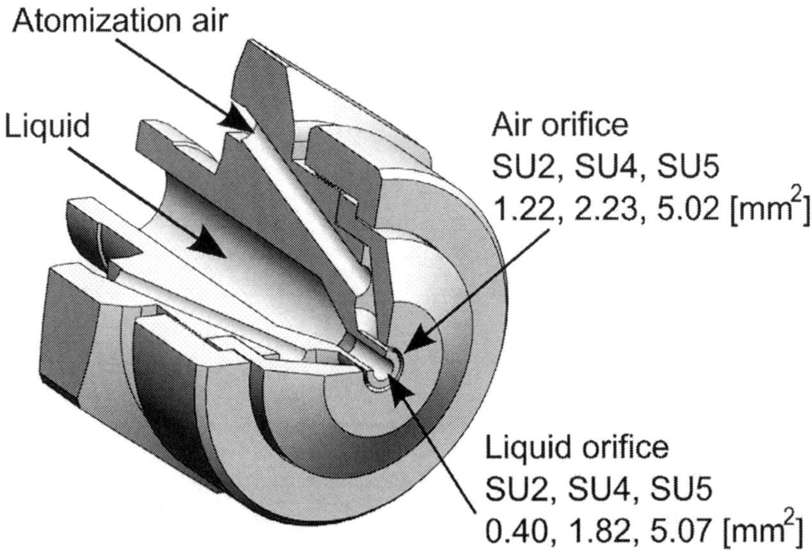

Figure 6.4 Cutaway view: externally mixed nozzle (Lujaji et al., 2016a,b).

6.2.2 Spray characteristics

Images of atomized spray droplets at set stoichiometric air flow rates displayed in Fig. 6.5 (see Appendix for complete set) show that it is possible to spray bio-oil/ethanol mixtures containing up to 40% bio-oil with water content as low as 12.6%.

These images show that the externally mixed nozzles could develop a solid, cone-shaped spray plume with a 40:60 bio-oil:EtOH blend. This follows the formation of a narrow liquid stream at the nozzle exit, as the liquid fuel exits the small orifice at the center of the nozzle tip. With this nozzle-arrangement atomization, air exits the nozzle concentrically around the liquid, thereby enveloping the liquid

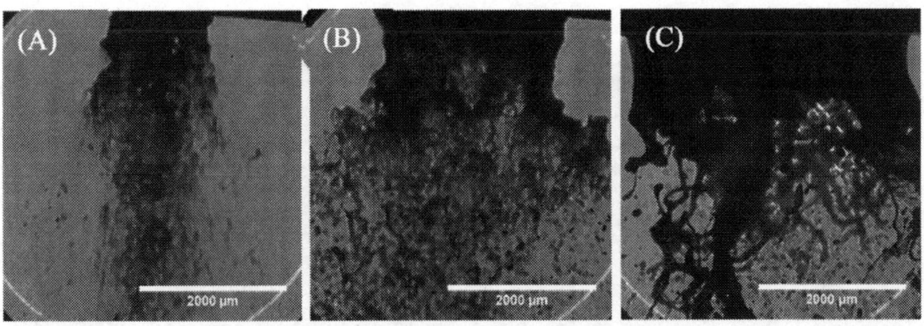

Figure 6.5 30 kW energy input spray of 40:60 bio-oil:EtOH blend at 20 SLPM atomization air flow with (A) SU2, (B) SU4, and (C) SU5 nozzles (Lujaji et al., 2016a,b).

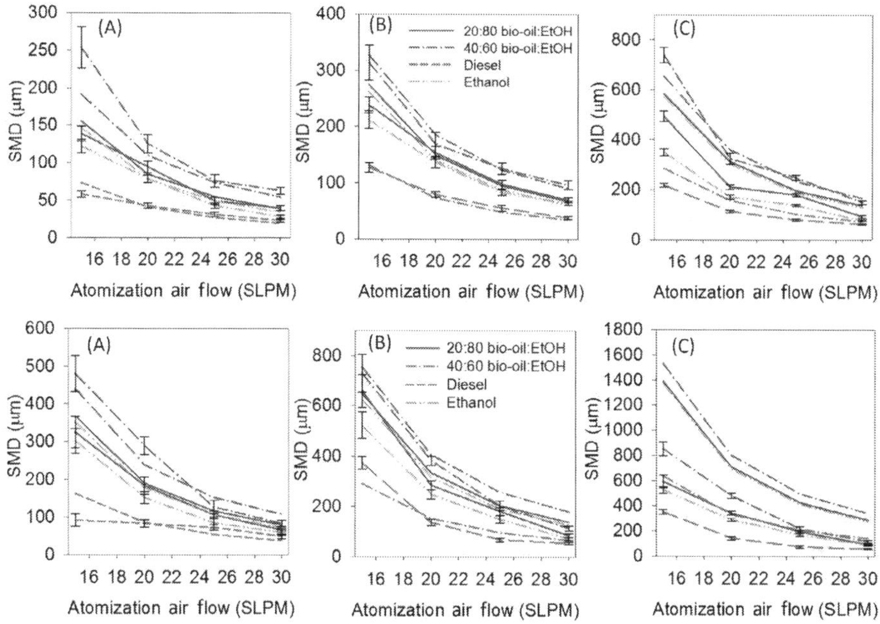

Figure 6.6 (A) Theoretical and experimental SMD of bio-oil blends, ethanol, and diesel spray at 30 kW energy input with (a) SU2, (b) SU4, and (c) SU5 nozzles. (B) Theoretical and experimental SMD of bio-oil blends, ethanol, and diesel spray at 50 kW energy input with (a) SU2, (b) SU4, and (c) SU5 nozzles (Lujaji et al., 2016a,b). *SMD*, Sauter mean diameter.

stream in a diffusion jet stream. The shearing of the liquid stream by the atomizing air results in the formation of liquid sheets that disintegrates into ligaments and later into droplets. As the orifice size increases (SU2, SU4, and SU5 nozzles), more interaction between air/fuel flow results. The atomization air required to meet the same energy input informs how the blended fuel properties, blend ratios, and air cross-sectional area affects droplet sizes and velocity of bio-oil/ethanol mixture.

Figs. A6.1–A6.7 (see Appendix) show that the sprays exhibit a consistent change of droplet diameters, decreasing sharply downstream. The interaction between the atomizing air stream and liquid surface tension is an important factor influencing different atomization characteristics such as different break up lengths and droplet sizes. The high viscosity blends such as 40% pyrolysis oil blend showed more resistance to the formation of droplets compared to number 2 diesel fuel at low atomization air flow rates.

6.2.3 Spray droplet sizes

There is generally a decrease in the droplet diameter for neat and blended liquids as atomization air flow rate increases (e.g., from 15 to 30 SLPM). This is attributable to

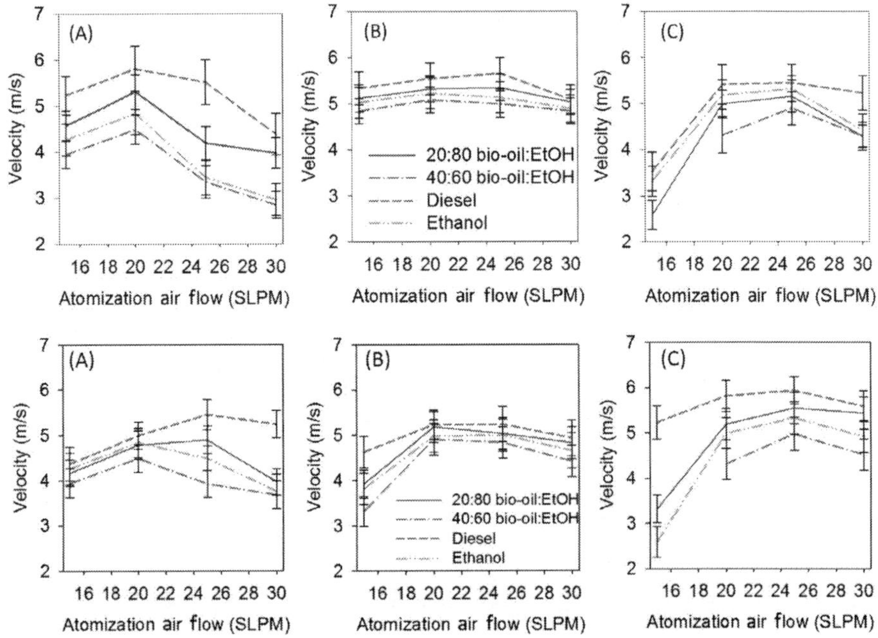

Figure 6.7 (A) Droplet velocity of bio-oil blends, ethanol, and diesel spray at 30 kW equivalent energy input with (a) SU2, (b) SU4, and (c) SU5 nozzles. (B) Droplet velocity of bio-oil blends, ethanol, and diesel spray at 50 kW equivalent energy input with (a) SU2, (b) SU4, and (c) SU5 nozzles (Lujaji et al., 2016a,b).

an increase in atomization air flow momentum that is responsible for strengthening the shear forces that break up the liquid stream (Lal et al., 2010). Calculated and measured SMD and tip velocities are shown in Figs. 6.6 and 6.7. These suggest that high pyrolysis oil viscosity results in larger droplet sizes than neat ethanol. SMD values for number 2 diesel fuel sprays are lowest for all nozzle sizes and all atomization air flow rates due to its high energy density, which, hence, results in lower quantities being swept by high stoichiometric, air flow rates.

The lower the viscosity of the liquid fuel, the higher the droplet velocity as low viscosity promotes faster rate of droplet formation, due to a quick liquid stream disintegration leading to fast moving droplets.

The key fuel properties of number 2 fuel oil, ethanol, pyrolysis oil, and three pyrolysis oil (py-oil)/ethanol blends are presented in Table 6.2. It is worth noting that all the blends have a lower heat content than neat diesel (the number 2 fuel oil), indicating the intended boiler/furnace will have a lower maximum heat output when fueled with each of these blends. While derating can be as much as 39% for ethanol combustion in a 40 kWth residential furnace, 50:50 ethanol/pyrolysis oil hovers around 26% and 21% for 20:80 ethanol/pyrolysis oil blend. This is small compared with the benefits of reducing the carbon footprint with bio-oil as a renewable fuel alternative.

Table 6.2 Comparison of thermal-fluid characteristics of bio-oil/ethanol blends (Martin and Boateng, 2014).

Fuel	Viscosity cP at 60°C	HHV (MJ/kg)	HHV (MJ/L)	Maximum boiler input (kWth)	% Decrease from fuel oil
Number 2 fuel oil	1.126	45.7	38.0	63.3	–
Ethanol	0.592	29.8	23.5	39.2	−39.2
Mixed py-oil	153.9	23.6	30.9	51.5	−18.7
20% py-oil/ 80% ethanol	0.963	28.6	25.5	42.6	−32.8
50% py-oil/ 50% ethanol	3.03	26.7	28.0	46.7	−26.2
80% py-oil/ 20% ethanol	18.05	24.8	29.9	49.9	−21.2

HHV, High heating value; *py-oil*, pyrolysis oil.

Table 6.2 presents the blending of pyrolysis oils with ethanol. Viscosity can be lowered from a high of 100 cP to as low as 18 cP with about 20% ethanol blend and 0.9 cP for 80% ethanol and 20% pyrolysis oil. As the table indicates, all the blends have lower heat content than the number 2 diesel fuel oil indicating that the furnace will have a lower maximum heat output when fueled with these blends. A derating of the said furnace is, therefore, eminent, since the maximum fuel flow is usually limited by nozzle size. For a typical residential boiler of 40 kWth operating on number 2 fuel oil, replacing it with neat pyrolysis oil can derate the boiler by as much as 18.7% if the neat pyrolysis oil can be atomized and burnt. This is compared with over 39% derating if the number 2 fuel oil is replaced with ethanol. For the 20:80 py-oil/ethanol blend that is easily atomized and can sustain combustion, the derating of the boiler is estimated to be just above 32% (Table 6.2).

6.3 Pyrolysis oil/diesel fuel emulsions

Because pyrolysis oil and hydrocarbon are immiscible, another promising way to promote the use of pyrolysis oil "as produced" is to form emulsions of diesel and pyrolysis oil blends that can sustain combustion in large volumes in industrial furnaces. The main drawback, however, has been the stability issues plaguing such mixtures as we discussed earlier. However, stable emulsions can be created by introducing the right surfactants in the mixture. Martin et al. (2014) produced and studied several emulsions consisting of biomass pyrolysis oil (bio-oil) in diesel fuel at the U.S. Department of Agriculture (USDA) and analyzed them for their stability over time. Using an ultrasonic probe microscopy, droplets of bio-oil suspended in diesel fuel were generated and then stabilized by surfactant chemicals. Potential surfactants tested include those listed in Table 6.3.

Table 6.3 Surfactant building blocks (Martin, et al., 2014).

Head types	
Polyethylene glycol (n) $C_{2n}H_{4n+2}O_{n+1}$	
Sorbitan $C_6H_{12}O_5$	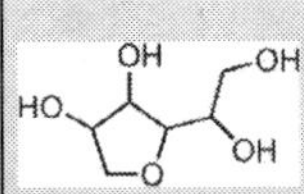
Tail types	
Oleic acid $C_{18}H_{34}O_2$	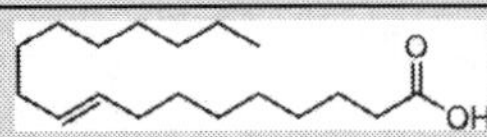
Stearic acid $C_{18}H_{36}O_2$	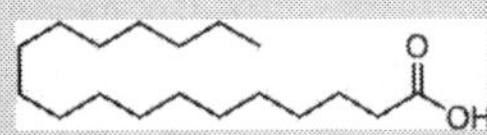
Stearyl alcohol $C_{18}H_{38}O$	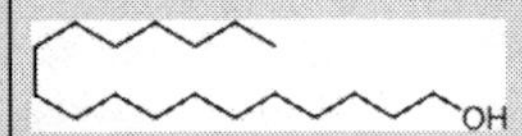
12-Hydroxystearic acid $C_{18}H_{36}O_3$	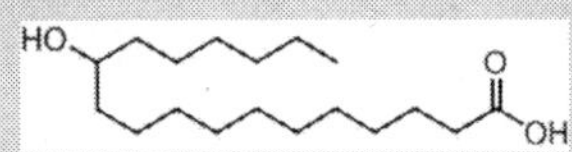

For emulsions, settling or phase separation, as we saw in the ternary mixtures earlier, does not permanently destabilize since droplets can be dispersed back into the emulsion through simple mechanical agitation without the need to perform the emulsification process again. The process that will permanently destabilize the emulsion is the coalescence of droplets. When two droplets collide, they will coalesce if there is not enough surfactant to surround the droplet, thus there must be enough surfactant to completely coat the droplets to prevent rapid coalescence. Even if excess surfactant is provided, the droplets can coalesce if they collide with sufficient force. As the surfactant-coated droplets encounter each other, they will initially resist deformation which, for a spherical object, will increase its surface area, but the surface tension of the surfactant film will likely resist such increase in surface area.

Due to the acidic nature of pyrolysis oils, only nonionic surfactants may be used to prevent reactions between the bio-oil components and the surfactant. For nonionic surfactants the hydrophilic–lipophilic balance (HLB) number informs the selection of the ideal surfactant mixture for emulsion application (Martin, et al., 2014). The HLB number is estimated as

$$HLB = \frac{MW_{head}}{MW_{head} + MW_{tail}} \times 20 \tag{6.3}$$

Table 6.4 Comparison of bio-oil:surfactant ratios under microscope.

Ratio	1 minute after emulsification		1 day after emulsification	
8:1 Bio-oil:surfactant	Mean diameter: 0.49 μm		Mean diameter: 0.48 μm	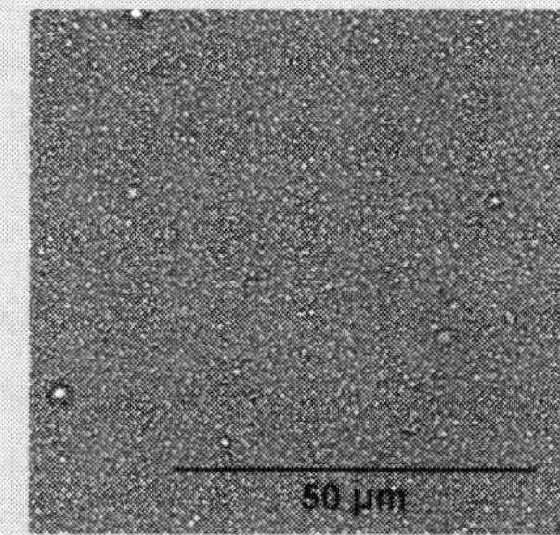
16:1 Bio-oil:surfactant	Mean diameter: 0.73 μm	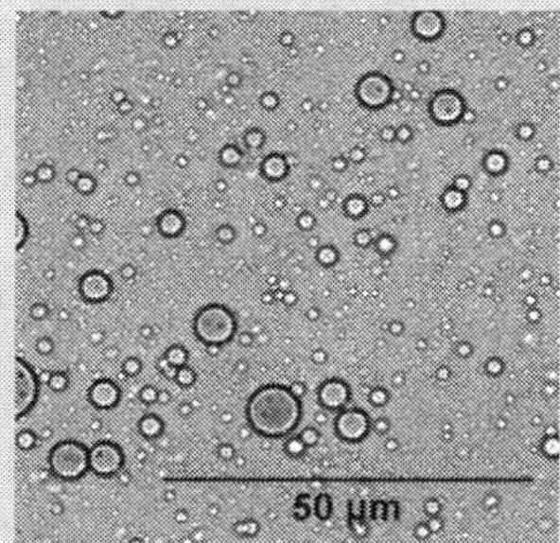	Mean diameter: 0.63 μm	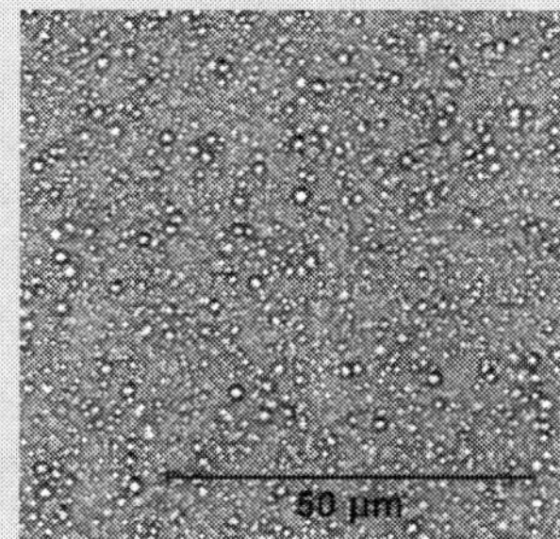

(Continued)

Table 6.4 (Continued)

Ratio	1 minute after emulsification		1 day after emulsification	
32:1 Bio-oil:surfactant	Mean diameter: 0.97 μm	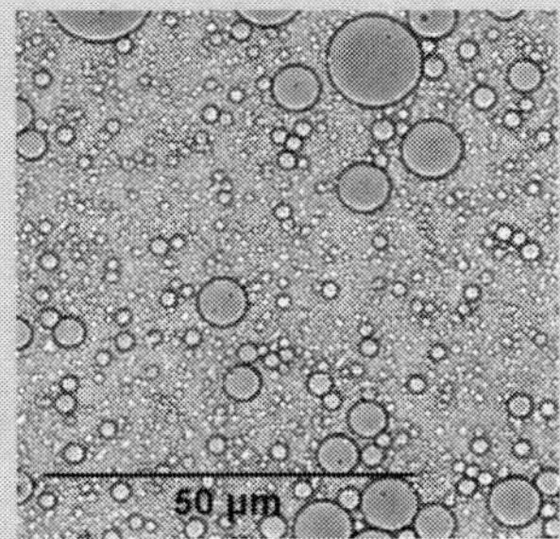	Mean diameter: 0.48 μm	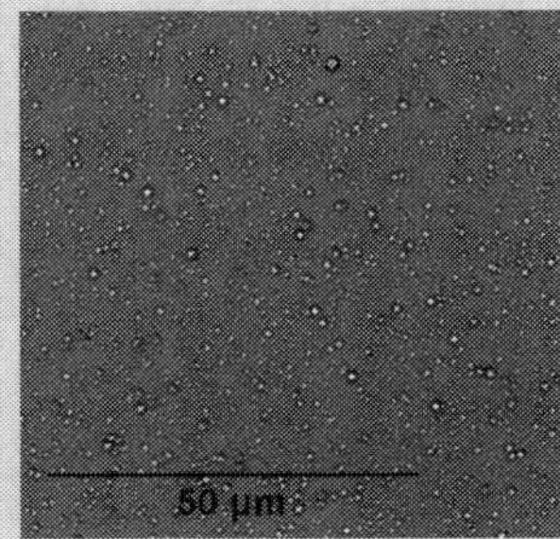
64:1 Bio-oil:surfactant	Mean diameter: 1.30 μm	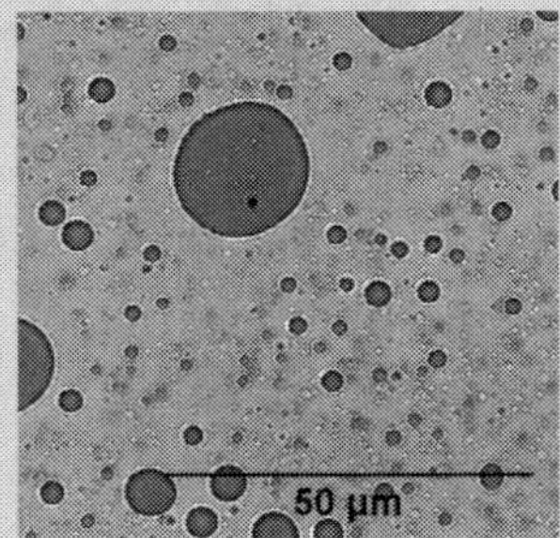	Mean diameter: N/A	

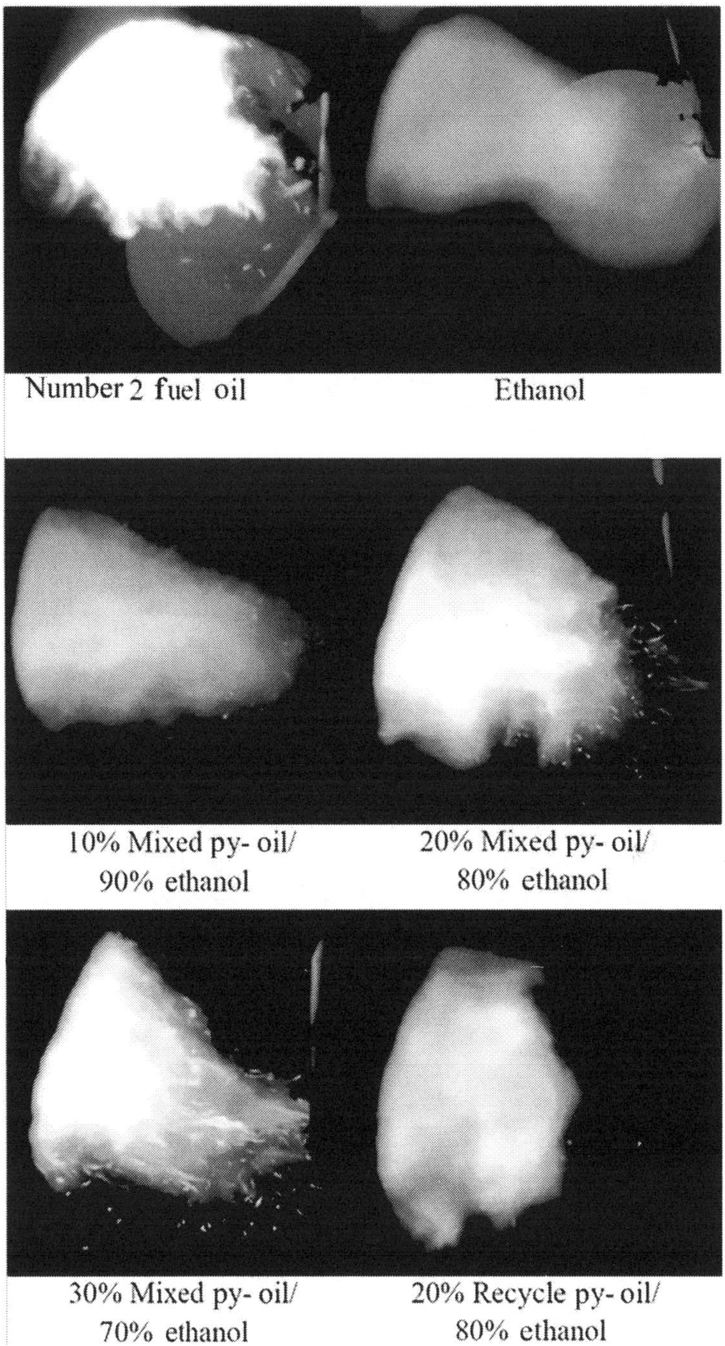

Figure 6.8 Flame structures in residential boiler (Martin and Boateng, 2014).

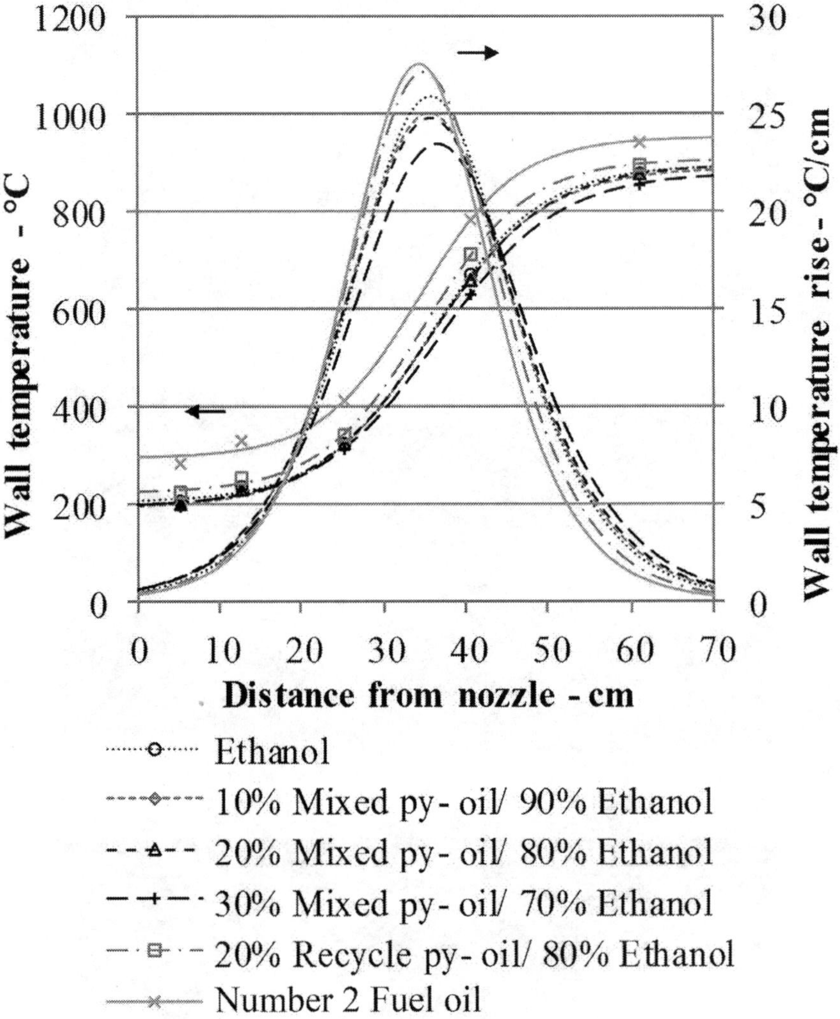

Figure 6.9 Chamber wall temperatures (Martin and Boateng, 2014).

where MW_{head} represents the molecular weight of the hydrophilic head section, and MW_{tail} represents the molecular weight of the lipophilic tail section. To select an ideal surfactant, one must first determine the ideal HLB of the emulsion and then compare different head/tail combinations that produce the ideal HLB. Although one surfactant may not have quite the ideal HLB, it can still be mixed with another surfactant to bring the HLB to the desired value. Typically, an HLB

Table 6.5 Boiler heat balance: mean value ± standard error from multiple test runs (Martin and Boateng, 2014).

	Number 2 fuel oil	**Ethanol**	**10% Mixed py-oil/90% ethanol**	**20% Mixed py-oil/80% ethanol**	**30% Mixed py-oil/70% ethanol**	**20% Recycle py-oil/80% ethanol**
Heat to water (boiler efficiency) (%)	62.78 ± 0.54	63.04 ± 1.15	62.34 ± 1.37	62.55 ± 0.51	61.67 ± 3.91	65.11 ± 0.10
Dry exhaust loss (%)	18.14 ± 0.27	16.18 ± 0.42	17.76 ± 0.02	17.23 ± 0.09	18.53 ± 0.16	15.88 ± 0.21
Vapor exhaust loss (%)	8.42 ± 0.01	12.12 ± 0.04	11.85 ± 0.01	11.24 ± 0.02	10.81 ± 0.04	10.97 ± 0.06
Fuel moisture loss (%)	0.00 ± 0.00	0.00 ± 0.00	0.10 ± 0.00	0.20 ± 0.00	0.30 ± 0.00	0.23 ± 0.00
Other losses (%)	10.65 ± 0.35	8.65 ± 1.16	7.94 ± 1.38	8.78 ± 0.62	8.68 ± 4.04	7.80 ± 0.25

Table 6.6 Boiler gas emissions corrected to 7% O_2: mean value ± standard error from multiple test runs (Martin and Boateng, 2014).

Fuel	CO_2(vol.%)	CO (ppm)	NO_x (ppm)	Percentage increase in NO_x versus number 2 fuel oil
Number 2 fuel oil	10.33% ± 0.00%	6.80 ± 0.58	113 ± 1	
Ethanol	10.19% ± 0.04%	3.19 ± 0.26	61 ± 3	−46
10% Mixed py-oil/ 90% ethanol	10.49% ± 0.02%	6.29 ± 0.27	138 ± 3	22
20% Mixed py-oil/ 80% ethanol	10.67% ± 0.01%	8.63 ± 0.93	188 ± 3	67
30% Mixed py-oil/ 70% ethanol	10.94% ± 0.03%	12.19 ± 0.51	245 ± 1	117
20% recycle py-oil/ 80% ethanol	10.60% ± 0.01%	5.38 ± 1.84	251 ± 4	122
20% Switchgrass py-oil/80% ethanol	10.61% ± 0.05%	3.59 ± 0.25	273 ± 1	142
20% Oak and pine py-oil/80% ethanol	10.57% ± 0.04%	6.91 ± 0.51	174 ± 1	54
20% Equine waste py-oil/80% ethanol	10.59% ± 0.02%	8.46 ± 0.35	294 ± 3	160
20% Eucalyptus py-oil/80% ethanol	10.72% ± 0.02%	10.10 ± 0.99	99 ± 5	−12
20% Miscanthus py-oil/80% ethanol	10.65% ± 0.00%	5.08 ± 1.02	150 ± 3	33
20% Pennycress py-oil/80% ethanol	10.42% ± 0.02%	5.40 ± 0.87	832 ± 13	636

Py-oil, Pyrolysis oil.

of 4−8 is considered best for emulsions created from bio-oil/diesel fuel mixtures of interest.

Martin et al. (2014) observed, under microscope, changes in mean diameter 1 minute and after 1 day after emulsification (Table 6.4) and showed that a polyethylene glycol dipolyhydroxystearate surfactant with a HLB number of about 4.75 can be most effective for diesel/bio-oil/surfactant blend of 32:8:1 ratio, that is, 20% utilization of bio-oil. This emulsion consisted of uniformly sized droplets with an average diameter of 0.48 μm, and exhibits no observed coalescence of droplets after 1 week. If left undisturbed, these droplets would slowly settle to the bottom of the mixture at a rate of only 2.4 mm/day and requires only a slight agitation to keep the droplets suspended indefinitely. This level of stability facilitates utilization of 20 wt.% of raw bio-oil in diesel as a renewable liquid fuel for spray combustion without the need for costly and energy-intensive upgrading.

6.4 Steady state combustion

In the USDA work reported by Martin and Boateng (2014) a 40 kWth oil-fired commercial boiler was fueled with some of the blends of biomass py-oil and ethanol discussed earlier to determine the feasibility of using these blends as alternative fuels to fuel oil in home heating applications. The combustion performance of blends of ethanol with different concentrations of pyrolysis oil (10%, 20%, and 30% py-oil by mass) produced from different biomass feedstocks, including

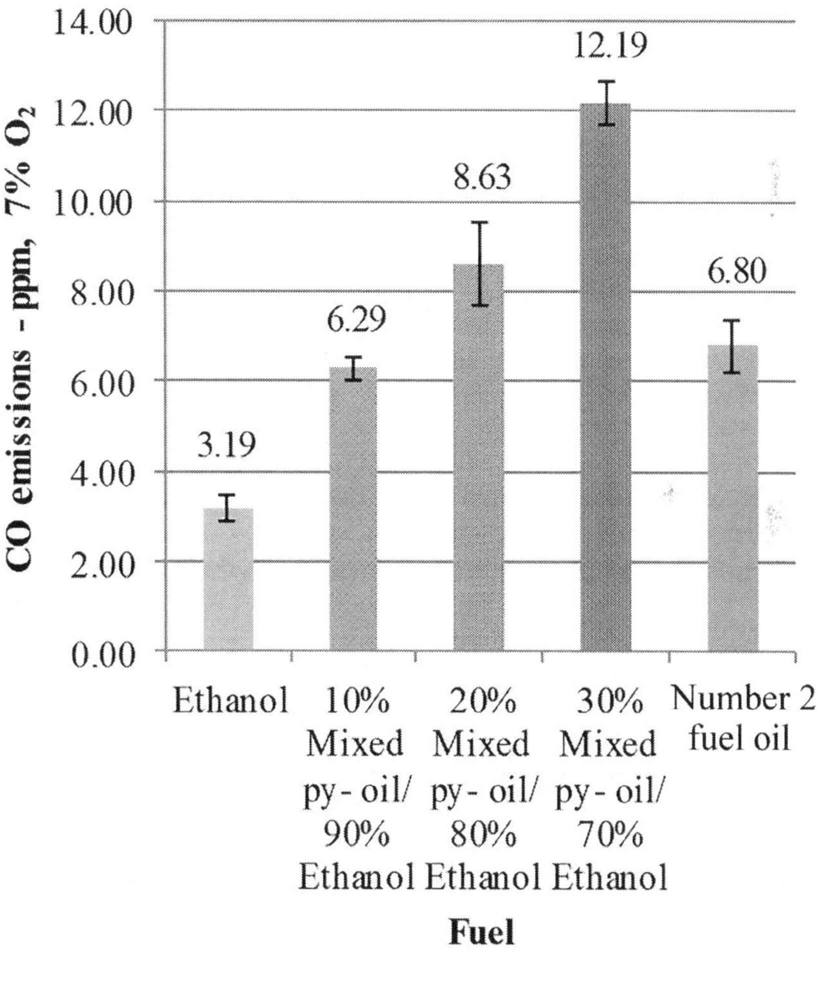

Figure 6.10 CO emissions at different pyrolysis oil/ethanol ratios (Martin and Boateng, 2014).

switchgrass, miscanthus, eucalyptus, pennycress, forest residues, and soiled animal bedding, was compared with number 2 fuel oil as a control.

Performance measures such as flame structure (Fig. 6.8) and combustion chamber axial temperature profile (Fig. 6.9) along with the total heat input, the gross heat output, heat losses to the flue gas, as well as exhaust gas emission concentrations are shown in Tables 6.5 and 6.6 to reveal some interesting characteristics. While they found that a blend of 20% pyrolysis oil/80% ethanol can be used as an alternative fuel in residential boilers with minimal retrofitting and derating as discussed earlier, there are subtle differences in emissions patterns. For example, while the 20:80 pyrolysis oil/ethanol blend ratio produced no detectable change in the CO and hydrocarbon emissions compared to number 2 fuel oil, NOx emissions can depend on the biomass from which the pyrolysis oil was created. When the pyrolysis oil fraction of the blend was produced from biomass with low amounts of nitrogen such as eucalyptus, NOx emissions reduced by as much as 12% compared to number 2 fuel oil, but pyrolysis oils produced from proteinaceous feedstocks led to an order of magnitude increase in NOx emissions.

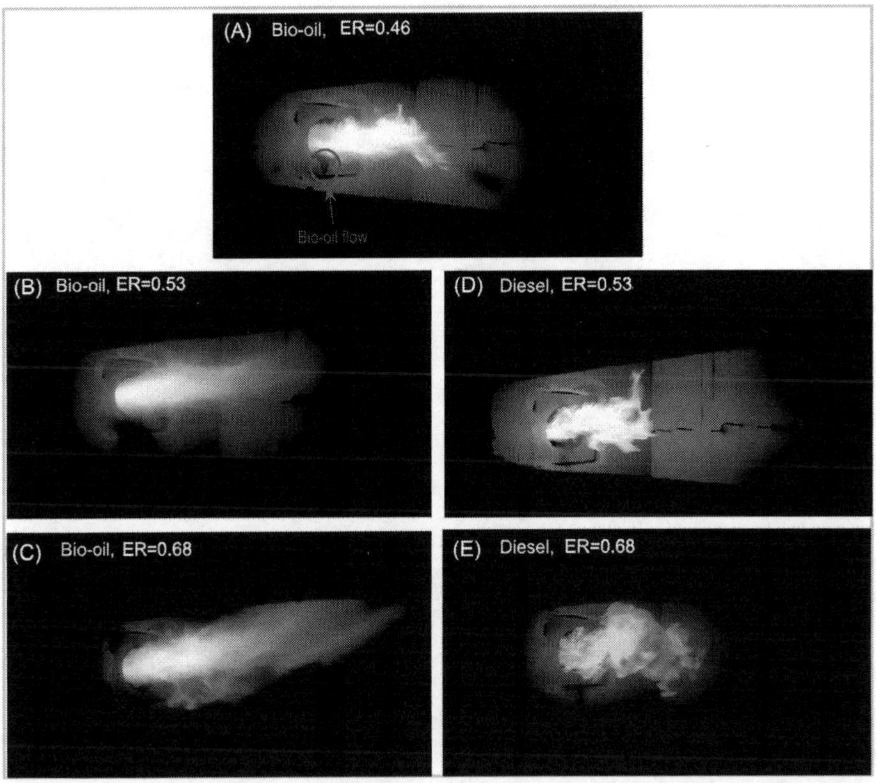

Figure 6.11 Flame structures of bio-oil and diesel at different air-fuel ER for 24 kW energy input (Lujaji et al., 2016a,b). *ER*, Equivalence ratio.

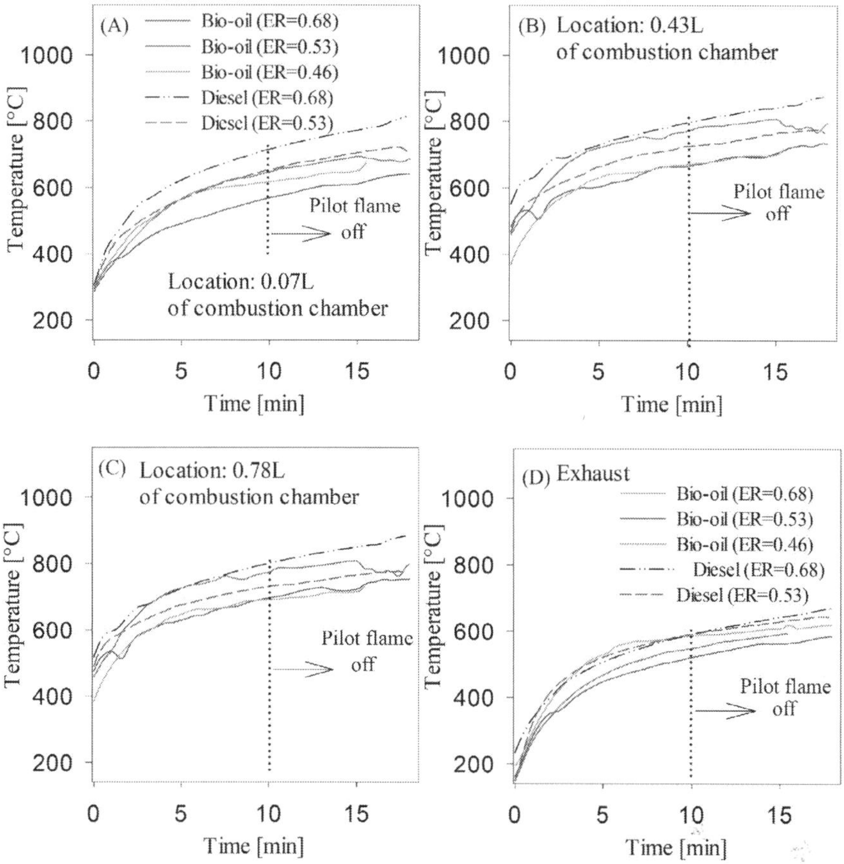

Figure 6.12 Temperature rise at three axial positions from the burner tile and the exhaust at different air/fuel ER: (A) 0.07 L (10 cm), (B) 0.43 L (60 cm), (C) 0.78 L (110 cm), and (D) exhaust. *ER*, Equivalence ratio.

The flame structures (Fig. 6.8) show, perhaps unsurprisingly, luminosity differences at the same heat input with the number 2 fuel oil flame being the most luminous, ethanol flame, the most transparent, while the luminosities of the flames of the fuel blends fall in-between the two extremes; the luminosity increasing with increased pyrolysis oil concentration. Pyrolysis oil introduces an increased number of incandescent particles from fine char particle residues that lead to unburned carbon in the exhaust as we learnt from earlier studies (Tzanetakis et al., 2011). The rate of temperature increase (Fig. 6.9) is an indication of the flame's heat release profile. Observed closely, Fig. 6.9 shows that as the pyrolysis oil concentration increases, the peak temperature drops, and the flame stretches indicating longer and increasingly lazy flame with less intensity compared with number 2 oil flame.

The heat balance (Table 6.5) indicates that the boiler efficiency was practically unchanged over all fuel blends (61.7%–65.1%) at the set heat duty, except that the combustion of ethanol/bio-oil blends produces more water as product due to the high mixture moisture content, thus more of the heat of combustion is lost due to the vaporization of this water.

CO emissions (Fig. 6.10) trend with increasing pyrolysis oil content of the blended mix fuel. This corresponds with the increasing viscosity of the fuel, which leads to poorer air/fuel mixing at the nozzle and also leads to incompleteness or delay in the oxidation of CO to CO_2 due, perhaps, to lower peak temperatures.

Similar studies, this time in a horizontal combustion chamber of a rectangular cross section with the air-assisted externally mixed nozzle described earlier mounted, were conducted to explore diffusion flame characteristics and performance. Combustion at several air/fuel equivalence ratios, including 0.46, 0.53, and 0.68 (116%, 88%, and 47% excess air, respectively) the externally mixed nozzle, could effectively atomize and ensure stable combustion of neat pyrolysis oil at the set heat rate even without ethanol blend. This, however, comes with a penalty associated with a lower peak flame temperature and, hence, lower heat flux compared with number 2 diesel fuel. The formation of carbon monoxide (CO) decreases with an increasing air/fuel equivalence ratio for bio-oil combustion. The levels of carbon dioxide (CO_2) and nitrogen oxides (NO_x) increase with an increasing air/fuel equivalence ratio for bio-oil combustion and were slightly higher than that generated by diesel. Hydrocarbon emissions do not follow any defined trend with an increasing air/fuel equivalence ratio for bio-oil, as typically observed for diesel fuels as a result of the oxygenated nature of bio-oil.

Captured digital images (Fig. 6.11) at 24 kWth energy input do not visually show any vast differences in flame intensity, except that the bio-oil flames appear to be longer and relatively more bushy than the diesel flames (Lujaji et al., 2016a,b). However, as Fig. 6.11B and C depict, the flame front tends to divert upward away from the burner horizontal axis for the pyrolysis oil as a result of a buoyancy effect exhibited by bio-oil combustion products despite the fact that the flames were equally

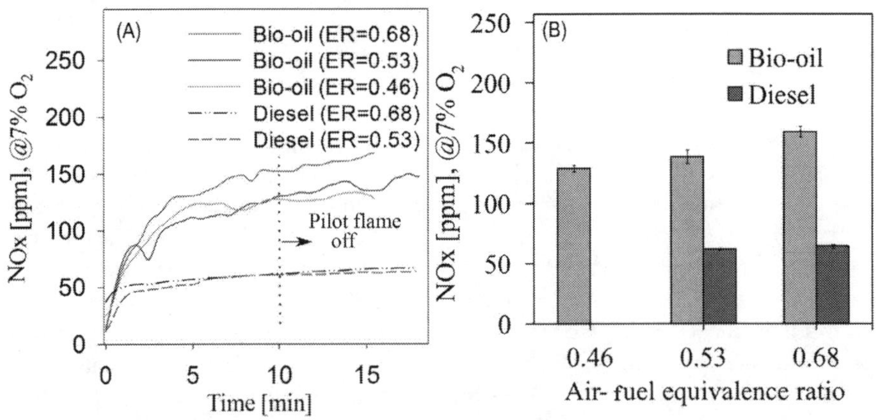

Figure 6.13 NO_x trending with neat pyrolysis oil combustion in a horizontal furnace.

intense. The high water content and the wide range of volatility in bio-oil lead to a slight decrease in flame temperature as reflected in the lowering of the combustion chamber wall temperature (Fig. 6.12). This has the tendency to reduce thermal NO_x, but fuel NO_x associated with inherent nitrogen of the biomass from which the pyrolysis oil is derived still persists (Fig. 6.13).

References

Chiaramonti, D., Oasmaa, A., Solantausta, Y., 2007. Power generation using fast pyrolysis liquids from biomass. Renew. Sustain. Energy Rev. 11, 1056–1086.

Czernik, S., Bridgwater, A.V., 2004. Overview of applications of biomass fast pyrolysis oil. Energy Fuels 18, 590–598.

Lal, S., Kushari, A., Gupta, M., Kapoor, J.C., Maji, S., 2010. Experimental study of an air assisted mist generator. Exp. Therm. Fluid Sci. 34, 1029–1035.

Lefebvre, A.H., 1980. Airblast atomization. Prog. Energy Combust. Sci. 6 (3), 233–261.

Li, Y., Barr, P.V., Watkinson, A.P., 2004. Bio-oil firing for industrial kiln operations. In: Proc. 6th Int. Symposium on Waste Processing and Recycling, COM 2004, Hamilton.

Lopez Juste, G.L., Salva Monfort, J.J., 2000. Preliminary test on combustion of wood derived fast pyrolysis oils in a gas turbine combustor. Biomass Bioenergy 19, 119–128.

Lujaji, F.C., Boateng, A.A., Schaffer, M.A., Mullen, C.A., Mtui, P.L., Mkilaha, I.S., 2016a. Pyrolysis oil combustion in a horizontal box furnace with an externally mixed nozzle. Energy Fuels 30, 4126–4136.

Lujaji, F.C., Boateng, A.A., Schaffer, M.A., Mtui, P.L., Mkilaha, I.S., 2016b. Spray Atomization of bio-oil/ethanol blends with externally mixed nozzles. Exp. Therm. Fluid Sci. 71, 146–153.

Martin, J.A., Boateng, A.A., 2014. Combustion performance of pyrolysis oil/ethanol blends in a residential-scale oil-fired boiler. Fuel 133, 34–44.

Martin, J.A., Mullen, C.A., Boateng, A.A., 2014. Maximizing the stability of pyrolysis oil/diesel fuel emulsions. Energy Fuels 28 (9), 5918–5929.

Naidu, P.R., Krishnan, V.R., 1996. Viscosities of binary liquid mixtures. Proc. Indian Acad. Sci. – Sect. A 64 (4), 229–236.

Schwietzke, S., Ladisch, M., Russo, L., Kwant, K., Makinen, T., Kavalov, B., et al., 2008. Analysis and identification of gaps in research for the production of second-generation liquid transportation biofuels. In: Report of the IEA Bioenergy Task 41, Project 2.

Shaddix, C.R., Hardesty, D.R., 1999. Combustion Properties of Biomass Flash Pyrolysis Oils: Final Project Report. Sandia National Laboratories, Albuguerque, New Mexico.

Shihadeh, A., Hochgreb, S., 2000. Diesel engine combustion of biomass pyrolysis oils. Energy Fuels 14, 260–274.

Shihadeh, A., Hochgreb, S., 2002. Impact of biomass pyrolysis oil process conditions on ignition delay in compression ignition engines. Energy Fuels 16 (3), 552–561.

Solantausta, Y., Nylund, N.-O., Westerholm, M., Koljonen, T., Oasmaa, A., 1993. Wood-pyrolysis oil fuel a diesel-power plant. Bioresour. Technol. 46, 177–188.

Tzanetakis, T., Moloodi, S., Farra, N., Nguyen, B., Thomson, M.J., 2011. Spray combustion and particulate matter emissions of a wood derived fast pyrolysis liquid ethanol blend in a pilot stabilized swirl burner. Energy Fuels 25 (4), 1405–1422.

Pyrolysis conversion technology systems and integration

7

7.1 Introduction

The synergy between petroleum refinery and thermo-catalytic biorefinery mentioned earlier on, offers a promising potential to engineer related unit operations for lignocellulosic biorefinery at equally high-efficiency rates. However, this promise has not been fully realized as the challenges encountered with the processing of lignocellulosic feedstocks using existing catalysts have been more than anticipated, leading to several false starts and outright closures. As such, technologies for the commercial production of pyrolysis oils followed by integrated upgrading systems are still in their developmental stages, although pyrolysis per se' is an old art. Therefore pyrolysis systems still fall under the classic definition of developing technologies as defined by sound engineering practice. According to the US Department of Energy (DOE) technology readiness assessment guidelines (Document DOE G 413.3-4A, 2015), a technology development is the process of developing and demonstrating new or unproven technology, the application of existing technology to new or different uses, or the combination of existing and proven technology to achieve a specific goal. This puts pyrolysis biorefinery, whether as a stand-alone system or its integration into existing petrochemical infrastructure, in this category. Technology development associated with a biorefinery project must, therefore, be identified early in the project life cycle, and its maturity level should have evolved to a confidence level that allows the project to establish a credible technical scope, schedule, and cost baseline. The US DOE employs a technology readiness level (TRL) assessment based on a scale used by the Department of Defense (DoD) and also NASA to rate bioenergy technology systems beginning from fundamental research through commercialization as described in Table 7.1.

In this chapter, efforts at various levels of pyrolysis systems technology development for the production of pyrolysis oil as well as their integration within existing petrochemical infrastructure are presented. Pyrolysis oil-production scales including laboratory or desk-top pyrolysis studies requiring a few grams of biomass per hour through pilot (5−10 kg/h) scale development (TRL = 2−3) to demonstration scale (TRL = 5−6) designed to process 1−2 tons/h have been developed within the thermo-catalytic biomass conversion community including the U.S. Department of Agriculture (USDA). Commercial production of neat pyrolysis oil for combustion fuel application is practiced by Ensyn, a Canadian entity, alone and together with Envergent, a joint venture with UOP-Honeywell (Des Plains, United States). Similar large-scale systems are currently being offered by Biomass Technology Group in the Netherlands. Systems integrating pyrolysis oil as a cofeed with

Pyrolysis of Biomass for Fuels and Chemicals. DOI: https://doi.org/10.1016/B978-0-12-818213-0.00007-2

Table 7.1 Referenced US Department of Energy technology readiness levels (TRL).

TRL-3	Analytical and experimental critical function and/or characteristic proof of concept: Active research and development is initiated. This includes analytical studies and laboratory studies to physically validate analytical predictions.
TRL-4	Component and/or breadboard validation in laboratory environment: Basic technological components are integrated to establish that they will work together.
TRL-5	Component and/or breadboard validation in relevant environment: Fidelity of breadboard technology increases significantly. The basic technological components are integrated with reasonably realistic supporting elements, so it can be tested in a simulated environment.
TRL-6	System/subsystem model or prototype demonstration in a relevant environment: Representative model or prototype system, which is well beyond that of TRL-5, is tested in a relevant environment. This represents a major step up in a technology's demonstrated readiness.
TRL-7	System prototype demonstration in an operational environment: This represents a major step up from TRL-6. It requires the demonstration of an actual system prototype in an operational environment, such as in a light duty vehicle on the road.

vacuum gas oil (VGO) in existing petrochemical processes are continuously being tested at Petrobras, Tesoro, WR Grace, etc.

7.2 Developmental scales

When it comes to the various scales of pyrolysis systems, units for the development of the technology at the microscale have played an important role in charting the course. Micropyrolyzers such as those offered by CDS Analytical (Oxford, Pennsylvania) and Frontier Laboratories (Fukushima, Japan) offer desktop analytical pyrolysis systems such as pyroprobe furnaces directly connected to gas-chromatograph/mass spectrometer (GC−MS) known as Py-GC/MS (Fig. 7.1) that give valuable data on the screening of large pools of biomass for their pyrolysis potential and efficacy (TRL = 1−2). These micro-pyrolyzers typically comprise an interface heater (100°C−400°C) and a pyrolysis furnace (40°C−800°C) in which milligrams of biomass samples or biomass/catalyst mixture sample may be instantaneously pyrolyzed and analyzed online. A carrier gas such as helium is typically used to purge air in the sample prior to pyrolysis and to convey the pyrolysis gas through the pyrolysis reactor, a filter, and then to a GC−MS. The fraction of the pyrolysis vapors that are not detected by GC may be analyzed by LC−MS, NMR, etc., when produced on larger-scale systems where all condensation products are collected. The GC analyzes high molecular weight compounds eluted in the pyrolysis vapor, that is, pyrolysis products that could be collected as condensates (bio-oil) in real

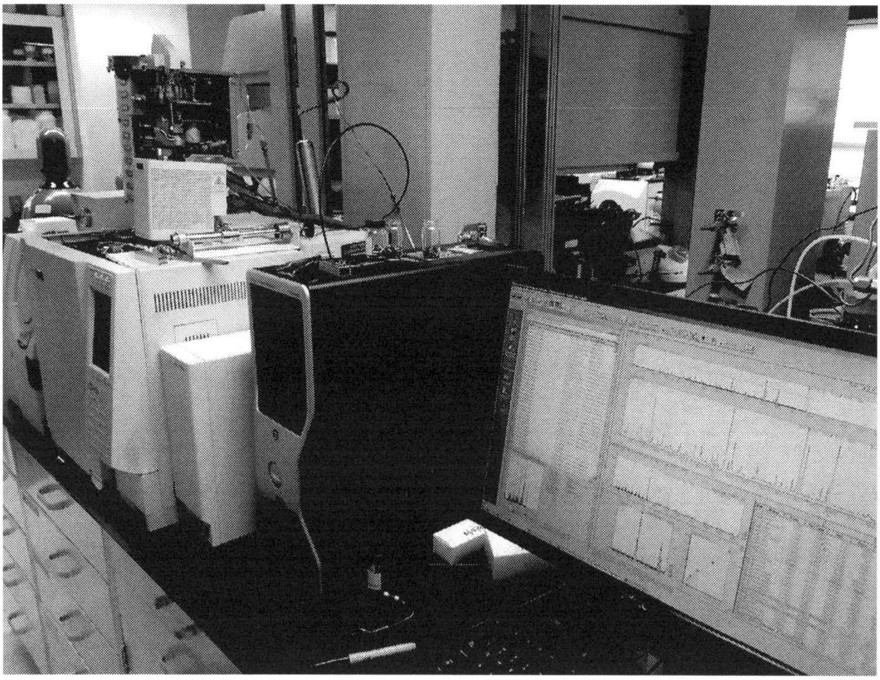

Figure 7.1 Pyrolysis coupled with gas chromatography and mass spectroscopy (Py-GC/MS), USDA.

systems. Among the limitations of such small-scale systems is that there is no actual collection of the liquid samples and neither is there a determination of the deactivation of the catalyst, which necessitates the use of large excesses of catalyst. However, through auto sampling that such systems offer, the composition, quantity, and quality of the pyrolysis products for a large sample pool of biomass can be evaluated, screened, and downselected for larger scale experiments.

Beyond this analytical scale are real systems for pyrolysis liquid production that may begin at a small scale (TRL = 2–3). For this, fluidized-bed systems with gram-scale capacities (Fig. 7.2) or pilot (kilogram-scale) capacities (Fig. 7.3) have been used as process development units (PDUs) to screen for small amounts of products that give a true heat and material balance needed for larger scale design. For these scales, up to 5 kg/h biomass such as switchgrass (SwG) can be uploaded into the unit and useful data such as energy requirements and product yields that it provides can be used as design parameters for larger systems based on the processing of select feedstock from Py-GC/MS. With these unit operations, bio-oil yields greater than 60% on a mass basis have been demonstrated for SwG and other perennial grasses, with energy conversion efficiencies ranging from 52% to 81%. The results are also used to show that char yielded would suffice in providing process heat for all the energy required for the endothermic pyrolysis reaction process; an

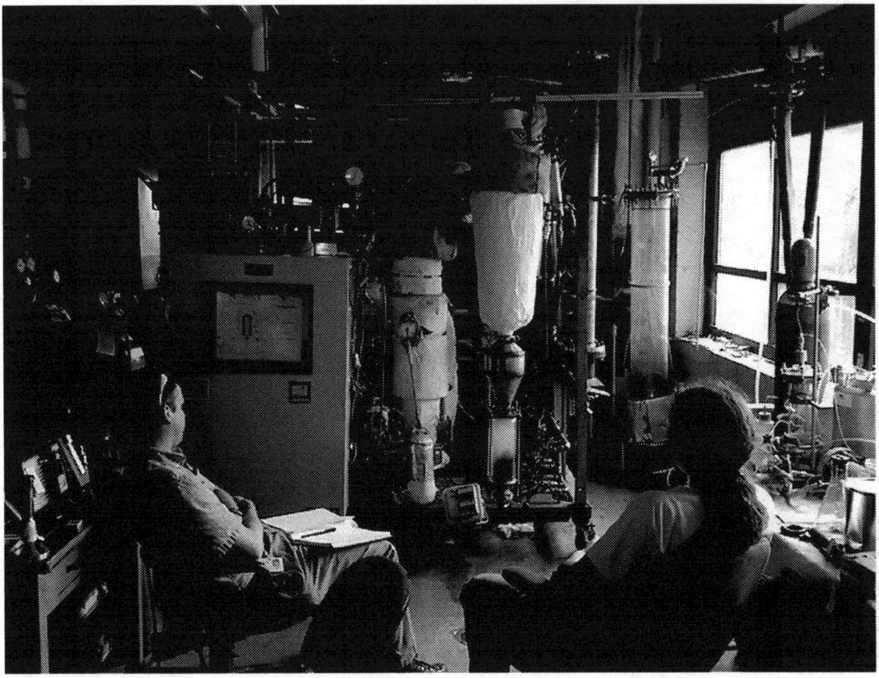

Figure 7.2 USDA's pyrolysis PDU, "The Kwesinator." *PDU*, Process development unit; *USDA*, U.S. Department of Agriculture.

important step for the design of energy self-sufficient in-forest or on-farm systems. Like all PDU's, these TRL's do not demonstrate long times on stream to assess their in-field potential. It requires a demonstration scale at TRL6/7 to do so.

Because stand-alone processes are needed to operate at the biomass source heat integration is a major part of the pyrolysis systems development. This can be addressed in several ways, one example being the patented combustion–reduction integrated pyrolysis system (CRIPS) with the layout shown in Fig. 7.4 that addresses heat integration by heating medium circulation. As shown, the CRIPS, a joint patent between USDA and the University of Pretoria, consists of two fluidized bed reactors in which one functions as a pyrolyzer, a reducing bed fluidized by exhaust gas stream; and the other as a combustor, an oxidizing bed fluidized by an air stream. The combustor provides the needed heat by the combustion of residual char that is returned along with the transfer of spend pyrolysis medium from the pyrolyzer to the combustor via an auger system. Such design arrangement not only generates its own heat for the endothermic pyrolysis process but also allows for the regeneration of spent catalysts when such medium is used by burning off carbon deposits in the combustion chamber.

Another effective integrated heat system is the circulating fluidized bed (CFB) that underlies the fluid catalytic cracking (FCC) technology used in the petroleum industry, first patented by UOP-Honeywell. The use of the CFB principle for heat integration and catalyst regeneration in bioconversion (Fig. 7.5) has been the

Figure 7.3 Pilot scale (10 kg/h) fluidized-bed pyrolysis system.
Source: Courtesy USDA.

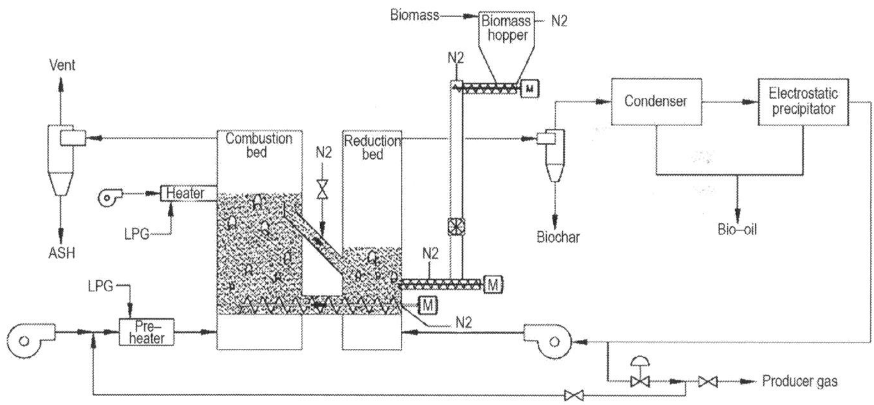

Figure 7.4 CRIPS (Boateng et al., 2019). *CRIPS*, Combustion–reduction integrated pyrolysis system.

cornerstone of Green Fuel Nordic's (Finland) biorefinery concept and Ensyn's Rapid Thermal Process (RTP) (Canada). This arrangement is employed in their full-scale, commercial design for their pyrolysis oil production. Like the CRIPS char and sand or catalyst is directed to a reheater for combustion, heat recuperation, and in the case of catalytic fast pyrolysis (CFP) regeneration of spent catalyst. Its ability to

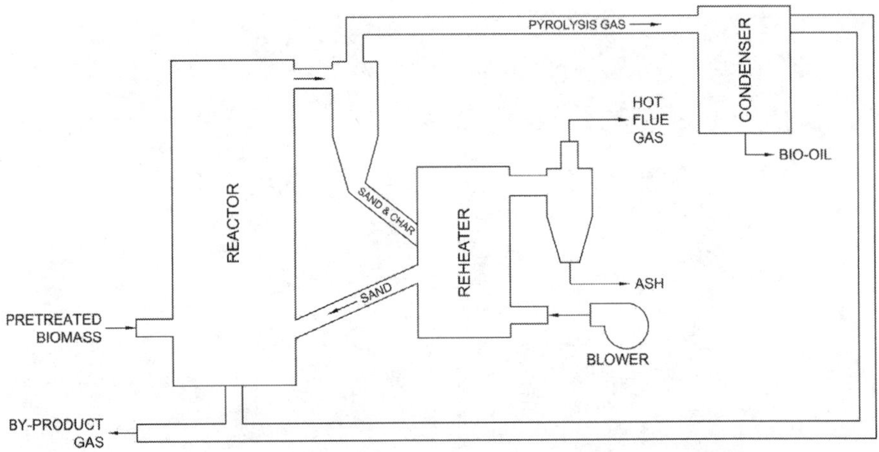

Figure 7.5 Generic circulating fluidized-bed based biomass pyrolysis process arrangement underlying Green Fuel Nordic and Ensyn's rapid thermal process (RTP) technologies.

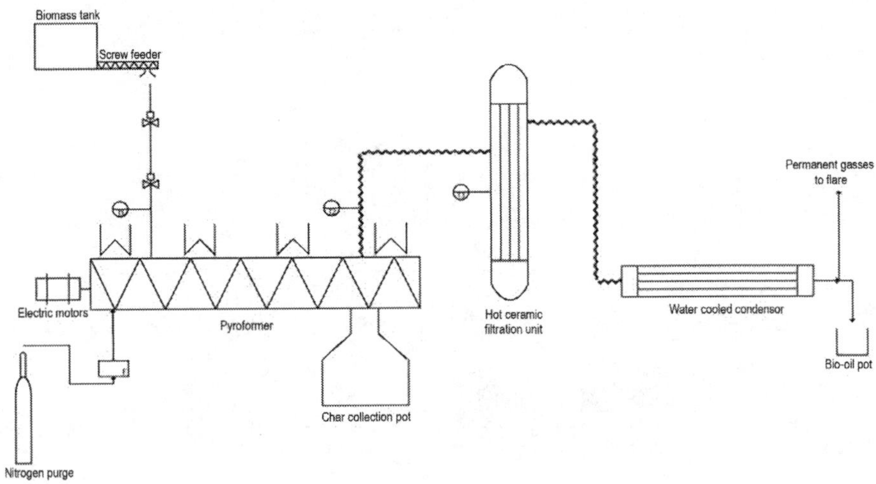

Figure 7.6 Auger pyrolysis system (Mahmood et al., 2013).

regenerate spent heterogeneous catalysts must have been one of the attractions for Ensyn's joint venture with UOP, the inventor of the FCC to form Envergent Technologies.

For the same heat-integration reasons the auger pyrolysis systems such as the Pyroformer patented by Aston University (Fig. 7.6) have found a meaningful application in the pyrolysis biorefinery space since pyrolysis is heat transfer limited. Constructed of carbon-steel, the twin horizontal rotary coaxial screw system offers the advantage of defined residence times. The outer screw transports a fraction of the char produced during pyrolysis back, so that there is an internal recycling of heat transfer material medium. The system's compactness favors small design footprints suitable for on-farm applications. These systems are currently at a TRL < 6 due to issues with scalability.

7.3 Distributed on-farm/in-forest biorefining

The distributed/satellite biorefinery concept (Fig. 7.7) whereby pyrolysis is carried out at the biomass source and the upgrading or utilization done at a centralized location is considered the most economic system for biofuels production since bulky biomass is uneconomical to hull beyond certain distances known as economic

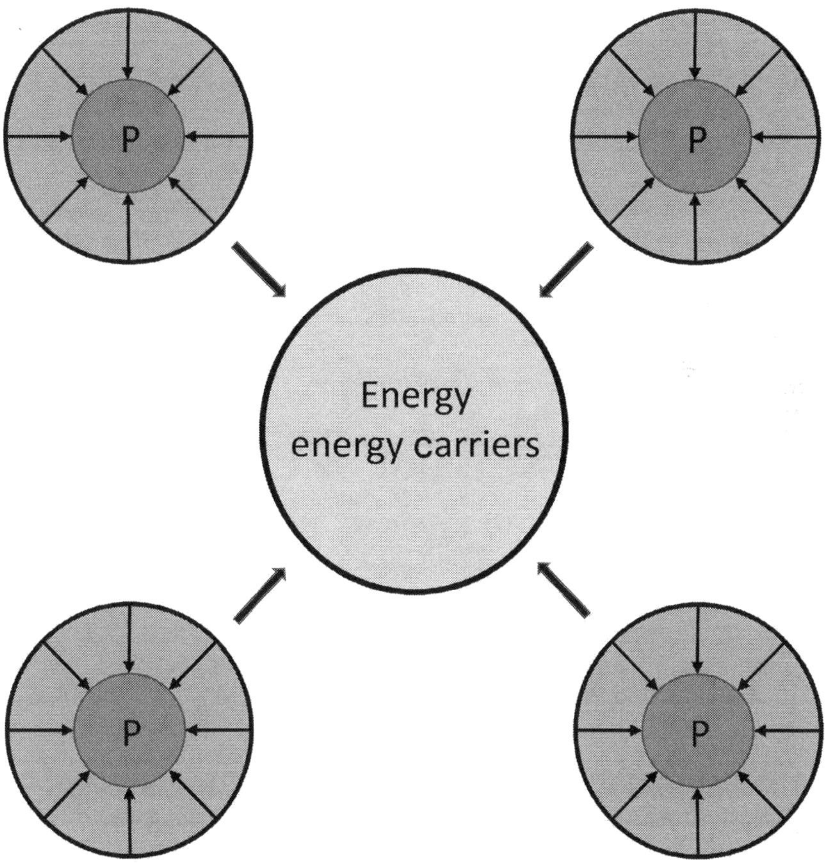

Figure 7.7 Concept for a distributed or satellite model using a centralized refinery.

radius. Such systems are most adaptable to the rural economies, including famer groups, perhaps, the most important stakeholders of the biofuels opportunities.

Distributed pyrolysis is a rural or cooperative concept at times called the village-scale system. The system (Fig. 7.8) has been a target by industrial groups and venture capitalists for a long time but has not had traction at the moment. Companies such as Google Inc., philanthropist organizations, and hedge-fund companies have touted distributed biorefineries as a model of importance to villages in poor countries, where small-scale pyrolysis can make a real difference in local energy production and provide a path for survival in the oil-constrained world. If properly deployed, distributed on-farm pyrolysis might play a major role in energy independence and improve the livelihood of rural America and overseas.

Most studies have consistently shown that the 2000 metric tons per day (MTPD) biomass processing plant should be the ideal economic capacity for the production of fungible fuels. These include DOE research facilities such as the National Renewable Energy Laboratory (NREL) and Pacific Northwest National Laboratory (PNNL). This means that if this plant is centralized, near-source biomass will quickly deplete and large amounts of biomass will need to be transported from far distances if the plant should continue to operate over several years of the product life cycle. Use of biomass feedstock such as perennial grasses will suffer economically due to their low bulk density. Densification of the biomass at site has,

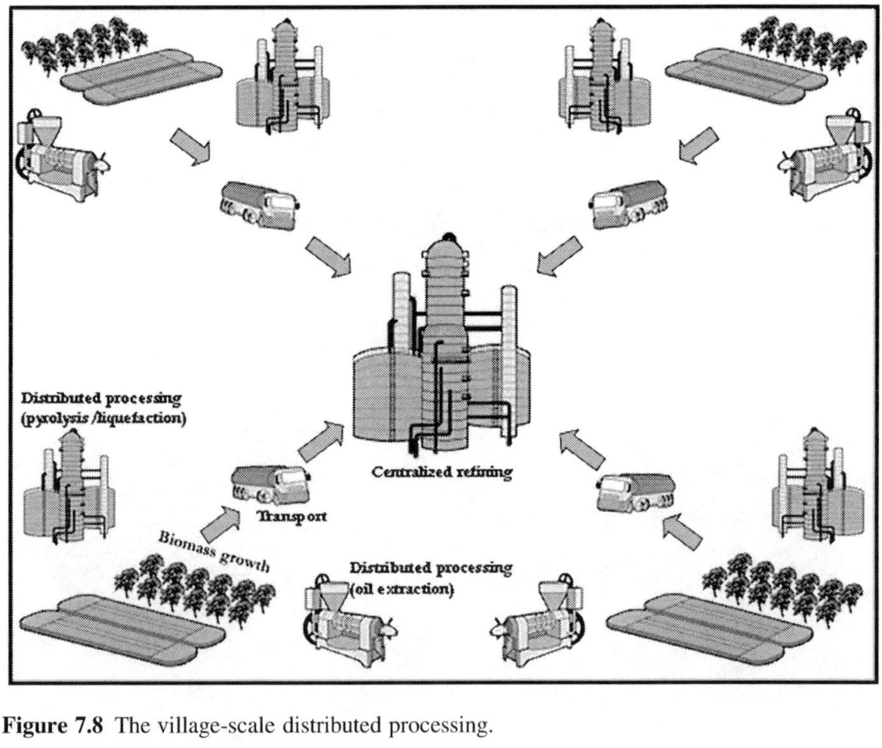

Figure 7.8 The village-scale distributed processing.

therefore, shown to be effective. Lin et al. (2016) showed that the transportation costs for biomass would generally follow the pattern of coal transportation. And that converting biomass to fuels, for example, ethanol locally, and shipping it over long distances is most economical, similar to the existing grain-based biofuel system. For pyrolysis biorefinery, this is equivalent to pyrolyzing the biomass at site and shipping the pyrolysate (bio-oil) to a centralized facility for direct use (combustion or gasification) or upgrading in pyrolysis biorefinery systems. The nominal 2000 MTPD facility could, therefore, be distributed into at least ten 200 MTPD processing plants distributed over a certain economic radius. However, the critical or minimum economic size is yet to be accurately defined. Wright et al. (2008) investigated three on-farm distributed processing of pyrolysis systems comprising 5.4 MTPD capacity; "small cooperative" size pyrolyzers of 55 MTPD capacity and "large cooperative" pyrolyzers of 550 MTPD capacity. Distributed processing was combined with very large centralized bio-oil processing plants that accept bio-oil for catalytic upgrading to transportation fuels (hydrodeoxygenation). They found the selling price to be competitive with gasoline. The bioliq-process developed by KIT (Karlsruhe Institute of Technology in Germany) together with the firm LURGI uses a twin-screw mixing reactor (sand cracker) as a pyrolysis reactor deployed at the biomass source. The produced bio-oil and part of the char is mixed to form a stable slurry which is transported to a centralized location to be transformed into syngas in an entrained pressurized flow gasifier with subsequent direct synthesis of dimethylether. Before we explore the effect of scale on the techno-economics of systems in the next chapter, suffice it to say that distributed systems are the economic models that pyrolysis biorefining technologies must be designed to address biomass transportation costs issues but only when the small units can be mass-produced as these are cost implications associated with engineering them individually.

Distributed systems must comprise a collection or satellite unit operations at a predetermined radius. If a mobile system is scale appropriate, it would reduce the set of unit operations to a selected few that could be deployed and redeployed at the biomass site when needed. To evaluate its strategic importance to sustainability a 2 MTPD skid-mounted and mobile CRIPS unit was designed and built at the USDA to test the distributed concept where, instead of building several local plants involving several capital expenditures, the mobile unit would visit each farm for an equal amount of time to simulate the rural distributed on-farm concept. System features are shown Fig. 7.9.

7.3.1 Mobile pyrolysis systems

The USDA trailer-mounted pyrolysis apparatus for on-farm or in-forest applications, shown in Fig. 7.10, is based on the CRIPS. The mobile system demonstrates efficacy of on-farm production and the coordination of station-to-station operation to simulate a collective system of several unit operations within a distributed/satellite system of pyrolysis biorefinery. Key system design features that provide utility for remote operations including in situ (on-trailer) generation of heat and electric

Figure 7.9 Picture and simplified schematic of the 2 MTPD CRIPS constructed at USDA. *CRIPS*, Combustion–reduction integrated pyrolysis system; *USDA*, U.S. Department of Agriculture.

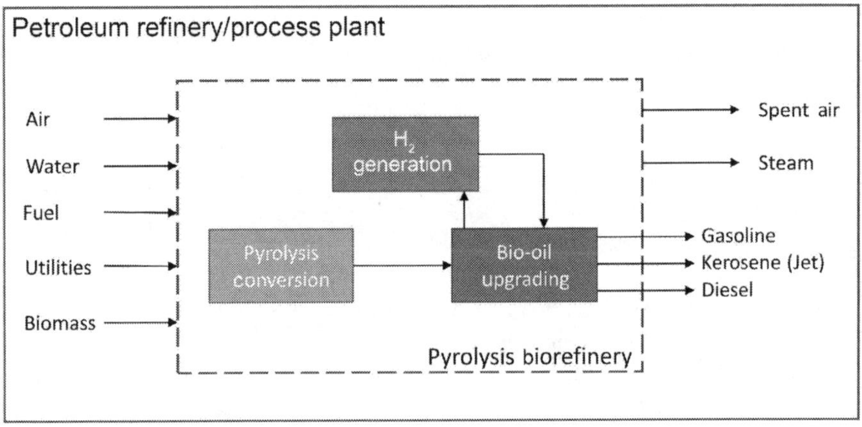

Figure 7.10 Colocation and integration of pyrolysis biorefinery into petroleum refinery.

power required for energy self-sufficiency have been successfully tested at this scale (TRL = 6–7). Extensive trials on three feedstocks important to the US agriculture, namely, woody biomass, switchgrass, and horse litter, were conducted with results showing that a processing rate of up to 40 kg/h (approximately 1 MTPD) is easily processed for the said biomass using the trailer. Beyond this throughput, however, operational problems such as pressure imbalance between the dual-bed reactors could hamper process control. The system's best operational feature is when the CRIPS is operated under CFP conditions during which catalyst regeneration is readily achievable with its heat recirculation design. Bio-oil yield for neat/traditional and catalytic pyrolysis were in the 45% and 5%−10% ranges, respectively, matching lab scale results and demonstrating production of large volumes of bio-oil on-farm. Bio-oil quality for catalytic pyrolysis is consistent over several hours on stream due to continuous catalyst regeneration that the system affords yielding high levels of BTEX compounds with the only limitation being some deactivation due to alkali contamination noticeable at cumulative biomass to catalyst ratios of >6/1. Compared with laboratory-scale results, noncatalytic pyrolysis product quality for the mobile, demonstration scale (TRL = 6) system wavered between that of regular pyrolysis oil produced under nitrogen atmosphere in the lab and that of the tail-gas reactive pyrolysis (TGRP) as partial recycle of effluent gas was possible with the mobile, skid-mounted system. The successful demonstration at 1 MTPD capacity qualifies the mobile system at the TRL for precommercial design suggested for blueprints (i.e., development of full-scale prototype).

7.4 Integrated pyrolysis biorefinery systems

Pyrolysis biorefinery integrated systems can be centralized, distributed, or collocated where the pyrolysis oil production and usage, for example, power generation or upgrading to fungible fuels, are collocated and coupled or decoupled. One technological challenge is how biofuel products can be seamlessly integrated into existing petroleum infrastructure. Given the fungibility of pyrolysis oils integration or insertion of pyrolysis products into existing petroleum processes is easier than biochemical products such as ethanol. Cofeeding of pyrolysis oils with VGO into an FCC cracker have been tested as a viable process that would integrate pyrolysis biorefinery and petroleum refinery in a near term. This is because bio-oil has viscosity and molecular weight range similar to that of VGO, and the hydrogen-rich nature of VGO can chemically complement the bio-oil's deficiency in hydrogen. Other systems whereby the pyrolysis oil production step is colocated within the petroleum refinery to take advantage of latter's hydrogen system (Fig. 7.10) was reported by PNNL (Elliott et al., 2009). These systems can share resources of the existing petroleum infrastructure such as hydrogen generation and many other unit operations including water and electrical utilities.

Many large petroleum companies have touted pyrolysis oil—VGO coprocessing as the most promising integration route for pyrolysis biorefinery in the nearer term although the bio-oil feed fractions are small. Bio-oil containing low oxygen content is used as a feedstock alongside VGO in FCC units to generate renewable gasoline and diesel at Ensyn. Here the VGO feedstock is displaced by up to approximately 5% bio-oil. Coprocessing of bio-oil and VGO in a FCC cracker has been tested in operating commercial refineries, FCC pilot facilities, and laboratory scale FCC test equipment around the world. It is envisioned that approximately 23 million gal/year of bio-oil at 5% bio-oil input usage rate would be used for FCC coprocessing. This opportunity has forced partnerships between bio-oil producers and the petrochemical industry. Tesoro partnered with Fulcrum BioEnergy and Ensyn to obtain bio-oils produced by Ensyn from municipal solid waste and tree residues for use as a feedstock for FCC coprocessing with traditional crude oil. Elsewhere, at W. R. Grace, a blend of 3 wt.% pyrolysis oil was coprocessed with 97 wt.% VGO using commercial equilibrium FCC catalyst in their pilot plant facility called DCR, a proprietary circulating riser pilot FCC unit and one of the leading, commercially available FCC technology for small-scale testing made possible thanks to innovations such as slide-valve, pressure balanced, and adiabatic operation. The DCR plant has provided data to aid in a better understanding of the behavior of bio-oil in large-scale reactors. In one test it was reported that the pyrolysis oil was not pretreated and contained 23 wt.% water and 53 wt.% oxygen as produced by thermal pyrolysis. Compared to 100% VGO processing, cofeeding pyrolysis oil at this initial oxygen content resulted in more coke (6.4–7.1 wt.%), less gasoline (49.1–47.5 wt.%), and production of CO (0.48 wt.%) and CO_2 (0.11 wt.%). Elsewhere, Petrobras in a collaboration with NREL coprocessed oak/sugarcane-bagasse–derived pyrolysis oil generated by NREL with VGO in a demonstration scale FCC unit (150 kg/h) developed at Petrobras (Fig. 7.11). Up to 20 wt.% bio-oil was coprocessed without pretreatment of the pyrolysis. The bio-oil and VGO were cracked into liquified petroleum gas (LPG) and gasoline with similar products yields as obtained for pure VGO when 10 wt.% bio-oil was used. The elemental analyses of the bio-oil used in various cofeeding tests (Table 7.2) show high water and high oxygen content compared with that produced with some of the recent pyrolysis technologies we have previously discussed including reactive pyrolysis or CFP which yields bio-oils with midlevel oxygen content indicating that there is tremendous room for product quality improvement in the foreseeable future should these become commercially available.

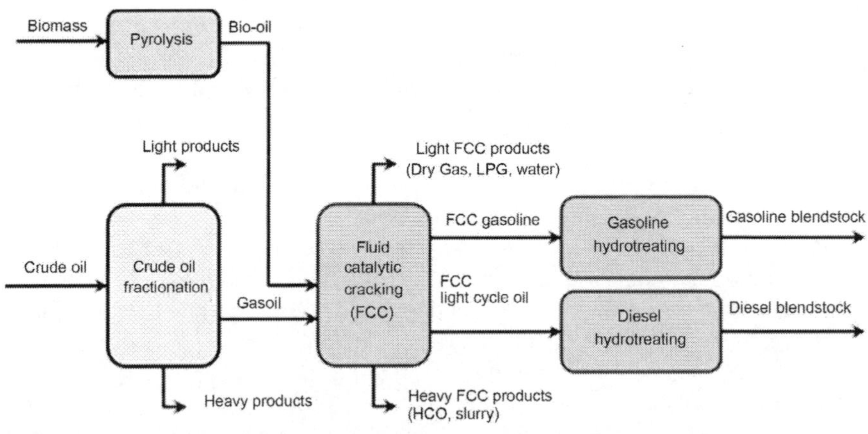

Figure 7.11 FCC unit operations; Petrobras FCC pilot unit (Pinho et al., 2017). *FCC*, Fluid catalytic cracking.

Table 7.2 Properties of the bio-oils.

	W.R. Grace pyrolysis oil	NREL pyrolysis oil	CRIPS bio-oil	TGRP bio-oil
Water content (wt.%)	23.0	25.5	20.3	7.2
%N (db)	–	–	0.3	1.1
%C (db)	55.5	56.8	58.9	71.2
%H (db)	6.5	4.9	6.0	6.3
%O (db)	38.0	38.3	34.8	21.3

CRIPS, Combustion–reduction integrated pyrolysis system; *NREL*, National Renewable Energy Laboratory; *TGRP*, tail-gas reactive pyrolysis.

Table 7.3 Elemental, high heating value (HHV), and proximate analysis of neat vacuum gas oil (VGO) and three bio-oil distillate bottoms from guayule bagasse (GB), a latex rubber plant biomass, spirulina, a microalgae [slow pyrolysis (SP)], and switchgrass (SwG), a herbaceous perennial grass (Choi et al., 2018).

	N (wt.%)	C (wt.%)	H (wt.%)	O (wt.%)	H/C (mol)	HHV (MJ/kg)	Volatiles (%)
VGO	–	88.0	12.1	–	1.65	46.8	100
GB bottoms	2.6	86.5	5.6	5.3	0.78	35.8	54
SP bottoms	11.9	75.9	5.5	6.7	0.87	30.5	59
SwG bottoms	1.0	77.8	4.8	16.4	0.74	30.3	46

At the USDA, distillate bottoms of bio-oils created by TGRP from three selected biomass with characteristics summarized in Table 7.3 were cofed and cracked with VGO over two heterogeneous catalysts, HZSM-5 and Y-zeolite, using the Py-GC/MS at 750°C–950°C. As the table shows, while the bio-oil solid residues have similar fuel contents, their oxygen contents range in an increasing order. Although the thermal system used possesses far less than FCC capabilities, the goal was to evaluate the cracking products individually and as a cofeed with VGO over the said FCC catalysts. Proton NMR was used to analyze grouped aromatics, olefins and alkanes produced when cofeeding and also to compare with what one would predict as arithmetic sum of such produced when cracked alone (i.e., theoretical product sum).

Tables 7.4 and 7.5 present the yields of the compound groups of olefins, aromatics, and alkanes from 1:1 VGO bio-oil bottom mixture cofeeds. Also shown in the tables are theoretical product expected based on percentages of the products each reactant is expected to contribute based on its own cracking and the percent change encountered between the actual (experimental) and calculated product. Carbon yield is also presented when cofed. For both temperatures at which the cofeeding experiments were conducted (850°C and 950°C), product yield patterns were similar for both catalysts employed (HZSM-5 and Y-Zeolite). Cofeeding

Table 7.4 Measured catalytic cocracking products of vacuum gas oil with bio-oil distillation bottoms (1:1 ratio) for HZSM-5 catalyst (Choi et al., 2018).

		850°C				950°C			
		Cofeed experimental	**Theoretical sum**	**Percentage of change**	**Percentage of carbon yield**	**Cofeed experimental**	**Calculated**	**Percentage of change**	**Percentage of carbon yield**
GB	Aromatics	4.1	3.03	35	4.34	5.59	5.13	9	5.91
	Olefins	6.18	4.97	24	6.09	8.4	7.96	6	8.27
	Alkanes	1.81	1.94	−7	1.62	3.14	2.95	6	2.78
SP	Aromatics	3.1	3.1	0	3.48	4.76	4.87	−3	5.34
	Olefins	6.66	5.63	18	6.96	12.04	8.55	41	12.58
	Alkanes	2.02	2.03	−1	1.92	3.59	2.99	20	3.39
SwG	Aromatics	4.91	3.14	56	5.44	9.12	5.23	74	10.11
	Olefins	6.23	4.9	27	6.43	11.58	7.92	46	11.96
	Alkanes	1.69	2.22	−24	1.58	3.28	3.24	1	3.05

GB, Guayule bagasse; *SP*, slow pyrolysis; *SwG*, switchgrass.

Table 7.5 Measured catalytic cocracking products of vacuum gas oil with bio-oil distillation bottoms (1:1 ratio) for Y-type zeolite catalyst (Choi et al., 2018).

		850°C				950°C			
		Experimental	**Calculated**	**Percentage of change**	**Percentage of carbon yield**	**Experimental**	**Calculated**	**Percentage of change**	**Percentage of carbon yield**
GB	Aromatics	4.17	3.52	18	4.41	5.31	4.33	23	5.62
	Olefins	5.64	3.79	49	5.56	11.13	7.73	44	10.96
	Alkanes	1.97	2.07	−5	1.76	3.58	3.61	−1	3.18
SP	Aromatics	3.84	3.18	21	4.31	5.43	3.85	41	6.09
	Olefins	6.23	4.28	46	6.51	11.91	8.46	41	12.45
	Alkanes	2.18	2.08	5	2.06	3.64	3.55	3	3.43
SwG	Aromatics	4.97	3.61	38	5.51	12.08	4.47	170	13.39
	Olefins	4.32	3.69	17	4.46	9.5	7.57	26	9.81
	Alkanes	1.65	2.31	−28	1.54	3.21	3.65	−12	2.98

GB, Guayule bagasse; *SP*, slow pyrolysis; *SwG*, switchgrass.

yielded more of the target products than expected except for a few cases, that is, alkanes from VGO-GB (guayule bagasse), VGO-SwG. A more fitting analyzes is directed toward Y-Zeolite because of its frequent use as FCC catalysts and its larger pore sizes that are suitable for bio-oil molecules.

Table 7.5 depicting the results for VGO/bottoms copyrolyzed over Y-type zeolite shows that overall, coprocessing reactions over Y-type zeolite showed more synergistic yield increases, primarily for aromatics and olefins. While both catalysts tend to suppress formation of saturated alkanes across the set temperatures, the characteristic of the starting materials is importantly significant. For spirulina and GB, with their low oxygen content and relatively higher H/C molar ratio, a nearly 50% increase in olefins yield is observed, with 12.5% carbon yield for olefins being the highest observable amongst the set. Over the Y-Zeolite the highest excess production of aromatics comes from SwG bottoms/VGO mixture at 950°C; yielding almost three times as much aromatic product as would be expected from individual reactants.

The data from the tables suggest that during cofeeding of VGO and bio-oil bottoms, VGO provides the availability of olefin compounds that enhance reactions particularly with oxygenated compounds toward the formation of mono-aromatics and the depression of coke formation that would be facilitated otherwise. SwG bottoms contained fewer volatiles, and the nonvolatiles therein possess a higher oxygen content. These nonvolatiles likely provide more opportunities for reactions of the oxygenated groups with VGO intermediates. It is also conceivable that radical intermediates would react more with fixed carbon that is already highly aromatic. Previous bio-oil/FCC studies (Gueudré et al., 2015; Ibarra et al., 2016b) have demonstrated the effects that oxygenated compounds have on coke formation during catalytic cracking. Furthermore, previous coprocessing studies (Wang et al., 2016; Ibarra et al., 2016a) showed that VGO cocracked with bio-oil suppressed the yields of light alkanes, while yields of other fractions such as gasoline, heavy oil, and coke remained the same. These studies suggest that the suppressed formation of alkanes occur at the expense of increased gas formation (CO, CO_2, and H_2O).

We have alluded to the fact that the system used in the USDA study is far different from FCC. While FCC tends to produce higher yields of lighter compounds, the system setup is more analogous to delayed coking than FCC, primarily due to the nature of the starting material and the yield distributions obtained. Delayed coking handles the lowest volatility residues with portions of fixed carbon that are polyaromatic in character.

In conclusion, distillate bottoms from mild-level oxygen bio-oils made by TGRP coprocessed with VGO over FCC catalysts in micropyrolyzers at elevated temperatures exhibit synergistic yields of key compound classes and inform their high probabilities in real FCC cofeeding. Specifically, yields of LPG-range olefins can be increased by nearly 50% for most bio-oil derived bottoms. For bottoms that are more hydrogen-deficient such as SwG-based bottoms (lignocellulosic biomass), cofeeding with VGO can generate as high as three times more aromatics than what would be expected without cofeeding. Y-type zeolite shows significantly better yields for aromatics compared to HZSM-5, likely due to the larger pore structure. For bottoms samples that originally possessed higher H:C ratios such as latex-sourced or cinobacteria, a microalgae, higher yields of olefins are readily realized. Each increased the production of olefins, and aromatics are almost always associated with a decreased production of light alkanes, compared with what is theoretically anticipated. We will examine cofeeding of pyrolysis oil and VGO in field trials carried out by oil companies, their technological challenges and their economic potentials in the chapters ahead when we discuss biorefinery performance measurements.

References

Boateng, A.A., Schaffer, M.A., Mullen, C.A., Goldberg, N.M., 2019. CRIPS Mobile demonstration unit for fast- and catalytic pyrolysis: the combustion reduction integrated pyrolysis system (CRIPS). J. Anal. Appl. Pyrolysis 137, 185–194.

Choi, Y.S., Elkasabi, Y., Tarves, P.C., Mullen, C.A., Boateng, A.A., 2018. Co-cracking of bio-oil distillate bottoms with vacuum gas oil for enhanced production of light compounds. J. Anal. Appl. Pyrolysis 132, 65–71.

Elliott, D.C., Hart, T., Neuenschwander, G.C., Rotness, L.J., Zacher, A.H., 2009. Catalytic hydroprocessing of biomass fast pyrolysis bio-oil to produce hydrocarbon products. Environ. Prog. Sustain. Energy 28, 441–449.

Gueudré, L., Thegarid, N., Burel, L., Jouguet, B., Meunier, F., Schuurman, Y., et al., 2015. Coke chemistry under vacuum gasoil/bio-oil FCC co-processing conditions. Catal. Today 257 (2), 200–212.

Ibarra, A., Rodríguez, E., Sedan, U., Arandes, J.M., Bilbao, J., 2016a. Synergy in the cracking of a blend of bio-oil and vacuum gasoil under fluid catalytic cracking conditions. 2016. Ind. Eng. Chem. Res. 55, 1872–1880.

Ibarra, A., Veloso, A., Bilbao, J., Arandes, J.M., Castaño, P., 2016b. Dual coke deactivation pathways during the catalytic cracking of raw bio-oil and vacuum gasoil in FCC conditions. Appl. Catal. B Environ. 182, 336–346.

Lin, T., Rodríguez, L.F., Davis, S., Khanna, M., Shastri, Y., Grift, T., et al., 2016. Biomass feedstock preprocessing and long-distance transportation logistics. GCB Bioenergy 8, 160–170.

Mahmood, A.S.N., Brammer, J.G., Hornung, A., Steele, A., Poulston, S., 2013. The intermediate pyrolysis and catalytic steam reforming of brewers spent grain. J. Anal. Appl. Pyrolysis 103, 328–342.

de Rezende Pinho, A., de Almeidaa, M.B.B., Mendesa, F.L., Casavechiab, L.C., Talmadgec, M.S., Kinchinc, C.M., et al., 2017. Fast pyrolysis oil from pinewood chips co-processing with vacuum gas oil in an FCC unit for second generation fuel production. Fuel 188, 462–473.

US Department of Energy. Technology Readiness Assessment Guide, 2015. Document DOE G 413.3-4A, Office of Project Management Oversight & Assessments. Washington, DC. Available at https://www.directives.doe.gov/directives-documents/400-series/0413.3-EGuide-04-admchg1/images/file

Wang, C., Li, M., Fang, Y., 2016. Coprocessing of catalytic-pyrolysis-derived bio-oil with VGO in a pilot-scale FCC riser. Ind. Eng. Chem. Res. 55 (12), 3525–3534.

Wright, M.M., Brown, R.C., Boateng, A.A., 2008. Economic analysis of distributed processing of biomass to bio-oil for subsequent production of Fischer-Tropsch liquids. Biofuels, Bioprod. Biorefin. 2, 229–238.

Biorefinery performance measurements

8

8.1 Introduction

The viability of any new technological development and its practice is demonstrated by its techno-economic feasibility as well as how its environmental sustainability measures up against the old practice. For that matter, bioenergy projects, including pyrolysis biorefining, must be assessed through this lens. The measurement thrust must aim at quantifying and analyzing the environmental and economic trade-offs along the full supply chain of the feedstocks converted and downstream technologies, some of which were discussed in the previous chapters. Material balance data along with techno-economic analysis established around potential feedstocks and their downstream transport logistics are critical components for building the foundation for life cycle inventory (LCI) needed for the estimation of green house gas (GHG) emissions as required under Energy Independence and Security Act (EISA) of 2007 and RFS2 (Schnepf and Yacobucci, 2012), that is, achieve 50% reduction in GHG. This must also include quantification of the 100-year global warming potential (GWP). Process flow evaluation based on credible experimental data may be gathered by thermodynamic models or chemical process simulation tools such as APEN + and Pro-II to establish a material and energy balance for which component equipment within the unit operations may be appropriately sized and cost analysis conducted in order to compare pyrolysis process conversion optimization trade-offs (McAloon et al., 2000; Kwiatkowski et al., 2006).

Life cycle assessment (LCA) is a systematic method for understanding the environmental consequences of products, processes, and activities. LCA models consider the energy and environmental emissions that occur in the unit operations of the product supply chains and the environmental credits of coproducts that can substitute for other products in the bioeconomy. For biofuel systems and transportation fuels, LCA models comprise the feedstock production, fuel conversion, distribution to end-use markets, and finally consumption, the so called "well-to-wheel (WTW)" models (Wang, 2011). Traditionally, process flow streams encompassing the system are evaluated according to the first law of thermodynamics where energy is conserved or by the second law where exergy is conserved. Exergy analysis, also known as availability, yields a more complete balance of resource use since it includes useful work that might be expended in the system. In that light, a more nuanced approach to evaluating life cycle energy that integrates exergetic transformations across the feedstock-to-refinery exit-gate supply chain may be created to represent changes to energy availability/usability (a measure of quality) of the biofuels and biochemicals as compared to existing petroleum-based fuels. This step

Pyrolysis of Biomass for Fuels and Chemicals. DOI: https://doi.org/10.1016/B978-0-12-818213-0.00008-4

will require the construction of exergy profiles for each pyrolysis route and biofuel pathway to be substituted with petroleum market alternatives such as diesel, gasoline, and jet fuel streams.

This chapter begins with a mass and energy or exergy balance principle and will follow a set of feedstocks and their pyrolysis oil production, their estimated yields, and the economics of scaled systems whether as stand-alone biorefinery or colocated with petroleum infrastructure followed by product life cycle in the contexts discussed. We will discuss some of the studies that use feedstocks that are important to US agriculture, some of which were mentioned earlier, including woody biomass, perennial grasses, and manure feeding into the pyrolysis technologies developed thus far to assess the economic viability and environmental sustainability. This will include some specific examples of when pyrolysis systems are operated as stand-alone, distributed, or integrated with existing infrastructure for production of energy or energy carriers. For these case studies, we begin with simple systems for the production and combustion of pyrolysis oil to generate energy in two ways, i.e., localized hot water heating and electricity/power generation then discuss more complicated systems for production of energy carriers (fungible fuels) exploring the cost and environmental benefits of coproducts, colocation, and other attributes of importance.

8.2 Mass balance, energy, and exergy analysis

We mentioned that establishing accurate material and energy balance for the process is critical to the design of the unit operations involved in the biorefinery and to establishing and conducting cost analysis which, hitherto for allow comparison of economic and environmental trade-offs of the product life cycle. The fundamental thermodynamic question that is often asked of bioconversion systems is whether the energy output matches or exceeds the energy input into the conversion system, that is, whether we are getting more energy out than we put into the system or vice versa. From the point of view of thermodynamics, the quick answer is that it depends on the system boundaries.

8.2.1 Energy balance

The first law of thermodynamics postulates that energy is neither created nor destroyed but transformed from one form to another; it asserts that the total energy is conserved in any process involving the exchange of heat and work between a system and its surroundings. To address the balance of energy flow, we consider the system boundary confined to the pyrolysis plant gate as shown in Fig. 8.1. If the pyrolysis system boundary is restrained to the reactor door step, then the balance is evaluated based on the energy/exergy flows associated with the material input and output. To establish an energy balance for this boundary condition, all material streams are accounted for and their heat of combustion or higher heating value (HHV) is established. The energy per unit mass is simply assigned the HHV,

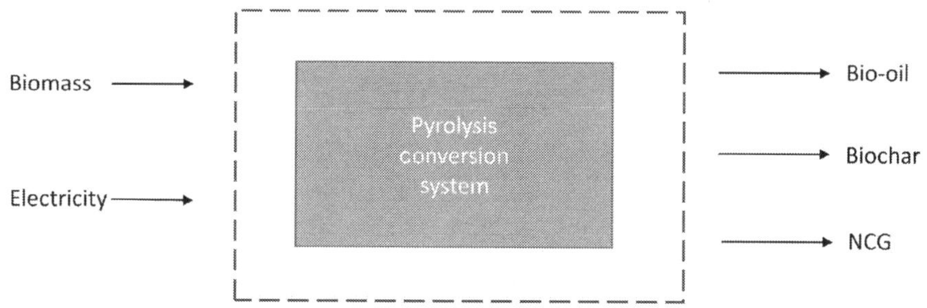

Figure 8.1 Energy and exergy flows for the system.

that is, E = HHV as the first law of thermodynamics teaches that the energy is conserved. Energy recovery may therefore be defined as:

$$\text{Energy recovery} = \frac{\text{Useful energy output}}{\text{Energy input}} \tag{8.1}$$

Per this definition, energy recovery will depend on the utilization of energy input and output streams for the pyrolysis conversion process. Because there are several possible ways that the internal energy of the by-products of pyrolysis can be recuperated, the energy recovery matrix may be based on several scenarios of product utilization. These include (1) targeting the bio-oil as the only energy carrier, for example, for use in vehicular application, (2) recognizing that bio-oil and biochar are both useful energy carriers, and (3) whether bio-oil, biochar, and the noncondensable gas (NCG) are all useful energy carriers that can be recycled as potential heat for the endothermic pyrolysis reactions. For example, in the last scenario, as much as 80% energy recovery can be achieved if all the products can be usefully harnessed.

However, if the system boundary is extended beyond the factory gate into the farms or forest where the biomass was cultivated, preprocessed, and transported to the plant gate then the associated energy flows must be accounted, which may tip the balance of the conversion efficiency.

8.2.2 Exergy balance

Unlike energy, exergy is characterized by the second law of thermodynamics, the classic definition of which states that a cyclic transformation, the only result of which is to transfer heat from a body at a given temperature to a body at a higher temperature, is impossible. This means that there must be some form of efficiency, central to which is entropy, a concept of a heat reservoir. Therefore exergy is an expression of the maximum theoretical work available from a substance if it were to achieve equilibrium with the environment. It is used in evaluating the potential use of energy resources and defined as:

$$e_x = (h - h_0) - T_0(s - s_0) + \frac{v^2}{2} + gz + e_{\text{xch}} \tag{8.2}$$

In this relationship the first two terms on the right-hand side of Eq. (8.2) represent the physical exergy (enthalpy and entropy changes), followed by the terms of kinetic and potential exergy and the chemical exergy. In the thermochemical conversion processes such as pyrolysis, one can assume the kinetic and potential exergies as negligible (Szargut and Morris, 1987). Again, since our system boundary is at the reference environment, exergy associated with heat transfer and the thermomechanical exergy of the cooling water used in the condensers may be neglected leaving the chemical exergy that, unlike energy, is a function of the lower heating value (LHV) rather than HHV.

$$E_{\text{xch}} = \beta(\text{LHV}) \tag{8.3}$$

where β is a function of the mass fraction of the chemical compound in the stream. There are empirical correlations one can use in estimating some of these, for example, for biomass and bio-oil (Eqs. 8.4 and 8.5) if their elemental compositions are measured (Keedy et al., 2014).

$$\beta_{\text{biomass}} = \frac{1.0412 + 0.2160(z_H/z_C) - 0.2499(z_O/z_C)\left[1 + 0.788(z_H/z_C)\right] + 0.0450(z_N/z_C)}{1 - 0.3035(z_O/z_C)} \tag{8.4}$$

$$\beta_{\text{bio-oil}} = 1.0401 + 0.1728\frac{z_H}{z_C} + 0.0432\frac{z_O}{z_C} + 2.169\frac{z_S}{z_C}\left(1 - 2.0628\frac{z_H}{z_C}\right) \tag{8.5}$$

The chemical exergies of the NCG are found by standard relationships (Szargut and Morris, 1987)

$$\bar{e} = \sum x_i \bar{e}_{\text{ch},i} + RT_0 \sum x_i \ln x_i \tag{8.6}$$

For the exergy of the water, only the first term of the above equation is necessary while the exergy associated with electrical power (considered work) is equivalent to the electrical energy (Keedy et al., 2014). Some pertinent sample values of exergy sourced from the literature were assembled by Keedy et al. (2014) as shown in Table 8.1.

Two exergy assessment metrices can be discerned when relating exergy to environmental assessment. The first is the global utilization of resources that is measured using the metric called the cumulative exergy demand (CExD). It measures the depletion of exergy associated with the conversion of material from its natural state to products. CExD values are available in the Ecoinvent Software or can be

Table 8.1 Chemical exergy of selected substances (Keedy et al., 2014).

Substance	Exergy (MJ/kg)
Nitrogen	0.026
Oxygen	0.124
Hydrogen	117
Carbon dioxide	0.451
Water (L)	0.050
Water (g)	0.527
Methane	52.0
Diesel fuel	43.0
Natural gas	49.0
Urea	11.5
Potassium chloride	0.259
Phosphate, P_2O_5	4.07
Forest residue biomass[a]	19.4
Forest residue bio-oil[a]	19.2
Equine waste biomass[a]	19.3
Equine waste bio-oil[a]	19.3
Switchgrass biomass[a]	18.0
Switchgrass bio-oil[a]	19.6

[a]Wet or "as is" basis, calculated from analysis provided in supporting information.

estimated following Eqs. (8.3)–(8.5). The second metric is the "breeding factor" (BF) or the exergy-based efficiency for the biomass conversion process, defined as (c. Eq. 8.1):

$$\text{Breeding factor} = \frac{\text{Exergy output}}{\text{Exergy inputs (not including biomass)}} \tag{8.7}$$

The BFs for bio-oil and useful products (after postprocessing) are considered if the system boundary extends to the upgrading of the produced pyrolysis oil to fungible renewable fuels. Else, the output is the chemical exergy of the bio-oil, the inputs of which include the CExD inputs to the pyrolysis process. If upgrading is considered, then output is the chemical exergy of useful products (e.g., representing gasoline, jet fuel, and diesel fuel) and inputs include the CExD of inputs to pyrolysis and postprocessing. If efficiency of BFs is greater than one (typically the case for bio-oil production), the process is producing exergy (i.e., output exceeds input).

The depletion of exergy associated with any input is traced through previous processes, including synthesis by nature associated with plant cultivation, harvest logistics, and preprocessing and identified by energy carriers as well as nonenergetic materials. The values of the CExD of energy carriers and materials are categorized by energy sources such as nonrenewable fossil, nuclear, primary forest and renewable hydro, direct solar, wind, biomass, and materials such as renewable water and nonrenewable metals and minerals. Quantitative values for CExD may be obtained from

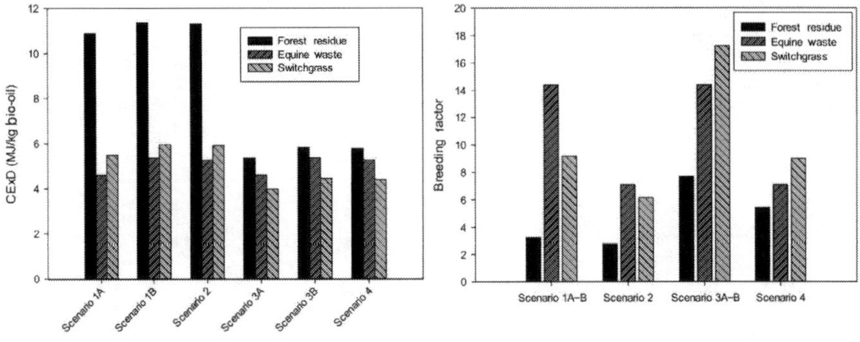

Figure 8.2 Comparison of CExD and exergy breeding factors for bio-oil production across various process scenarios used in producing bio-oil. (Scenario 1A–B): Grid electricity is used for postprocessing with pyrolysis gases and natural gas for pyrolysis heat. (Scenario 2): Local diesel generated electricity used for postprocessing and with pyrolysis gases and natural gas for pyrolysis heat. (Scenario 3A–B): Grid electricity for postprocessing with NCG and biochar coproduct utilization for pyrolysis. (Scenario 4): Local diesel generated electricity for postprocessing with NCG and biochar coproduct utilization for pyrolysis. Note: Exergy of electricity is the same regardless of source grid. *CExD*, Cumulative exergy demand; *NCG*, noncondensable gas.

databases such as the Ecoinvent database found in SimaPro (Ecoinvent 3.0/3.1, 2013). Values are selected that suitably fit the location and availability of resources. Electricity values are selected typical of the southwestern United States. Material resources, that is, water, nitrogen, hydrogen, and natural gas may be locally obtained, for example, at a refinery. These metrices are useful equivalents in assessing the true sustainability of resource use and their depletion. Fig. 8.2 shows CExD and BFs for three feedstocks and four bio-oil production process scenarios. Several conclusions can be drawn from the figure concerning the exergy-based sustainability of the processes examined. It is demonstrated that an evaluation of cumulative exergy depletion produces a significantly different and more comprehensive evaluation of sustainability than exergy. There are significantly different results for the three feedstocks examined because of the nature of their cultivation and preprocessing prior to the pyrolysis factory gate. There is an advantage in exergy production and less exergy depletion in converting biomass resources that do not require planting, fertilization, mowing, etc. These agricultural inputs, particularly fertilization, contribute substantially to exergy depletion. The choice of energy sources, for example, the selection of grid electricity or local electric power generation that uses diesel generators, has a significant sustainability impact. The CExD of grid electricity is influenced by the mix of energy carriers. The utilization of coproducts such as NCGs and biochar to recuperate heat and reduce or eliminate the use of additional fuel in the fast pyrolysis process has a very significant benefit in reducing cumulative exergy depletion and improving BFs, that is, exergy efficiency.

In the case study presented exergy values for dry bio-oil produced ranged from 24 to 27 MJ/kg and cumulative exergy depletion ranged from approximately 4 to 11 MJ/kg. Thus the exergy of the bio-oil is much less than traditional fuels, but the cumulative exergy depletion is also much less. BFs (exergy efficiencies) based on cumulative exergy depletion varied from approximately 2 to 5, indicating the potential exergy gain in production of bio-oil from the three feedstocks using fast pyrolysis.

In the next sections, we explore some case studies pertaining to the economic feasibility and/or environmental sustainability of select bioenergy systems as they pertain to biomass pyrolysis conversion and utilization.

8.3 Economics of production and combustion of pyrolysis oil

8.3.1 Case scenario 1—*equine waste for localized hot water heating*

We begin with small and simple application of pyrolysis oil derived from equine waste for use as combustion fuel at an equine rehabilitation facility (Morrisville State College, New York) where the waste is generated (Hammer et al., 2013). The rehabilitation center is surrounded by a collection of equine farms that can add to the availability of the waste. The goal of the study is to explore to see if it is possible and if so whether the facility can benefit from converting its waste and that from the surrounding area to energy by pyrolysis using equine waste as feedstock. It also explores subsequent use of the bio-oil as a renewable fuel in an existing boiler for localized hot water heating to the benefit of the animals. Necessary data on availability and pertinent characterization of the equine waste are exemplified in Tables 8.2 and 8.3 (Hammer et al., 2013). Equine waste is not only manure itself but also includes bedding that can be any lignocellulosic biomass such as hey and sawdust. In this scenario, there are two streams of waste material: one comes from the rehabilitation facility where there are four hot water boilers and the other comes from the farms in the surrounding community. The difference can range from bedding type and moisture content with associated disposal cost estimated as around $125/dumpster. The elemental analysis shows varying chemical composition of the organic matter. Ash can be higher for manure than bedding as expected, but the fixed carbon on dry ash-free basis is about the same making their energy contents similar.

Table 8.2 Manure data—Morrisville State College (MSC) equine rehabilitation facility and college barn (Hammer et al., 2013).

	Equine facility	MSC barn
Bulk density "as received" (kg/m^3)	283	283
Moisture % "as received" (wt.%)	40	48.7
Dry litter mass (kg/day)	1159	4240
Ratio of manure to bedding dry basis (kg:kg)	0.05	0.18
Volume of waste disposed (m^3/week)	7.6	27.6
Disposal costs ($/dumpster)	125	125
Number of boilers	4	
Duty of boilers (kW)	100.2	

Table 8.3 Analysis including high heating value of equine waste at Morrisville State College (Hammer et al., 2013).

	Manure			**Bedding**		
	As received	**Dry basis**	**DAF basis**	**As received**	**Dry basis**	**DAF basis**
Proximate analysis (wt.%)						
Moisture	1.75			1.27		
Ash	5.96	6.06		4.61	4.67	
Volatile matter	74.28	75.58	80.49	75.34	76.30	80.05
Fixed carbon	18.01	18.42	19.52	18.78	19.02	19.95
Ultimate analysis (wt.%)						
Moisture	2.13			1.89		
H	5.85	5.98	6.35	5.69	5.80	6.05
C	47.41	48.43	51.46	49.49	50.43	52.63
N	1.19	1.21	1.29	0.70	0.71	0.75
S	0.07	0.07	0.08	0.07	0.07	0.07
O	37.61	38.43	40.83	38.09	38.83	40.51
Ash	5.76	5.88		4.08	4.16	
High heating value (MJ/kg)	18.175	18.498	19.548	19.062	19.055	20.091

DAF, Dry ash free.

8.3.1.1 Process

The pyrolysis of the equine waste, that is, manure and bedding, was carried out in a fluidized bed reactor at two feed rates of 2 and 5 ton/day to establish feasibility, yields, and product distribution (Table 8.4). With this data the economic analyses were conducted for two process scales modeled at 6 oven-dry metric ton per day (MTPD) and 15 oven-dry MTPD. Using the mass and energy balances obtained from Aspen Plus model (Fig. 8.2 and Tables 8.5 and 8.6), each major piece of equipment was sized resulting in capital and operating costs of installing and operating the USDA's Combustion Reduction Integrated Pyrolysis System (CRIPS) at the prescribed scales shown in Table 8.7.

The results established that utilization of all the available waste from the site's 41 horses requires a 6 oven dry MTPD pyrolysis system, but it would require a 15-MTPD system to meet their hot water demand. Additional waste would have to be generated requiring 150 horses from the vicinity surrounding the rehabilitation center. The model results show the potential for the rehabilitation facility to displace diesel fuel (fossil) with renewable pyrolysis oil and at the same time alleviate a costly waste disposal problem (Table 8.5). While the estimates show that all the heat required to operate the CRIPS pyrolyzer could be supplied by the NCG supplemented by about 40% of the biochar coproduct, the techno-economic analysis shows neither scale

Table 8.4 Yield distribution of components for pyrolysis of horse manure and bedding (Hammer et al., 2013).

Yield distribution (wt.%)		
Components	Manure	Bedding
Biochar	20.69	29.74
Water	12.16	12.08
Organic pyrolysis oil	47.36	28.52
Total liquids	59.52	40.60
CO_2	12.37	19.68
CO	6.82	9.02
H_2	0.01	0.01
CH_4	0.59	0.94

Table 8.5 Mass balance of 2 ton per day (TPD) and 5 TPD system from Aspen (Hammer et al., 2013).

	2TPD	5TPD
Biomass drying and feeding		
Total flow rate in (kg/h)	429	1215
Particle diameter (mm)	3	3
Moisture in (%)	40	48.6
Moisture out (%)	10	10
Bio-oil fuel used (kg/h)	23 (as is)	75 (as is)
Combustion Reduction Integrated Pyrolysis System (CRIPS)		
Recycle flue gas flow rate (kg/h)	636	1574
Bio-oil produced (kg/h)	77.2 (DAF)	189 (DAF)
Char produced (kg/h)	73 (DAF)	169 (DAF)
NCG + H_2O produced (kg/h)	723	1821
Excess recycle gas to combustor (kg/h)	74	181
Char as combustion fuel (kg/h)	27.5 (DAF)	71 (DAF)
Excess char (kg/h)	41.2 (DAF)	97.5 (DAF)
Quench		
Bio-oil off condenser 1 (kg/h)	25 (as is)	64 (as is)
Bio-oil off condenser 2 (kg/h)	20 (as is)	50 (as is)
Bio-oil off condenser 3 (kg/h)	15 (as is)	38 (as is)
Bio-oil off ESP (kg/h)	32 (as is)	83 (as is)
Hot water generation		
Diesel flow into each chamber (kg/h)	7	7
Bio-oil flow into each chamber (kg/h)	1.67	1.67
Hot water flow rate (L/min)	265	265

DAF, Dry ash free; *NCG*, noncondensable gas.

Table 8.6 Energy balance of 2 Tons per day (TPD) and 5 TPD system from Aspen (Hammer et al., 2013).

	2TPD	5TPD
Biomass drying and feeding		
Grinder power requirement (kW)	0.66	1.65
Dryer (kW)	117.5	399
Combustion Reduction Integrated Pyrolysis System (CRIPS)		
Pyrolyzer (kW)	260	753
Quench		
Condenser 1 cooling required (kW)	77.1	192
Condenser 2 cooling required (kW)	29.7	73.9
Condenser 3 cooling required (kW)	41	102
Hot water generation		
Single boiler duty (kW)	97.2	97.2

Table 8.7 Annual cost and saving distribution for 6 metric ton per day (MTPD) and 15 MTPD (oven dry basis) Combustion Reduction Integrated Pyrolysis System (CRIPS) (Hammer et al., 2013).

	6 MTPD	15 MTPD
Total installed cost	$2,260,000	$3,463,000
Operating costs ($/year)		
Feedstock	$0	$0
Utilities	$17,000	$35,000
Labor, supplies, and overheads	$201,000	$253,000
Administration	$33,000	$50,000
Depreciation	$226,000	$346,000
Annual production cost	$477,000	$684,000
Annual savings ($/year)		
Manure disposal	$22,000	$61,000
Diesel displacement	$82,000	$191,000
Coproduct credits	$6000	$16,000
Annual revenue	$110,000	$268,000

design is economical at current market conditions. However, the 15-MTPD CRIPS design could break even when diesel prices reach $11.40/gal. This can be further improved to $7.50/gal if the design capacity is maintained at 6 MTPD but operated at 4950 hours per annum.

8.3.2 Case scenario 2—*electricity production with* Eucalyptus-*derived bio-oil in Brazil*

In a second scenario on power generation, pyrolysis oil is produced from *Eucalyptus* field in Brazil and either processed in a stand-alone, or in a distributed bio-oil production facility for postprocess power generation at a centralized location. Both systems employ the CRIPS as the primary pyrolysis technology. The generation of bioelectricity from pyrolysis oil proposed in this study consists of four main sections of the design layout as depicted in Fig. 8.4 (Pighinelli et al., 2018). Section components include (1) biomass cultivation, U1; (2) biomass pretreatment, U2; (3) biomass fast pyrolysis, U3; and (4) electricity generation using the produced bio-oil as combustion fuel, U4. Process flow sheet was developed and simulated in the SimSci PRO/II software, in which unit operation blocks are interconnected with material and energy stream functions. Fluid and energy flows are simulated to size the actual process equipment in the pyrolysis and electricity generation sections. With the technical information from the basic design of the plant and with input assumptions specified, the Pro-II software is used to generate the mass and energy balances that were used as basis for scaled design and the techno-economic analysis. For the distributed biomass processing plant, it was assumed that 10 biomass processing facilities of 200-MTPD capacity each are distributed over a predefined radius that supply bio-oil to a centralized power generation plant to generate bioelectricity using the bio-oil to fuel the boiler at a capacity of 950-MTPD bio-oil.

Calculations pertinent to distributed systems include the biomass hauling distances and scaling of capital expenditure from prototype designs. In the distributed approach, one important parameter to estimate is the economic radius to locate the satellite units. The average biomass transportation distance (r_{biom}) to place a centralized power station may be calculated using Eq. 8.8 following (Wright et al., 2008). This assumes that the power plant is stationed at the center of a square grid along which the satellite biomass–bio-oil conversion mini plants are located. For a rectangular grid road layout, tortuosity factor of $\tau = 1.5$ is considered for such designs, where $\tau = 1$ corresponds to a straight-line trajectory between two given points.

$$r_{\text{biom}} = \frac{1}{6}\tau\sqrt{\frac{F}{Yf}}\left(\sqrt{2} + \ln\left(1 + \sqrt{2}\right)\right) \quad (8.8)$$

The average bio-oil transportation distance ($r_{\text{bio-oil}}$) is a function of the amount of biomass that must be converted to bio-oil and the capacity of the distributed pyrolysis plants, expressed by the following equation:

$$r_{\text{bio-oil}} = 0.423\tau\left(\frac{F}{F_{\text{plant}}}\right)^{0.476}\sqrt{\frac{F_{\text{plant}}}{Yf}} \quad (8.9)$$

where τ is the tortuosity factor, F is the total biomass input (ton/year), Y is the biomass yield (ton/ha), the factor f is the fraction of land surrounding the plant that is devoted to biomass crops (%), and F_{plant} is the bio-oil input (ton/year).

Unit costs of equipment are typically scaled from base equipment cost (Eq. 8.10) following Peters et al. (2003).

$$\text{Cost}_{\text{EqNew}} = \text{Cost}_{\text{EqBase}} * \left(\frac{\text{CAP}_{\text{New}}}{\text{CAP}_{\text{Base}}}\right)^{n} \tag{8.10}$$

where $\text{Cost}_{\text{EqNew}}$ is the scaled new equipment cost (US$), $\text{Cost}_{\text{EqBase}}$ is the cost of the base equipment or prototype, n is the specific scaling factor for a particular equipment (ranging from 0.6 to 0.8), and CAP_{New} and CAP_{Base} are the sizes of the new and base equipment, respectively.

Table 8.8 presents the characterization of the feedstock, pyrolysis oil and biochar for this study and obtained from fast pyrolysis of *Eucalyptus benthamii* using the reactive pyrolysis system tail gas reactive pyrolysis (TGRP) as such the bio-oil obtained had a HHV of 30.76 MJ/kg (dry basis, db), which is higher than that for pyrolysis oil obtained from wood via traditional fast pyrolysis (Table 8.8) (Pighinelli et al., 2018).

Table 8.8 Properties of *Eucalyptus benthamii* biomass and tail gas reactive pyrolysis products (Pighinelli et al., 2018).

Variable	Biomass (wt.%, db)	Bio-oil (wt.%, db)	Biochar (wt.%, DAF)
Proximate analysis			
Water	11.56	2.5	3.79
Volatile matter	84.75	–	15.91
Fixed carbon	14.73	–	84.09
Ash content	0.50	–	7.43
Ultimate analysis			
Carbon (C)	50.14	73.54	87.09
Hydrogen (H)	5.40	6.17	2.59
Nitrogen (N)	0.08	0.56	0.28
Oxygen (O)	43.87	20.09	9.88
C/O (mol)	1.52	4.87	11.74
O/C (mol)	0.66	0.21	0.09
H/C (mol)	0.08	0.06	0.02
Other properties			
HHV (MJ/kg)	20.26	30.76	31.89
Viscosity (cP@40°C)	–	32.35	–
TAN (mg KOH/g)	–	19–41	–
Density (kg/L)	–	1.2	–
Specific heat (kJ/kg K)	–	3.00	–
Thermal conductivity (W/m K)	–	0.39	–

DAF, Dry ash free; *db*, dry basis, *HHV*, high heating value.

With this information the total purchase equipment cost (TPEC) for the various production units is presented in Table 8.9. From the dataset the fast pyrolysis unit (U3, Fig. 8.3) contributes the highest TPEC of about 71% (or US$ 32.4 million) of the 2000-MTPD single facility followed by the electricity generation section (about 15.1%). These cost factors result in the plant's total capital investment (TCI) of approximately US$ 137.4 million for the 2000-MTPD facility mainly driven by downstream unit operations such as pyrolysis and combustion reactors, boiler, turbines, generators, and dryers. For the distributed arrangement the fast pyrolysis equipment accounts for 78% of the TPEC while the electricity generation accounts for 6.5% of the TPEC. The TCI for the distributed case is more than double the

Table 8.9 Installed equipment costs (Pighinelli et al., 2018).

Unit	2000 MTPD (M$)	10 × 200 MTPD distributed (M$)
Feedstock handling and pretreatment	5.601	14.069
Pyrolysis	12.696	31.891
Quench	7.434	18.673
Heat recovery	12.257	30.788
Product recovery and storage	0.661	1.660
Electricity production	6.856	6.856
TPEC	45.510	103.937
TIC	137.425	313.890

MTPD; Metric ton per day; *TIC*, total installation cost; *TPEC*, total purchased equipment cost.

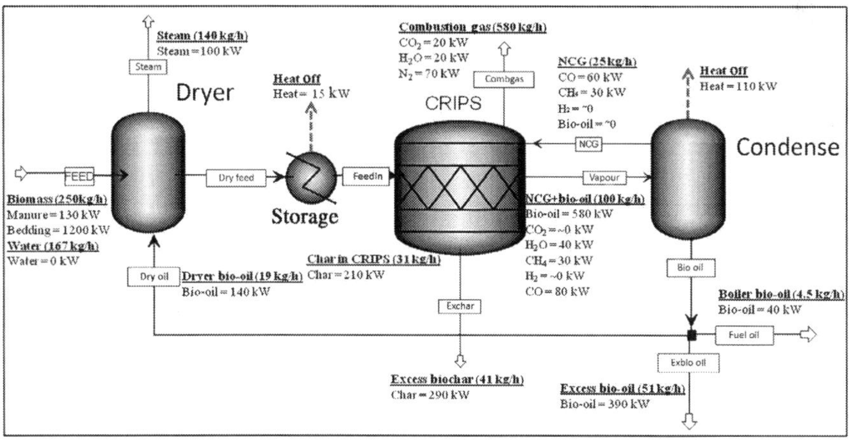

Figure 8.3 Distribution of energy within pyrolysis products for the 6 oven-dry MTPD system (ash-free basis) (Hammer et al., 2013). *MTPD*, Metric ton per day.

Table 8.10 Summary of economic analysis for bioelectricity generation (Pighinelli et al., 2018).

Economic indicator	2000 MTPD	10 × 200 MTPD-distributed
FCI (M$)	213.51	487.68
WC (M$)	32.03	73.16
TCI (M$)	248.27	567.07
Project contingency (M$)	35.6	81.28
Unit selling price ($/kWh)	0.23	0.37
PBP (year)	4.4	4.9
DCFROR (%/year	12.6	10.8

DCFROR, Discounted cash flow rate of return; *FCI*, fixed capital investment; *MTPD*, metric ton per day, *PBP*, payback period; *TCI*, total capital investment; *WC*, working capital.

single plant at US$ 313.9 million in this scenario. In fact, others have reported higher TCI value of US$ 438 million for a 2000-MTPD fast pyrolysis and gasification facility for bioelectricity production (Li et al., 2015) lending credibility to the use of the CRIPS.

Based on the estimated TCI and TCP, an internal rate of return on investment (IRR) of 12.6%/year can be obtained for a single unit operation using a discounted cash flow rate of return (DCFROR) analysis while the distributed facility stands at 10.8%/year IRR (Table 8.10). Fig. 8.5 shows the discounted cash flow diagram for the 2000 MTPD and distributed facilities for a 10-year plant life span.

Performance measurement: The selling price at the breakeven point of the electricity generated when the pyrolysis facility is designed around the TGRP technology with the bio-oil quality presented in Table 8.8 is estimated to be US$0.34 and US$0.62/kWh electricity for the single and the distributed scenarios, respectively, considering a 10-year payback period. The single capacity pyrolysis-to-electricity generation facility was found in this case study to have a better economic benefit over the distributed plants of small sizes under the 2017 conditions in Brazil, at least, for bioelectricity generation despite the favorability of satellite/distributed systems as discussed earlier. The results therefore indicate that pyrolysis of *Eucalyptus* wood for electricity in a single facility cannot be competitive with the 2017 electricity cost in Brazil, which stood at US$0.08−0.13/kWh. However, considering the forces of auxiliary benefits such as climate change and carbon credits, plus the continuous increases in the electricity market price in Brazil, the study showed both scenarios could be competitive if climate policies are monetized. This is also true if the smaller units in the satellite system are mass produced.

8.4 Economics of colocated pyrolysis biorefinery

Case scenario: Colocating a pyrolysis plant at a rubber-producing plant (Sabaini et al., 2018).

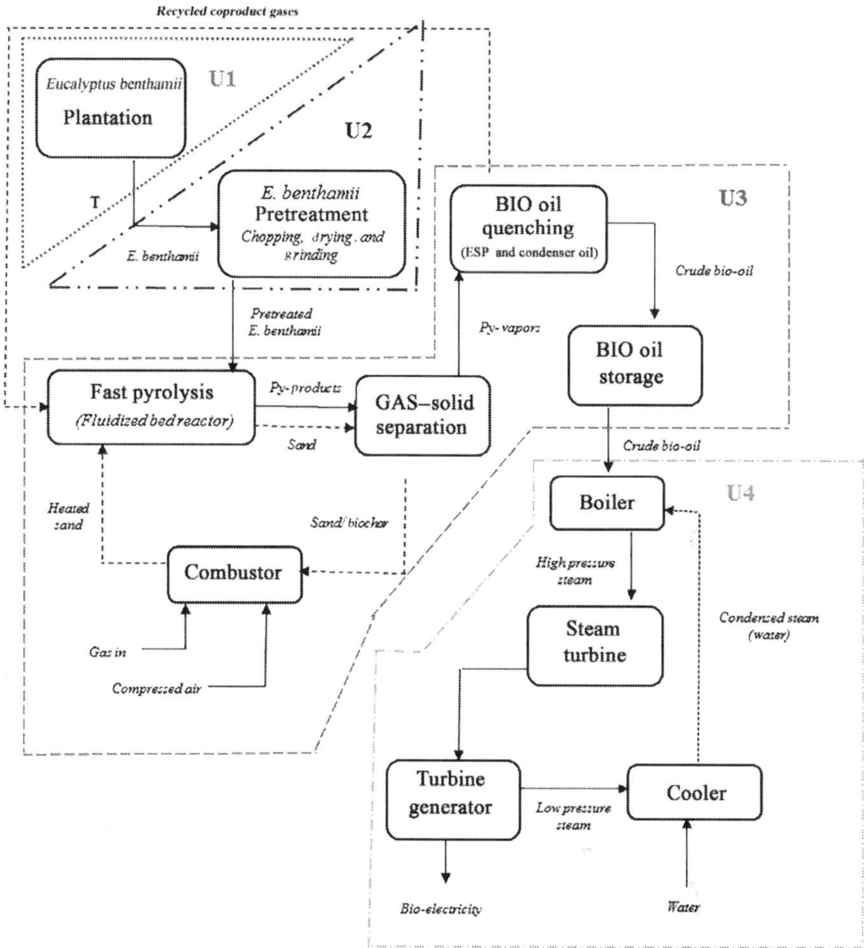

Figure 8.4 Process flow diagram of bioelectricity generation from fast pyrolysis oil (Pighinelli et al., 2018).

The tire industry is currently considering natural rubber from guayule (*Parthenium argentatum Gray*) as a viable alternative to imported Hevea natural rubber, or petroleum-based synthetics, to meet the expanding material needs of the industry. However, only 5%–10% of the harvested guayule plant is converted into rubber latex. For economic sustainability to prevail the industry must identify the viable uses for the residual biomass, termed bagasse. Bioenergy production has been considered, but conversion facilities must be colocated to avoid additional costs in transportation of the bagasse. Since the bagasse is already accounted for as a coproduct of the core rubber business, it is an opportunity fuel resource that needs not to be transported if a biorefinery is colocated at the rubber plant for its use as feedstock. This case study investigated the economics of processing a minimum of 200 MTPD of guayule

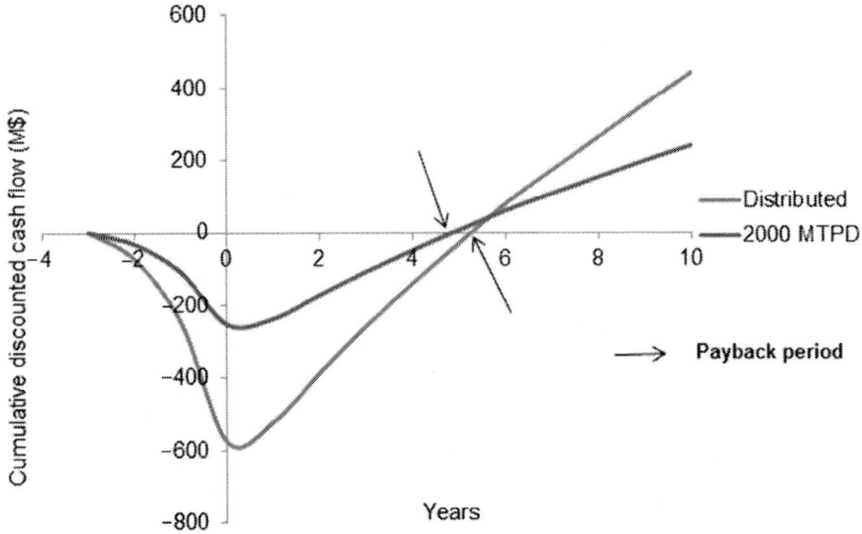

Figure 8.5 Discounted cash flow diagram for bioelectricity production via fast pyrolysis (Pighinelli et al., 2018).

bagasse, the current availability at US plants, to produce biofuels in a pyrolysis biorefinery colocated within a guayule latex processing facility (Fig. 8.6).

Process: A unique aspect of the simulated process was the use of the TGRP technology that formulates an intermediate bio-oil with less oxygenates and therefore requires only mild upgrading to fuel products. In fact, attempts at the USDA to produce pyrolysis oil from guayule latex bagasse using conventional thermal-only pyrolysis process encountered fouling problems due to their low boiling points, a problem that could only be resolved with the TGRP process. The TGRP process particularly suits this feedstock given the high concentration of resins and residual latex, the pyrolysis of which is prone to fouling in conventional pyrolysis units. The TGRP also yields less oxygenates with product quality that is readily distilled in high yield, pre- or posthydrogenation, while traditional bio-oils present some challenges. Herein, guayule bagasse, the residual biomass after latex extraction, or guayule plant leaves was used as TGRP feedstock to formulate a special intermediate bio-oil product that allows the use of conventional hydrotreating via hydrodeoxygenation (HDO) with conventional noble metal catalysts and a simple distillation process to synthesize hydrocarbon (drop-in) fuels (Boateng et al., 2015, 2016). Physical separation of the starting bio-oil by simple centrifugation results in 85 wt.% liquid yield and further 50–65 wt.% product following a continuous hydrotreatment over a common noble metal (Pt, Ru, or Pd) on a carbon support (Fig. 8.7).

Atmospheric distillation of the HDO product can yield as high as >95 wt.% of a hydrocarbon liquid fuel mixture comprising 30.4% gasoline (C_5–C_7), 37% jet (C_8–C_{12}), and 24% diesel (C_{13}–C_{22}). Analysis of a composite mixture of the

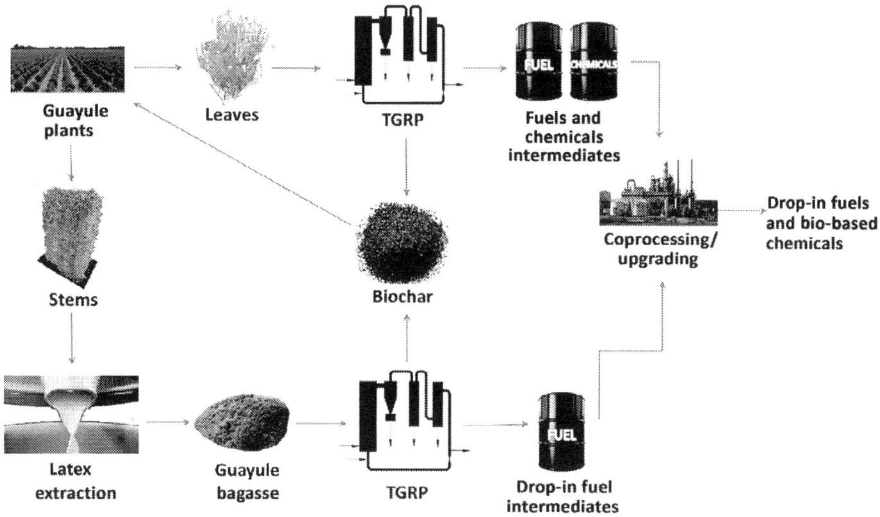

Figure 8.6 Colocated guayule pyrolysis biorefinery (Boateng et al., 2015).

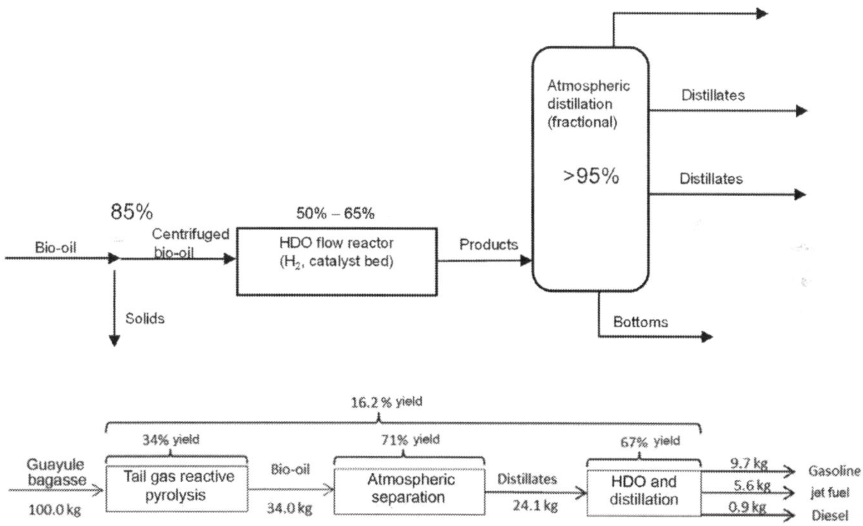

Figure 8.7 Process flow for upgrading of the organic phase of guayule bagasse bio-oil derived from TGRP. The yields are typical over Pt/C, Ru/C, and Pd/C. (Top) Volume basis and (bottom) Overall yield of the process, mass basis. *TGRP*, Tail gas reactive pyrolysis.

hydrotreated product from the bagasse showed that most of the sample ($\sim$66%) was C_{12} and below, which falls within the gasoline (naphtha) range with the greatest fraction of naphtha falling within the C_8–C_{10} range. Beyond C_{12}, the molecular weights increased through the diesel range (34%), with C_{37} being the highest observable

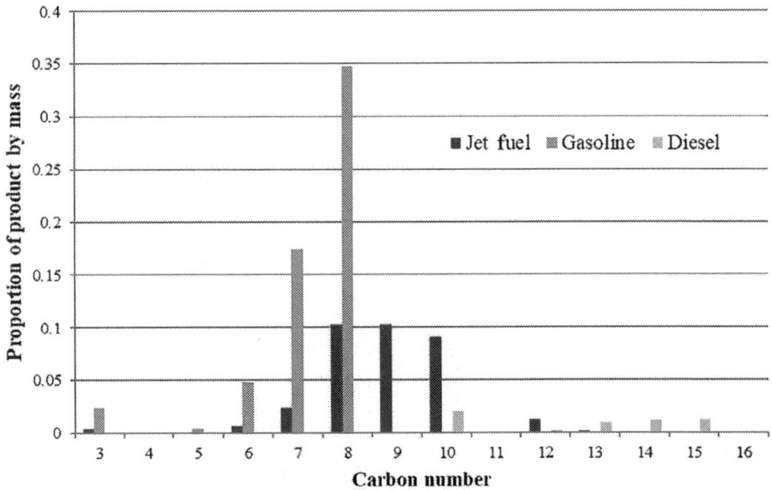

Figure 8.8 Molecular weight distribution by carbon number of the upgraded guayule TGRP bio-oil (Sabaini et al., 2018). *TGRP*, Tail gas reactive pyrolysis.

molecular weight (Fig. 8.8). The product meets several ASTM standards for drop-in fuels, but the sulfur content (primarily due to latex extraction additives) was relatively high at around 300 ppm, indicating that hydrodesulfurization may be required to be infrastructure ready. A total hydrocarbon yield of the entire biorefinery was estimated to be 16.2 wt.% of the input biomass equivalent to about 50 gal of gasoline per ton of biomass. The techno-economic study was to assess the economic viability of colocated pyrolysis biorefinery for renewable fungible fuels.

Performance measurement: The hydrocarbon yield of 16.2% (50 gal/ton biomass) was distributed in the gasoline (9.7%), jet fuel (5.6%), and diesel (0.9%) carbon ranges (Sabaini et al., 2018). The capital cost for the 200-MTPD colocated plant was estimated at $58.74 million, and the annual operating cost was estimated at $14.19 million for the plant. ADCFROR analysis was conducted to evaluate the economic feasibility based on a 30-year plant life and 10% internal rate of return. The minimum fuel selling price (MFSP) calculated was $1.88/L for gasoline, $1.84/L for jet fuel, and $1.91/L for diesel fuel, clearly showing the limitations imposed by economies of scale of the current guayule bagasse availability. However, the potential exists to reduce the MFSP by increasing the capacity of the facility and by utilizing the valuable coproducts that accompany guayule pyrolysis biorefining. Sensitivity analysis indicates the MFSP of gasoline can be lowered to $0.96/L considering the most optimistic scenario, comprising an integrated large facility of 2000 MTPD, a lower cost of hydrogen, and the sale of a premium-quality residual guayule biorefinery coke residue as green coke for anode making.

Key conclusions: The techno-economics of a colocated guayule bagasse pyrolysis biorefinery at a guayule latex extraction facility that produces hydrocarbon fuels were evaluated and the investment was deemed to be feasible (Sabaini et al., 2018). A processing capacity of 200 MTPD was an appropriate size for meeting the current bagasse availability at guayule latex plants that are currently located in Arizona. For this a capital cost was estimated as \$58.7 MM and annual operating cost was established at \$14.3 MM. MFSP based on a DCFROR analysis considering 30-year life plant and 10% internal rate of return predicted a MFSP of \$1.88/L for gasoline, \$1.84/L for jet fuel, and \$1.91/L for diesel fuel. Sensitivity analysis showed that the economic concerns are related to pyrolysis product yield and capital investment. It is demonstrated that at this scale the technology is not competitive with the current market price for fossil fuel at current subsidies, but the potential exists to become more cost competitive. Leverage points are found in improvements in the biorefinery technologies to facilitate higher yields, making coproducts with high value, scaling up to larger facilities and reduction of hydrogen price. In the light that this industry is already anticipating an expansion to 1000 MTPD in the near term, larger scale pyrolysis units could be proposed as the lowest hanging fruit. It is also envisioned that a full nominal industrial size of 2000-MTPD bagasse availability could be possible as guayule natural rubber successfully becomes an alternative material to imported Hevea and petroleum-based rubbers in the tire market. This could improve the economic feasibility of a colocated pyrolysis biorefinery. In an optimistic scenario encompassing the nominal capacity along with a reduced hydrogen cost of \$1.5/kg and the potential to sell biorefinery green coke at \$250/kg for use as carbon anodes, the MFSP of gasoline equivalent could further reduce to as low as \$0.96/L. The 2000-MTPD analysis was found to be most economically viable when the pyrolysis oil production plants are distributed at 10 satellite latex facilities at 200 MTPD each with a capability to transport the bio-oil to a larger, centralized facility dedicated for condensed-phase HDO upgrading to finished fuels (Fig. 8.9). Predicted MFSP for the

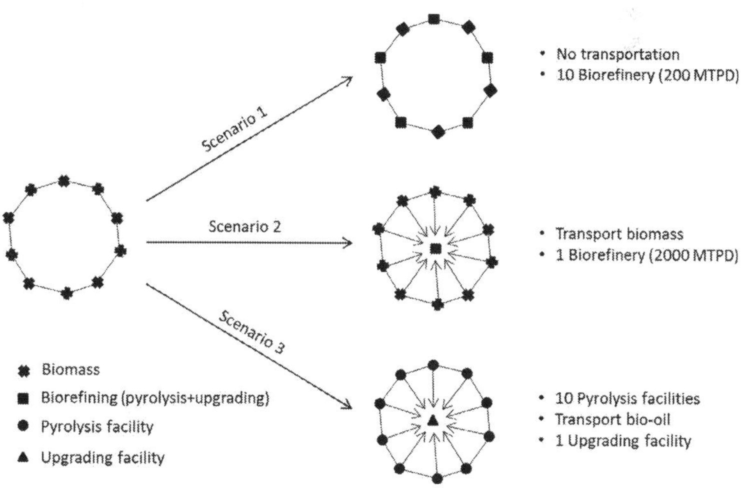

Figure 8.9 Distributed versus centralized scenario for guayule pyrolysis biorefinery (Sabaini et al., 2018).

Table 8.11 Minimum fuel selling price (MFSP) for different scenarios and scales for guayule pyrolysis biorefinery (Sabaini et al., 2018).

Scale/ distribution	10 × 200			4 × 500			2 × 1000		
Scenario	1	2	3	1	2	3	1	2	3
100 km	7.12	5.82	5.99	5.87	5.82	5.52	5.18	5.82	5.15
150 km	7.12	6.30	6.03	5.87	6.30	5.56	5.18	6.30	5.19
200 km	7.12	6.79	6.07	5.87	6.79	5.59	5.18	6.79	5.22

three various distributed scales at 10 × 200, 4 × 500, and 2 × 100 MTPD at set distances is presented in Table 8.11.

8.5 Techno-economic and exergetic life cycle assessment

Case scenario: Jet fuels from horse manure–derived pyrolysis oil (Sorunmu et al., 2017).

Horse manure, improper disposal of which may impose considerable environmental costs but constitutes an apt feedstock for conversion to renewable fuels and chemicals when appropriate technologies such as reactive pyrolysis, for example, TGRP is employed. TGRP data (Elkasabi et al., 2014) show that as high as ~36 wt.% product yield on feedstock basis with high concentrations of hydrocarbon and valuable chemicals such as phenol molecules can be produced using the TGRP system. For this reason a WTW LCA was carried out for a distributed processing of horse manure in a pyrolysis biorefinery to target aviation fuel and phenol as final products and to explore associated GWP and product life cycle (Fig. 8.10).

A Geographic Information System (GIS) map of the supply logistics of horse manure within the New York State (Fig. 8.11) indicates that in most of the New York area, there will be enough horses within any 50-mi radius to generate significant quantities of manure that amount to the availability required for the distributed design capacity of 200 MTPD previously modeled for guayule in Arizona.

8.5.1 The process

Herein, the coproduction of phenol was evaluated as a value-added renewable chemical for which chemical credits can be earned, alongside jet-range fuels within distributed TGRP systems for its techno-economic and LCAs. This case elucidated the metrics of GWP, CExD, that is, resource depletion, and cost for the conversion of 200 dry MTPD of horse manure to bio-oil and its subsequent upgrade to hydrocarbon fuel and phenolic chemicals. A process flow encompassing unit operations presented in Fig. 8.12 was modeled in Aspen Plus with TGRP oil yield assumed to be 36.9% (~100 gal/ton biomass) of the horse manure, based on previous experimental work (Elkasabi et al., 2014). Major processing steps include biomass

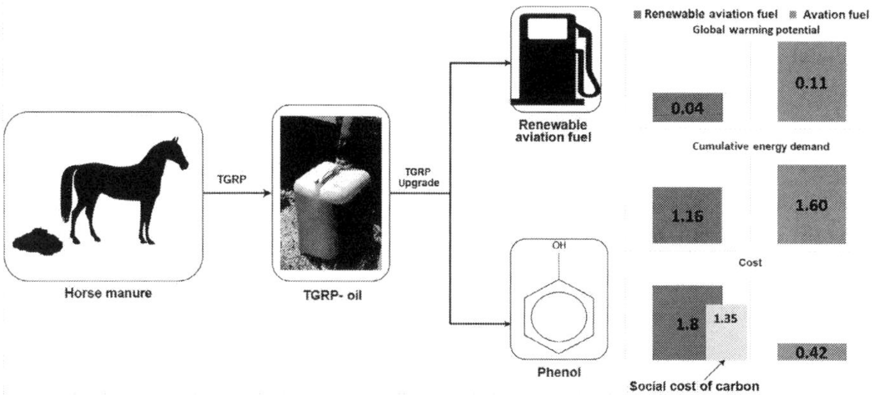

Figure 8.10 Horse manure pyrolysis biorefining (Sorunmu et al., 2017).

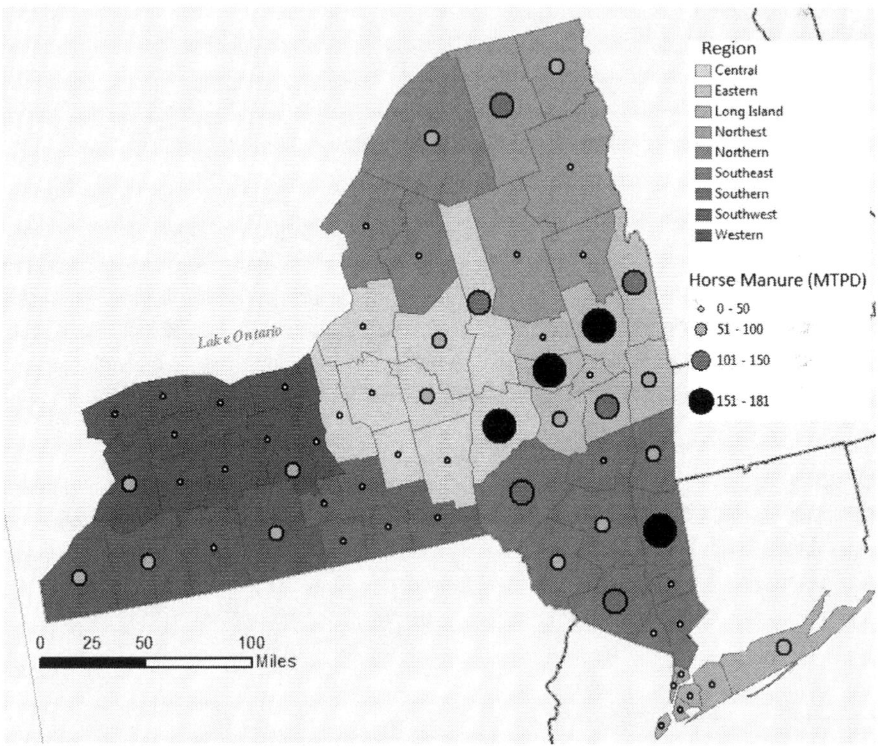

Figure 8.11 New York State map divided into counties. Color blocks represent the regions in New York State and graduated symbols represent horse manure availability in MTPD. Distributed pyrolysis systems scaled up to 200 MTPD would be feasible in multiple locations in New York State (Sorunmu et al., 2017). *MTPD*, Metric ton per day.

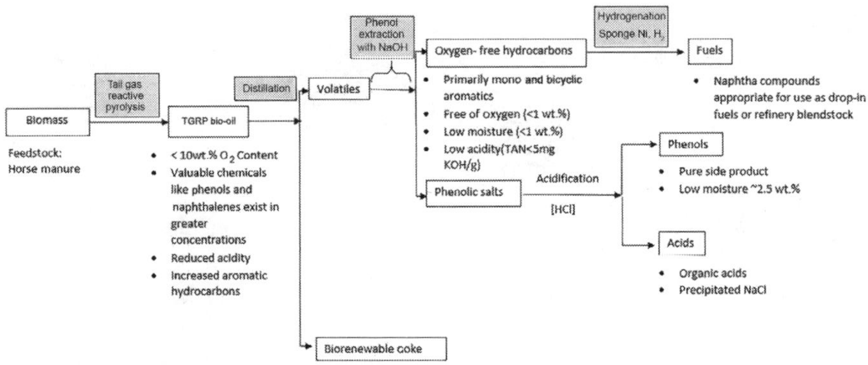

Figure 8.12 Process sheet for tail gas reactive pyrolysis of horse manure to bio-oil and further upgrade through distillation and isolation of the bio-oil distillates, forming value-added chemicals and fuels (Sorunmu et al., 2017).

preparation, TGRP and TGRP oil upgrade, consisting of distillation, extraction of phenolics from the distillates and hydrogenation of oxygen-free hydrocarbons to produce renewable fungible fuel products (Fig. 8.12).

The LCA model was carried out using the Simapro software. Data inputs include parameters specified in the feedstock harvest and collection, and the mass and energy balances derived from Aspen Plus (Fig. 8.13). The metrics evaluated under life cycle impact assessment include the 100-year GWP for CO_2, CH_4, and N_2O, as required according to Intergovernmental Panel on Climate Change (IPCC) guidelines, and CExD (Goedkoop et al., 2008). These metrics were used to describe the target products (jet-range fuels and phenol) postseparation and upgrading of the condensed-phase TGRP oil. The GWP-100 metric benchmark for new biofuel standards, as in accordance with the RFS-2 protocols, was applied, which states that advanced biofuels must reduce life cycle GHG emissions by 50% (Schnepf and Yacobucci, 2010). This was complimented by CExD that has not been as widely used but captures the exergy pertaining to resource depletion measure described earlier. This measure of "exergy" under the second law of thermodynamic defines the maximum theoretical available work from a system if it were to achieve equilibrium with the environment. Since exergy is a measure of available work, it represents a more complete indicator of resource use, compared to cumulative energy consumption. Energy can be converted to different forms, but exergy is consumed in all processes. CExD also accounts for the consumption of nonenergetic raw materials. However, for nonrenewable energy−intensive products, the two results are similar (Sorunmu et al., 2017).

8.5.2 Performance measurement

Assigning credits to offset the coproducts such as phenol and green coke, the net GWP and CExD of jet fuel derived from TGRP system were estimated at 10 g of

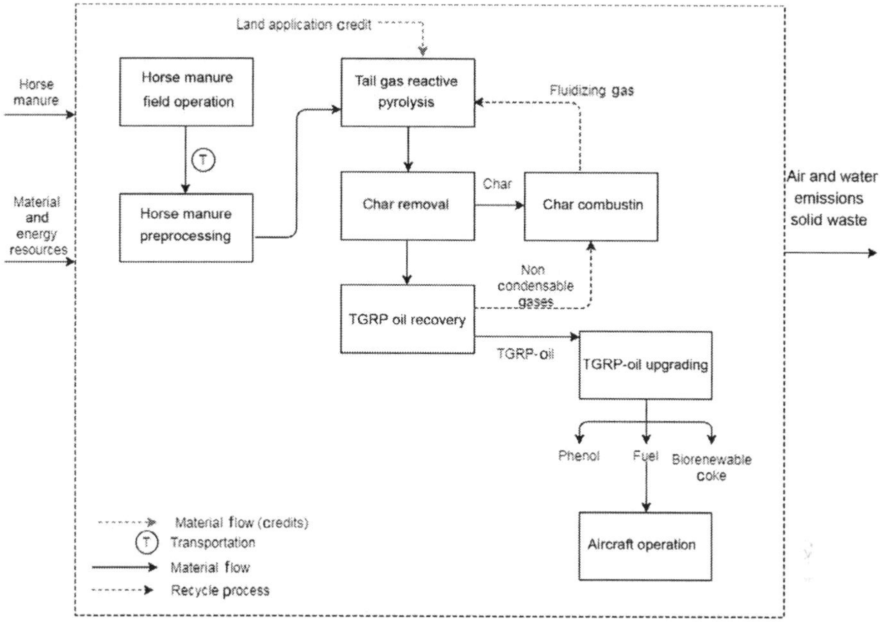

Figure 8.13 Life cycle system boundary to produce fuel and phenol from horse manure via TGRP and upgrading through isolation and extraction of TGRP oil distillates (Sorunmu et al., 2017). *TGRP*, Tail gas reactive pyrolysis.

CO_2 equiv. and 0.4 MJ per passenger kilometer (PKM) distance traveled, respectively (Fig. 8.14). These values are considerably lower than the GWP and CExD of petroleum-based aviation fuel. The GWP of petroleum-based aviation fuel (110 g of CO2 equiv.) combined with phenol from the cumene process (30 g of CO_2 equiv.), per PKM of distance traveled. The MFSP of the TGRP bio-oil–derived jet fuel was estimated at \$1.35–\$1.80 L^{-1}, which is much greater than that of current petroleum-based aviation fuel at \$0.42 L^{-1}. However, under optimized fuel conversion and coproduct market conditions, this can be \$0.53–\$0.79 L^{-1} when we include a market price for carbon. The key cost overlays for TGRP over conventional thermal-only pyrolysis shown in Table 8.12 indicates the capital cost for this modest capacity is quite significant compared with thermal pyrolysis only (Carrasco et al., 2017; Pourhashem et al., 2013) resulting in high aviation fuel selling price. However, a sensitivity analysis shows that the MFSP is most sensitive to feedstock cost and yield and, to a lesser extent, on the market prices of phenol and green coke for this integrated biorefinery while costs for hydrogen and hydrogenation catalyst required for the upgrading steps do not exhibit any significant effect on the MFSP because they are used in small quantity, compared to the other raw materials in the process. This means that the MFSP value could be significantly reduced from \$1.8 to 1.1 L^{-1} when there is no monetary value (\$0 kg^{-1}) associated with it (option A, Table 8.12). It is further anticipated that MFSP reductions will depend on the market selling prices of phenol and green coke. This implies that increasing the pyrolysis process yields and

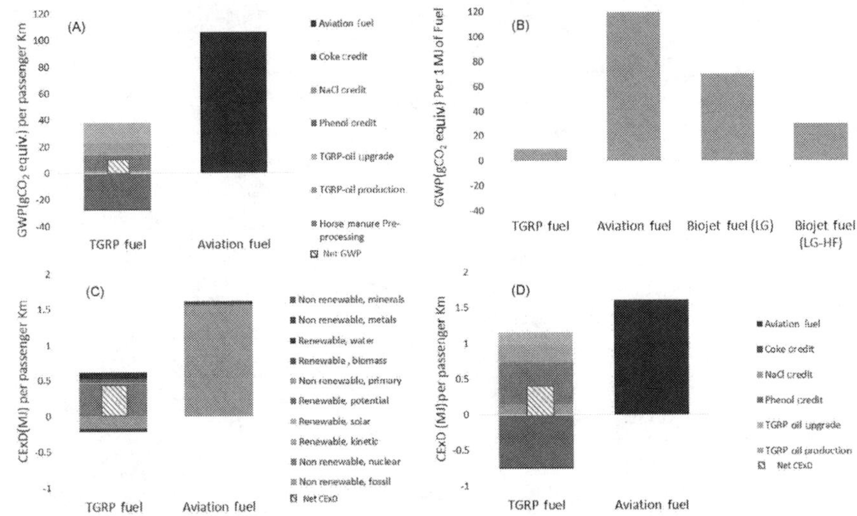

Figure 8.14 Life cycle environmental impact of TGRP fuel. (A) GWP comparing TGRP fuel to aviation fuel per passenger km. (B) GWP comparing TGRP fuel to aviation fuel and other biojet fuels produced from the bioconversion of poplar biomass per 1 MJ of fuel produced. (C) CExD comparing TGRP fuel to aviation fuel per PKM by natural resource contribution. (D) CExD comparing TGRP fuel to aviation fuel per PKM by process type. LG refers to lignin gasification and LG-HF refers to lignin gasification and hog fuel (Carrasco, et al., 2017). *CExD*, Cumulative exergy demand; *GWP*, global warming potential; *PKM*, passenger kilometer; *TGRP*, tail gas reactive pyrolysis.

Table 8.12 Capital and operating costs of fast pyrolysis and the incremental cost of tail gas reactive pyrolysis (TGRP) upgrading (Sorunmu et al., 2017).

	Fast pyrolysis	**TGRP upgrade**
Total project capital costs ($)	24,700,000	7,680,000
Annual operating costs ($/year)	9,360,000	4,510,000
Raw material ($/year)	4,280,000	1,950,000
Utilities ($/year)	768,000	313,000
Total operating labor and maintenance costs ($/year)	1,130,000	1,060,000
Total additional fixed operating cost ($/year)	712,000	423,000
Depreciation—linear (10 years) ($/year)	2,470,000	768,000
Total revenue ($/year)		5,620,000
Coproduct sales ($/year)		3,020,000
Sales from fuels ($/year)		2,600,000
MFSP ($/L)	1.1[a]–1.8[b]	

[a]MFSP for option B (free feedstock).
[b]MFSP for option A (feedstock price $0.055/kg).

the price of coproducts and reducing the price of feedstocks can render the process economically feasible, but current research does not indicate the potential to do so. Overall, an investment analysis over a project life of 10 years results in a net present value (NPV) of \$1.6 million/year; hence, the project is economically feasible under the assumptions made. Jones et al. (2009) evaluated the production of fuels via pyrolysis and HDO upgrade and found a positive net present with a MFSP of \$0.54 L^{-1}, which is far below the MFSP for the TGRP oil upgrade process, \$1.1L^{-1}.

Both (C) and (D) of Fig. 8.14 indicate the cumulative exergy destruction per PKM is much higher for aviation fuel, compared to TGRP fuel; this is true, even without phenol and coke credits. The positive TGRP area shown on the figure represents the CExD for total production. Of the individual process contributions shown on Fig. 8.14D, the results indicate that the majority of the CExD in the overall process is attributed to horse manure preprocessing. The bulk of the energy consumed in the preprocessing step is from electricity used to operate milling, drying, and conveying equipment. This indicates that the material and energy needs of the preprocessing step could be targeted for CExD reduction.

8.6 Economics of cofeeding pyrolysis oil with vacuum gas oil in petro/biorefinery

Fluid catalytic cracking (FCC) is one of the most flexible processes in a petroleum refinery that can readily adjust to changes in feed quality through modifications to catalyst and operating conditions. Hence, this is the insertion point of the petrochemical infrastructure that pyrolysis oils could be integrated. We discussed earlier that companies such as WR Grace, Petrobras, and Teroso have explored such opportunity (Fig. 8.15). While technical challenges prevail, and the environmental benefits are obvious, the economics are yet to be deciphered.

Scenario: In a case study, coprocessing of typical FCC feed stream with raw and/or mildly hydrotreated bio-oils was carried out by companies such as CPERI, IRCELON, WR Grace, and other institutions of learning. WR Grace coprocessed a blend of 3 wt.% pine-based pyrolysis oil with 97 wt.% mid-continent vacuum gas oil (VGO) over a low-metal commercial-grade equilibrium catalyst (Bryden et al., 2013).

Process conditions: While others used HDO products as cofeed, Grace used raw pyrolysis oil with composition reported as 39.5 wt.% carbon, 7.5 wt.% hydrogen, and 53 wt.% oxygen. A percentage of 100 mid-continent VGO was cracked as a control case. Riser outlet temperature was 521°C (970°F) for both feeds.

Results: Yields at insertion into the refinery at the point of the FCC reactor was summarized in Table 8.13 (Zacher et al., 2014). As the table shows coprocessing the mild HDO oil at 20 wt.% blend resulted in gasoline and LCO yields comparable to VGO alone, but higher dry gas yields and coke along with evidence of hydrogen consumption occurred according to Zacher et al. (2014) because of additional, but incomplete deoxygenation of the bio-oil. WR Grace has reported that cofeeding of only 3% raw pyrolysis oil resulted in more coke, less gasoline, and production of

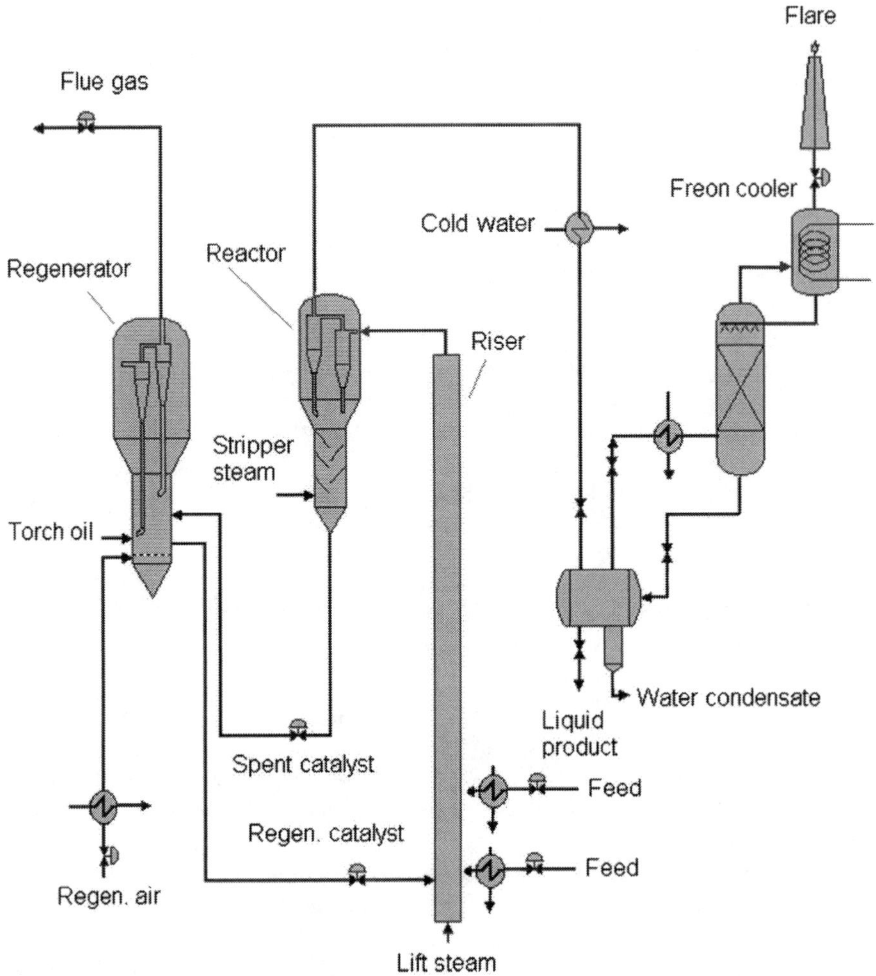

Figure 8.15 Schematic diagram of fluid catalytic cracking for coprocessing (Pinho et al., 2017).

CO and CO_2 in the product gas as informed by others who processed high oxygen content pyrolysis oils (Bryden et al., 2013). It would appear from an untrained eye that the differences in the product distribution for 3% replacement with pyrolysis oil are insignificant but there are several qualities of product and process challenges associated with this minute change. We are informed based on WR Grace's DCR analysis (Bryden et al., 2013), that at the same feed preheat and catalyst temperature, the blend of pyrolysis oil and VGO in this case study required slightly (~0.3) less catalyst-to-oil ratio to maintain the set 520°C (970°F) riser outlet temperature with the DCR operated in adiabatic mode. They further show that the exothermic reactions of the oxygen in the pyrolysis oil reduce the heat requirements for

Table 8.13 Fluid catalytic cracking (FCC) coprocessing of pyrolysis oils efforts (Zacher et al., 2014).

Org	**CPERI**				**IRCELYON**		**Twente**		**Grace**		
Year	**1998**		**2009**		**2010–11**		**2010–11**		**2013**		
Reactor	MAT, batch		Pilot, Circ.		MAT, batch		MAT, batch			Pilot, Circ.	
Cofeed	LCO	LCO	LCO/ VGO		VGO		Long res.		None	VGO	VGO
wt.% HDO bio-oil	0%	15%	0%	2%	0%	20%	0%	20%	100%	0%	3% raw
FCC conversion (%)			57–74	63–73	75	75	60	60	60	81.6	81.7
T (°C)	550	550	520	520	500	500	520	520	520	521	521
Catalyst/oil (g/g)	2–6	2–6			1–6	1–6	3.1	3–4	12–20	9.9	9.8
Dry gas (C_1, C_2) (wt.%)					1.5	2	1.5	1.9–2.5	6–11	3.2	3
LPG (C_3, C_4) (wt.%)					24	20	8.5	9–11	10–12	8.5	8.1
Gasoline (C_5) <221°C (wt.%)	17–22	20–25	38–46	42–47	46	44	44	44–45	22–36	49.1	47.5
LCO 221°C–370°C (wt.%)			17–22	18–21	20	20	25.2	23–25	11–19	14.1	14.2
HCO > 370°C (wt.%)					4	3	14.8	12–13	7–8	4.4	4.2
Coke (wt.%)	2–3	3–5	2–5.5	3.7–5.7	3.2	4.6	5.9	5.5–7.8	22–38	6.4	7.1

LCO, Light cycle oil; *HCO*, heavy cycle oil; *HDO*, hydrodeoxygenation; VGO, vacuum gas oil.

coprocessing pyrolysis oil with VGO. However, only trace amounts of oxygenates were found in the liquid product. While running pyrolysis oil, CO and CO_2 were detected in the product gas, amounting to a total of $\sim$22% of the oxygen in the pyrolysis oil. By difference, $\sim$78% of the oxygen in the pyrolysis oil reacted to water according to the Grace report. They conclude that cofeeding VGO with pyrolysis oil, like all or most nonpetroleum-based feedstock components, can result in significant yield shifts, even at small addition quantities. This seems marginal given the huge promise that the synergy between pyrolysis and petrochemical refining offers, but others, including Petrobras, claim to have shown better results.

8.6.1 Performance measurements

While no economic benefits were offered by the test companies in Table 8.11 besides the obvious environmental impacts, TEA studies have been carried out on similar processes under the Petrobras–NREL Cooperative Research and Development Agreement (CRADA) formed to elucidate the techno-economic benefits (Pinho et al., 2017). NREL/Petrobras has demonstrated up to 20 wt.% pyrolysis oil cofeed and shown that using the JMP statistical analysis software, that profit can be realized when cost is less than the breakeven values. For a 400 dry MTPD pyrolysis plant colocated in the refinery, the pyrolysis oil breakeven values can be \$84–\$88/barrel pyrolysis oil and a \$65/barrel crude oil price. For a larger pyrolysis oil plant at the nominal capacity of 2000 dry MTPD, these are \$55 and \$55, respectively.

Asmaa et al. (2018) developed a model in Microsoft Excel of a petroleum refinery along with a colocated biomass pyrolysis plant and reported the economic impact on the NPV, relative to a base case, in addition to a sensitivity analysis of key parameters such as the impacts of changes in crude oil prices, refinery capacity, biomass plant capacity, and bio-oil yield are examined. It is clear from their results (Fig. 8.16)

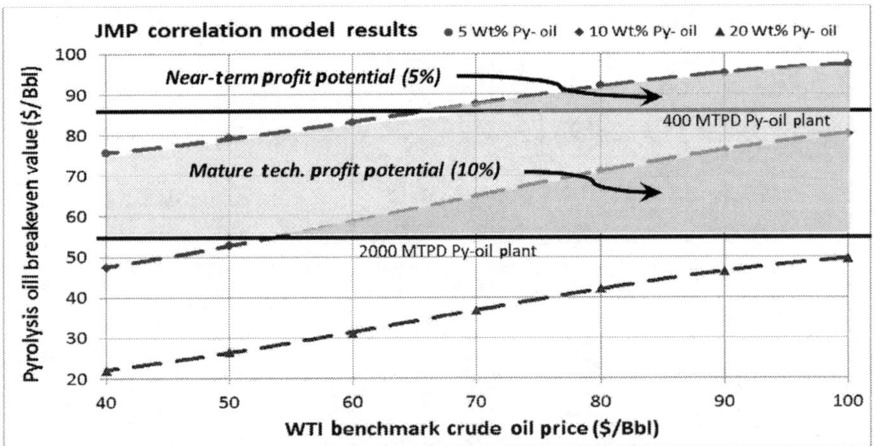

Figure 8.16 Pyrolysis oil breakeven value analysis National renewable energy lab (NREL) (Pinho et al., 2017).

that coprocessing is highly sensitive to changes in prices of crude oil and petroleum products. Coprocessing is shown to be economically attractive at crude oil prices at below 60$/bbl for raw bio-oil and above $120/bbl for HDO bio-oil. Although the coprocessing option is favorable to bio-oil producers, technological advancements are still required in order to improve the quality of the bio-oil produced.

References

Asmaa, A.M.A., Mustafa, A., Yassin, Kamal E., 2018. A techno-economic evaluation of bio-oil co-processing within a petroleum refinery. Biofuels . Available from: https://doi.org/10.1080/17597269.2018.1519758.

Boateng, A.A., Elkasabi, Y., Mullen, C.A., 2016. Guayule (*Parthenium argentatum*) pyrolysis biorefining: fuels and chemicals contributed from guayule leaves via tail gas reactive pyrolysis. Fuel 163, 240–247.

Boateng, A.A., Mullen, C.A., Elkasabi, Y., McMahan, C., 2015. Guayule (*Parthenium argentatum)* pyrolysis biorefining: production of hydrocarbon compatible bio-oils from guayule bagasse via tail-gas reactive pyrolysis. Fuel 158, 948–956.

Bryden, K., Weatherbee, G., Habib Jr., E.T., 2013. Grace catalysts technologies. Flexible pilot plant technology for evaluation of unconventional feedstocks and processes. In: Catalagram, A Catalysts Technologies Publication, No. 113/Spring 2013/grace.com.

Carrasco, J.L., Gunukula, S., Boateng, A.A., Mullen, C.A., DeSisto, W.J., Wheeler, M.C., 2017. Pyrolysis of forest residues: an approach to techno-economics for bio-fuel production. Fuel 193, 477–484.

Elkasabi, Y., Mullen, C.A., Boateng, A.A., 2014. Distillation and isolation of commodity chemicals from bio-oil made by tail-gas reactive pyrolysis. ACS Sustainable Chem. Eng. 2 (8), 2042–2052.

Goedkoop, M., Oele, M., De Schryver, A., Vieira, M., 2008. Simapro Database Manual: Methods Library. Pre′ Consultants, Amersfoort, The Netherlands. Available via the Internet from: <http://www.presustainability.com/manuals>.

Hammer, N.L., Boateng, A.A., Mullen, C.A., Wheeler, M.C., 2013. Aspen Plus® and economic modeling of equine waste utilization for localized hot water heating via fast pyrolysis. J. Environ. Manage. 128, 594–601.

Jones, S., Holladay, J., Valkenburg, C., Stevens, D., Walton, C., Kinchin, C., et al., 2009. Production of Gasoline and Diesel from Biomass via Fast Pyrolysis, Hydrotreating and Hydrocracking: A Design Case PNNL-18284, p. 76.

Keedy, J., Prymak, E., Macken, N., Pourhashem, G., Spatari, S., Mullen, C.A., et al., 2014. Exergy based assessment of the production and conversion of switchgrass, equine waste, and forest residue to bio-oil using fast pyrolysis. Ind. Eng. Chem. Res. 54 (1), 529–539.

Kwiatkowski, J.R., McAloon, A.J., Taylor, F., Johnston, D.B., 2006. Modeling the process and costs of fuel ethanol production by the corn dry-grind process. Ind. Crop. Prod. 23, 288–296.

Li, Q., Zhang, Y., Hu, G., 2015. Techno-economic analysis of advanced biofuel production based on bio-oil gasification. Bioresour. Technol. 191 (2015), 88–96.

McAloon, A.J., Taylor, F., Yee, W., 2000. Determining the cost of producing ethanol from corn starch and lignocellulosic feedstocks. In: NREL Report TP-580-28893. National Renewable Energy Laboratory, Golden, CO.

Peters, M.S., Timmerhaus, K.D., West, R.E., 2003. Plant Design and Economics for Chemical Engineers. McGraw-Hill, New York.

Pighinelli, A.L.M.T., Boateng, A.A., Schaffer, M.A., 2018. Utilization of eucalyptus for electricity production in Brazil via fast pyrolysis: a techno-economic analysis. Renew. Energy 119, 590–597.

Pinho, A., de, R., de Almeidaa, M.B.B., Mendesa, F.L., Casavechiab, L.C., Talmadgec, M.S., et al., 2017. Fast pyrolysis oil from pinewood chips co-processing with vacuum gas oil in an FCC unit for second generation fuel production. Fuel 188, 462–473.

Pourhashem, G., Spatari, S., Boateng, A.A., McAloon, A.J., Mullen, C.A., 2013. Life cycle environmental and economic tradeoffs of using fast pyrolysis products for power generation. Energy Fuels 27 (5), 2578–2587.

Sabaini, P.S., Boateng, A.A., Schaffer, M., Mullen, C.A., Elkasabi, Y., McMahan, C., et al., 2018. Techno-economic analysis of guayule (*Parthenium argentatum*) pyrolysis biorefining: production of biofuels from guayule bagasse from via tail-gas reactive pyrolysis. Ind. Crop. Prod. 112, 82–89.

Schnepf, R., Yacobucci, B.D., 2010. In renewable fuel standard (RFS): overview and issues; CRS report for congress. <https://www.ifdaonline.org/IFDA/media/IFDA/GR/CRS-RFS-Overview-Issues.pdf>.

Schnepf, R., Yacobucci, B.D., 2012. Renewable Fuel Standard: Overview and Issues. p. 31.

Sorunmu, Y., Billen, P., Elkasabi, Y., Mullen, C.A., Macken, N., Boateng, A.A., et al., 2017. Fuels and chemicals from equine waste derived tail gas reactive pyrolysis oil: techno-economic analysis, environmental and exergetic life cycle assessment. ACS Sustain. Chem. Eng. 5 (10), 8804–8814.

Szargut, J., Morris, D.R., 1987. Cumulative exergy consumption and cumulative degree of perfection of chemical processes. Energy Res. 11, 245–261.

Wang, M.Q., 2011 GREET 1. Center for Transportation Research, Argonne National Laboratory.

Wright, M.M., Brown, R.C., Boateng, A.A., 2008. Distributed processing of biomass to bio-oil for subsequent production of Fischer-Tropsch liquids. Biofuels Bioprod. Biorefin. 2 (3), 229–238.

Zacher, A.H., Olarte, M.V., Santosa, D.M., Elliot, D.C., Jones, S.B., 2014. A review and perspective of recent bio-oil hydrotreating research. Green Chem. 16, 491–515.

Energy crops—biomass resources and traits

9

9.1 Introduction

The development of biomass resources as feedstock for bioenergy and bioproducts is a big agricultural undertaking and energy crops are at the center of the industry. Feedstock development, including pre- and postharvest processing, supply chain logistics, and quality, is among the critical components in evaluating the economic viability and environmental sustainability of the lignocellulosic biorefinery. Ensuring continuous and reliable supply of quality and economically affordable biomass feedstock to the factory gate is critical to the longevity of operation and reliable supply of fungible fuels. In 2016 the US Department of Energy (DOE) released a report estimating that the United States has the potential to produce at least 1 billion dry tons of biomass resources annually by 2040 (Billion-ton report, 2011). Of this potential, roughly 365 million dry tons are currently used in the existing US bioeconomy. However, untapped resources in the form of agricultural residues, wastes, and forest residues are available now, while energy crops, algae, and other additional waste streams offer growth potential in the coming years. For example, quantitative assessment indicates that utilizing 1 billion dry tons of biomass could amount to about 50 billion gallons of biofuels, 50 billion pounds of biobased chemicals and bioproducts, 75 billion kWh of electricity, and 990 trillion BTU of thermal energy according to the US Bioeconomic Initiative Implementation Framework working group.

Based on the "billion-ton" study, agricultural residues and wastes are about 244 million dry tons currently and expected to increase to 404 million dry tons by 2030 at a farmgate price of $60 per dry ton. Their models show that in the near term, that is, by 2022 when the RFS2 is supposed to be fully enacted, the total agricultural resources (crop residues and energy crops) will reach 910 million dry tons at the $60 price. Energy crops are the largest potential source of biomass feedstock, but its potential varies considerably depending on what is assumed about productivity. At a 2% annual growth rate, energy crop potential is estimated at 540 million dry tons by 2030 and 658 million dry tons if an annual increase in productivity of 3% is assumed. Increasing yield growth to 4% pushes the energy crop potential to nearly 800 million dry tons. At lower farmgate prices of, for example, $40 and $50 per dry ton, it is predicted that total land-use change will be between 33 and 44 million acres, respectively, for energy crop production. Hence, tremendous R&D investment has gone into developing knowledge base and understanding of upstream processes such as feedstock genetic improvement, feedstock production and management, and feedstock logistics for energy crops. The U.S. Department of

Pyrolysis of Biomass for Fuels and Chemicals. DOI: https://doi.org/10.1016/B978-0-12-818213-0.00009-6

Agriculture (USDA)–DOE through the Biomass R&D Initiative (BRDi) funding allocations have focused on the development of new technologies from molecular mapping of genes to landscape models to properly design and place biomass energy crops within agricultural and forestry landscapes. While focusing on the increasing yields of dedicated energy crops will inspire producers to meet feedstock demand, it is not enough. Life cycle analysis surrounding the sustainability of energy crops is important and so is feedstock quality, that is, compositional traits that affect conversion. Feedstock compositional traits must be compatible with the conversion process as they affect biofuels product yields, distribution, and quality. For example, while lignin is a recalcitrant to the biochemical conversion of lignocellulosic biomass, it may be a potential attribute to thermochemical conversion such as pyrolysis.

Energy crops that are deemed to have high potential to succeed as biofeedstocks include perennial grasses such as switchgrass (*Panicum virgatum* L.), big bluestem (*Andropogon gerardii*), and Indian grass (*Sorghastrum nutans*), annual crops such as high-yielding sorghum, and short rotation woody crops, managed either as a single rotation (i.e., harvest before replanting) or managed as a multirotation (i.e., coppicing) crop. Switchgrass is a prairie grass native to North America adapted for a wide range of environments. Breeding and selection research on native perennial grasses such as switchgrass started in 1936 when the USDA at Lincoln, Nebraska, began breeding native grasses to revegetate land damaged by the drought of the 1930s. Since that time, scientists at the USDA biomass centers and other institutions have evaluated native collections and selected and bred improved cultivars for most areas of the United States (Figs. 9.1 and 9.2). Several characteristics make perennial grasses such as switchgrass desirable biofeedstock energy crops. They are broadly adapted reducing the concerns for becoming invasive species. Other

Figure 9.1 Switchgrass harvesting: (A) hand harvest of ripe seed and (B) standing biomass of switchgrass at senescence.
Source: Photo by Michael Casler, USDA, Madison, WI.

Figure 9.2 The Plant genome 4:3 (Casler, 2012).

important attributes include its water- and nitrogen-use efficiency and low fertilizer requirements and many more, some of which are depicted in Fig. 9.3. Yields can be anywhere between 2 and 10 tons/acre/year, establishment cost of \$210–\$410, and harvest cost of about \$41–46 per acre (~\$3/dry ton). Long-term plot trials and farm-scale studies have indicated that switchgrass is productive as well as enhances and protects environmental quality and can be potentially profitable in the cellulosic biofuels market. USDA's goal established for their biomass centers has been to develop new and improved perennial grass cultivars for marginal cropland in the Central United States to produce biomass for bioenergy and bioproducts. Critical aspect to this effort is to identify the key factors that impact biomass production and conversion, including the effects of harvest time, across multiple cultivars, growing locations, and harvest years.

This chapter will highlight some energy crops specifically switchgrass developed using USDA resources. The chapter will address specific cultivars, harvest yield, and track their compositional traits as they affect pyrolysis product yield, distribution, and quality. The information is drawn from studies conducted by several USDA supported consortia such as CenUSA funded under the USDA-National Food and Agriculture (NIFA)'s Coordinated Agricultural Project (CAP) grants and

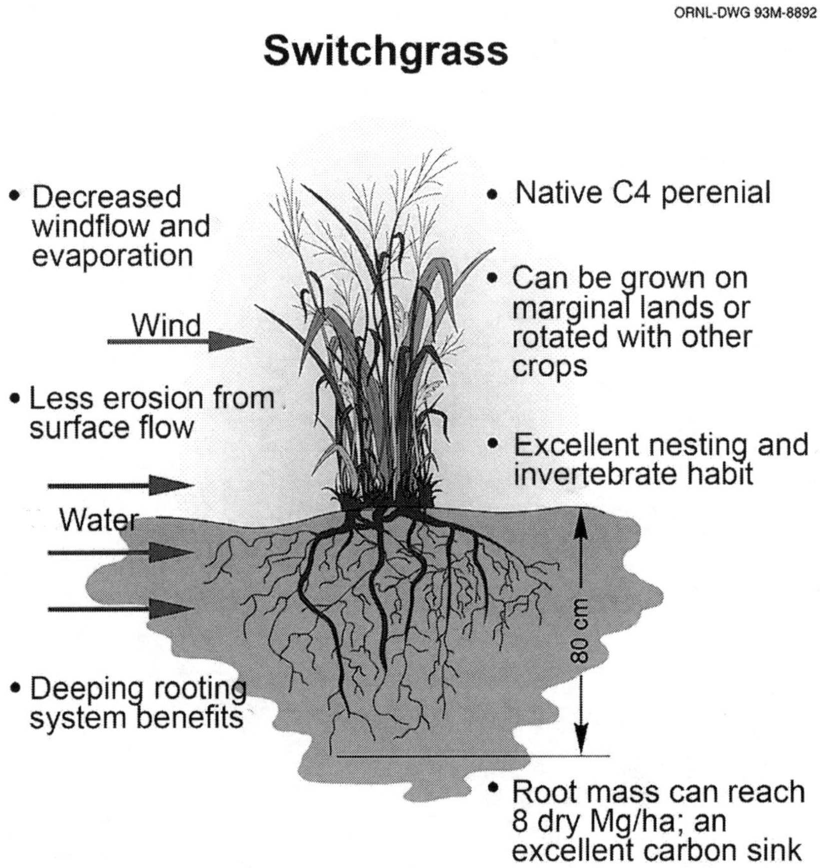

Figure 9.3 Beneficial switchgrass traits (courtesy of Oak Ridge National Lab, ORNL).

others who are involved in breeding upland and lowland ecotypes and their hybrids that are wind pollinated and self-incompatible natural accessions. Lowland ecotypes are typically taller with larger leaves, thicker stems that flower later than upland ecotypes. The CenUSA Bioenergy partnership was led by Iowa State University that investigated a Midwestern system for producing biofuels and bioproducts using perennial grasses grown on marginal lands.

9.2 Harvest time and cultivar on fast pyrolysis

A large study conducted by CenUSA over a 5-year period (2012–17) on the breeding, cultivation, and harvest logistics for upland and lowland switchgrass was expanded to evaluate the effects of harvest time on bioenergy production (CenUSA consortium, 2012). The large field trials evaluated the harvest of six switchgrass cultivars grown at three northern US locations over 3 years and harvested at upland peak crop (anthesis), postfrost, and postwinter (Serapiglia et al., 2016, 2017a,b). The six cultivars that were examined in the CenUSA study include releases such as "Blackwell," "Cave-in-Rock," "Hiawatha," "St. Croix," "Sunburst," and "Kanlow." In the study, designated plots were harvested at three separate times to evaluate harvest date effects, including (1) upland peak, approximately anthesis for Cave-in-Rock (August); (2) approximately 2 weeks following killing frost (October); and (3) postwinter (April), that is, as soon as fields were sufficiently dry to support harvesting equipment (Casler et al., 2018).

The biofuels goals for the study were to explore whether the harvested biomass, with cell wall polymer composed of hemicellulose, cellulose, and lignin, can be converted to liquid fuels via various biochemical and thermochemical processes at what yields and product distribution based on the time of harvest. Within the harvest time frame the biomass composition can vary widely depending on several factors, including genetics, the growth environment, crop management strategies, harvesting methods, and storage practices of the biomass postharvest.

When the target technology is biochemical conversion by enzymatic hydrolysis followed by fermentation of sugars, Serapiglia et al. (2017b) found that harvesting after frost is recommended as it allows for enrichment of the biomass for structural carbohydrates, which increases its value to the processor on a per mass basis. Delaying the time of harvest until after frost or postwinter increased the concentration of structural carbohydrates by 14 wt.% (from 500 to over 570 g/kg) in the biomass and increased lignin content by 25 wt.% (160 to over 200 g/kg). This is because the delay allows for the translocation of valuable plant nutrients and non-structural carbohydrate back into below ground tissues, avoiding processing issues incurred by the presence of simple sugars upfront of pretreatment. They observed that delaying harvest does lead not only to increased sugar yields largely because of higher carbohydrate concentration in the later harvested material but also because the later harvested material had reduced biomass recalcitrance.

In the case of pyrolysis, both catalytic (e.g., over HZSM-5) and noncatalytic pyrolysis product yields can be significantly affected by the field trial location, year of harvest, cultivar, and harvest time also. Delaying harvest time of the switchgrass crop leads to greater production of deoxygenated aromatics, thereby improving the efficiency of catalytic fast pyrolysis and bio-oil quality. The changes in the pyrolysis product yield are related to biomass compositional changes, and key relationships between cell wall polymers, concentration of minerals nutrients such as potassium in the biomass, and pyrolysis products have been identified and well documented (Serapiglia et al., 2017a). The study has shown that the loss of minerals in the biomass when harvest time is delayed combined with the greater proportion in cellulose and lignin in the biomass has significantly positive influences on fast pyrolysis conversion. Some of the results are interesting and can serve as guide to the bioenergy crop grower supplying biomass to the pyrolysis biorefinery.

9.2.1 Harvest time affects pyrolysis yield production

We have seen that delaying harvest changes compositional traits of the biomass crop. To see the effect of harvest time on pyrolysis yield and distribution, a three-way interaction of year × location × harvest allows one to establish harvest time effects on product distribution and the pyrolysis of switchgrass cultivar (Fig. 9.4;

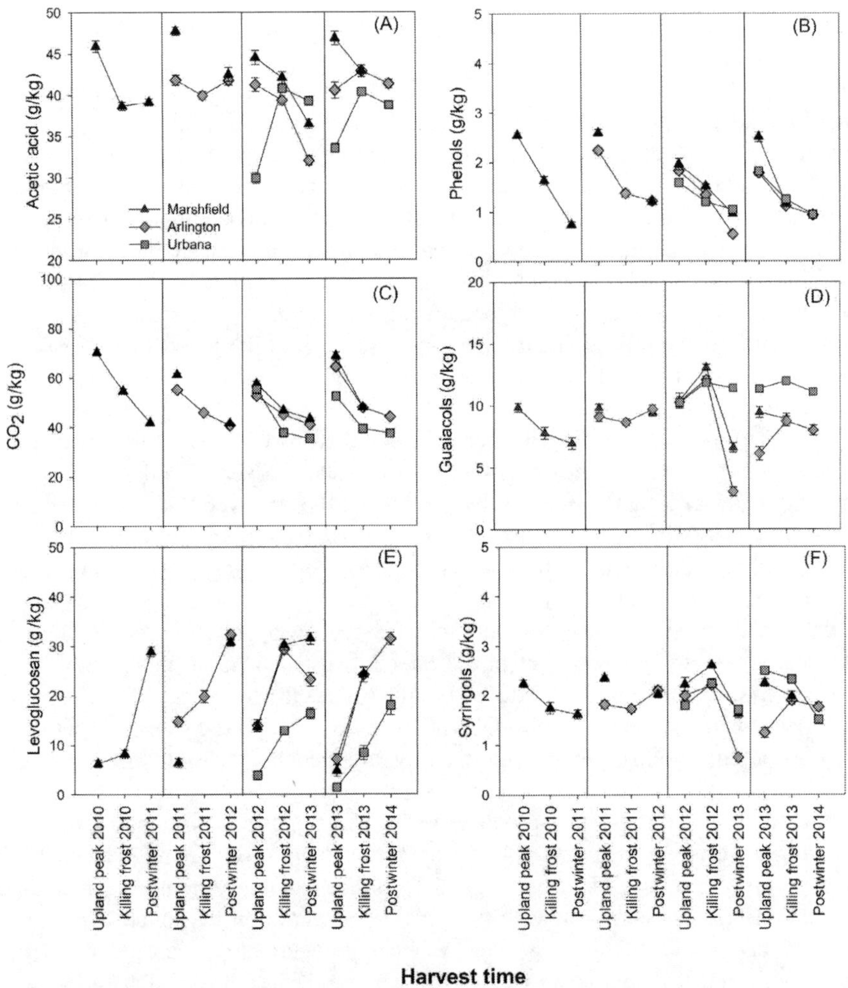

Figure 9.4 Pyrolysis product yield from switchgrass biomass at each of the three harvests, upland peak, after frost, and postwinter at three locations (Arlington and Marshfield, WI and Urbana, IL). Means were calculated from six cultivars and four field replicates ($n = 24$) and error bars are ± 1 standard error of the mean. (A) Acetic acid, (B) phenols, (C) CO_2, (D) guaiacols, (E) levoglucosan, and (F) syringols (Serapiglia et al., 2017a).

Table 9.1 ANOVA for selected noncatalytic pyrolysis products showing main effects and interactions, with significant effects indicated by *P*-values in bold (Serapiglia et al., 2017a).

						P-value		
Source of variation	**CO**	**CO_2**	**Acetic acid**	**Acetol**	**Phenols**	**Guaiacols**	**Syingols**	**Levoglucosan**
Year	**<.001**	**<.001**	**<.001**	**<.001**	**<.001**	**<.001**	**<.001**	**<.001**
Location	**<.001**	**<.001**	**<.001**	**<.001**	**<.001**	**<.001**	**<.001**	**<.001**
Year × location	**<.001**	**<.001**	**<.001**	**<.001**	**.043**	**<.001**	**<.001**	**<.001**
Harvest	**<.001**	**<.001**	**<.001**	**<.001**	**<.001**	**<.001**	**<.001**	**<.001**
Year × harvest	**<.001**	**<.001**	**<.001**	**<.001**	**<.001**	**<.001**	**<.001**	**<.001**
Location × harvest	**<.001**	**<.001**	**<.001**	**<.001**	**<.001**	**<.001**	**<.001**	**<.001**
Year × location × harvest	**<.001**	**<.001**	**<.001**	.081	**<.001**	**<.001**	**<.001**	**<.001**
Cultivar	**<.001**	**<.001**	**<.001**	**<.001**	**<.001**	**<.001**	**<.001**	**<.001**
Year × cultivar	**.042**	**.007**	**.018**	**.005**	.110	**.001**	**.010**	.805
Location × cultivar	**.040**	**.019**	.086	.061	.443	**.003**	**.040**	**.004**
Year × location × cultivar	.155	.073	**.008**	.120	.137	**.004**	**<.001**	**.021**
Harvest × cultivar	**<.001**	**<.001**	**.004**	**.021**	.096	**.024**	.087	**<.001**
Year × harvest × cultivar	**.003**	**<.001**	.167	.097	**.036**	.709	.494	.327
Location × harvest × cultivar	**<.001**	**.009**	.362	.270	.450	.906	.863	.274
Year × location × harvest × cultivar	.382	.059	**.004**	.114	**.040**	.510	.094	.064

Table 9.1). We have previously seen that levoglucosan, an anhydrous sugar formed from cellulose depolymerization and dehydration during pyrolysis, is a major source of oxygenated compounds that destabilize bio-oil. It turns out that levoglucosan yield increases as harvest time is delayed, and this is a consistent trend across most cultivars (Figs. 9.4 and 9.5). In the reported USDA study, pyrolysis of the switchgrass cultivar "Kanlow" produced significantly less levoglucosan when harvested at the killing frost harvest compared to the other upland ecotypes. The production of CO_2 and phenols is observed to be decreased over harvest time. The yield of acetic acid, guaiacols, and syringols, and other bad actors that affect bio-oil stability was rather inconsistent across locations and years, indicating that there are multiple factors affecting the production of these compounds. While the yield of nonmethoxylated phenols decreased over harvest time, there was less of an effect on the

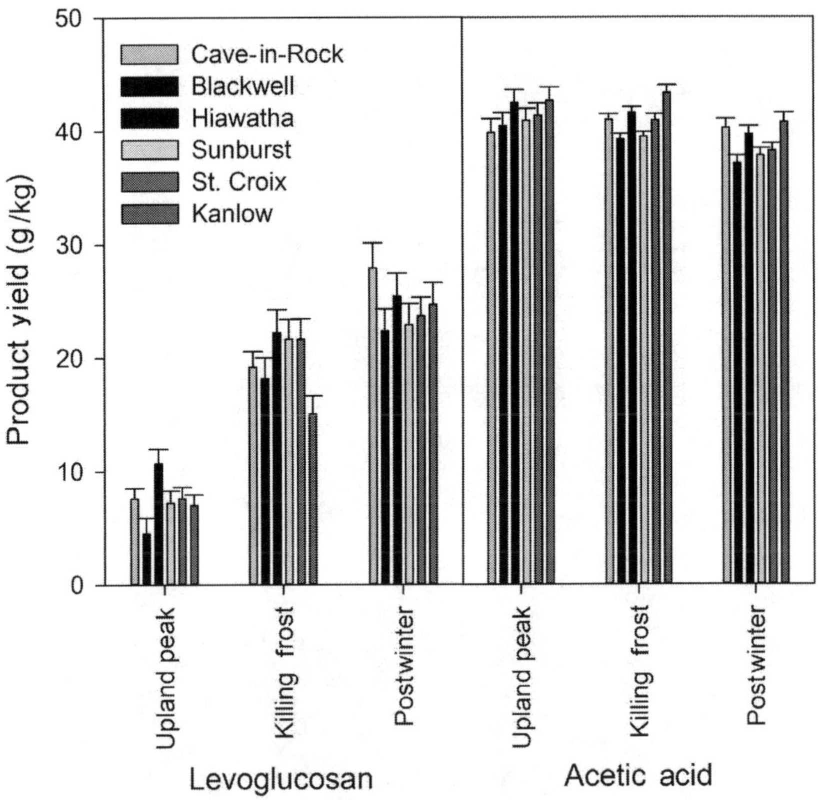

Figure 9.5 Levoglucosan and acetic acid production from six switchgrass cultivars over the three harvest times. Means were calculated from all years, all locations, and all for blocks ($n = 36$) and error bars are ± 1 standard error of the mean (Serapiglia et al., 2017a).

guaiacols, a consistent trend across the cultivars (Figs. 9.4 and 9.6). Guaiacols and syringols are known to be derived directly from the guaiacyl and syringyl units present in the lignin polymer; however, the study could not identify any relationship between these pyrolysis products and total lignin content in the biomass. Both the guaiacols and syringols do increase slightly with transition from the upland peak harvest to the killing frost harvest, and then decrease in the spring harvest. Fig. 9.7 shows that aromatic hydrocarbons from catalytic pyrolysis over HZSM-5, that is, benzene, toluene, *p*-xylene, naphthalene, and methyl-naphthalene increase as harvest time is delayed. While aromatic hydrocarbon yield is the highest at spring, harvest selectivity can be different across cultivars.

9.3 Mineral compositional effects on pyrolysis products

Several studies have consistently demonstrated that ash in the biomass (particularly alkali metals such as K) impacts the pyrolysis of biomass, reducing overall bio-oil liquid yield and reducing the bio-oil quality (Nowakowski and Jones, 2008). However, the biological range of mineral concentration that affects pyrolysis yield and to what extent has alluded producers. Answers to these questions can inform growers involved with crop management strategies to gauge the appropriate time to

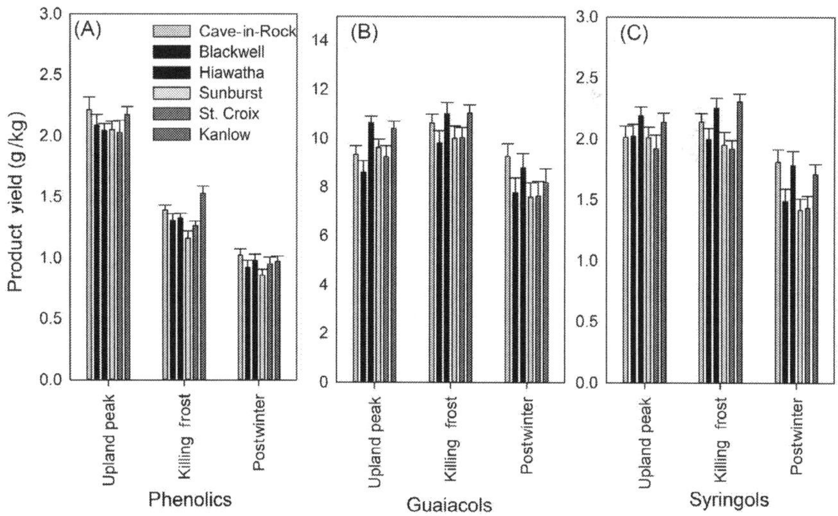

Figure 9.6 Phenolic production from six switchgrass cultivars over the three harvest times. Means were calculated from all years, all locations, and all for blocks ($n = 36$) and error bars are ± 1 standard error of the mean. (A) phenolics, (B) guaiacols, and (C) syringols (Serapiglia et al., 2017a).

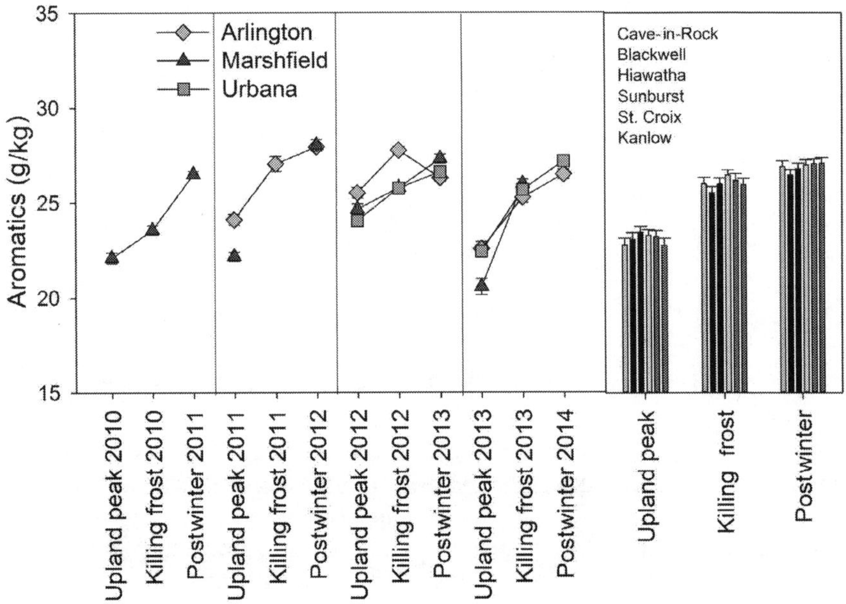

Figure 9.7 Aromatic hydrocarbon yield from catalytic fast pyrolysis in response to harvest time across the sites and cultivars (aromatics = benzene, toluene, xylene, naphthalene, and methyl-naphthalene) (Serapiglia et al., 2017a).

harvest their biomass given that delaying harvest allows for translocation of valuable plant nutrients back into below ground tissues.

There is a consensus that the presence of K in the biomass can catalytically change the path of pyrolytic cellulose breakdown from the production of levoglucosan to favor compounds such as acetic and propanoic acids, cyclopentanone derivatives, hydroxyacetaldehyde, acetol, and phenol (Serapiglia et al., 2017a,b). The large data analysis afforded by the USDA study confirms this for switchgrass across cultivars (Fig. 9.8). The figure shows that anhydrous sugar such as levoglucosan production will increase as K builds up to a maximum of around 1.2 g/kg, then decreases thereafter. The low K values are observed after winter harvest indicating depletion of minerals back to the soil.

As K content in the biomass declines with the delay of harvest time, the deoxygenation reactions that occur during the pyrolysis process also decline. This is further supported by the decrease in CO_2 and nonmethoxylated phenolics production over harvest time (Fig. 9.4) and the positive correlation with K (Serapiglia et al., 2017a).

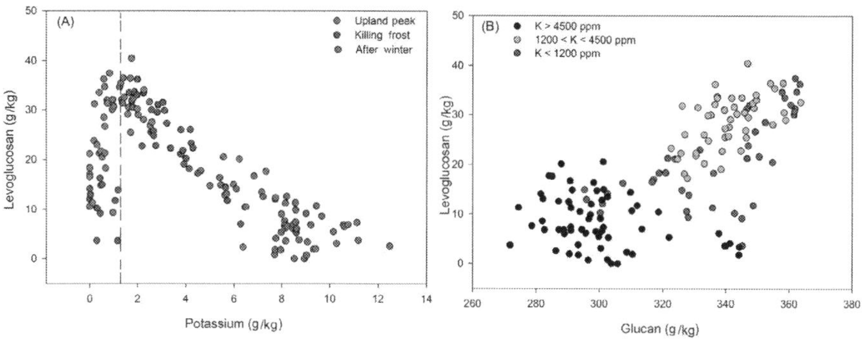

Figure 9.8 Relationship of levoglucosan with K content (A) and glucan content (B) in the biomass. Dotted line in panel A is at 1.2 g/kg of K. (A) Regression statistics for above 1.2 g/kg are $Y = -3.44X + 36.6$, $P < .001$, $R^2 = 0.86$. Regression statistics for below 1.2 g/kg are $Y = 7.3 + 16.88$, $P = .14$, $R^2 = 0.09$. (B) Regression statistics for above 4500 ppm K are $Y = -0.1X + 38.93$, $P = .01$, $R^2 = 0.10$. Regression statistics for below 4500 ppm K are $Y = 0.32X - 83.36$, $P = < .001$, $R^2 = 0.36$ (Serapiglia et al., 2017a).

Aromatic hydrocarbons from catalytic pyrolysis over HZSM-5 increased with harvest time, mainly due to the cleaner initial depolymerization of cellulose with decreasing K, as noted by the correlation of noncatalytic levoglucosan production with catalytic conversion to aromatic hydrocarbons (Fig. 9.9). Catalytic fast pyrolysis over HZSM-5 converts the mixture of primary pyrolysis products to deoxygenated aromatic hydrocarbons, with the majority of those coming from the depolymerization of cellulose (Mullen et al., 2011).

The product distribution effects observed is not restricted to potassium but holds true for related minerals such as phosphorous and iron. A study of herbaceous bioenergy crops, including sorghum, switchgrass, and miscanthus, evaluated for their potential as phytoremediators for the uptake of phosphorus in the Delmarva Peninsula, and their subsequent conversion to biofuel intermediates (bio-oil) showed a related influence of P and Fe on catalytic and noncatalytic pyrolysis product yield and distribution (Boateng et al., 2015, Mullen et al., 2014). The composition of the bio-oil obtained from the pyrolysis of biomass with high P, K, and ash was altered by the lower yields of levoglucosan and furfural and higher yields of phenol. Mullen et al. (2014) found a positive correlation between ash content and carbon conversion to aromatic hydrocarbons, which was particularly strong for switchgrass samples in a catalytic pyrolysis. The foregoing leads to a conclusion that the effects of compositional traits such as K and Fe on the pyrolysis chemistry particularly catalytic pyrolysis to aromatic hydrocarbons follow the chemical pathways as suggested in Fig. 9.10. The scheme shows that on one hand, increasing the K content leads to fragmentation reactions during pyrolysis which, as intermediates, in turn, decrease the carbon yield of aromatic hydrocarbon products. Meanwhile, the aromatics, which are produced from this pathway, tend to be alkylbenzenes,

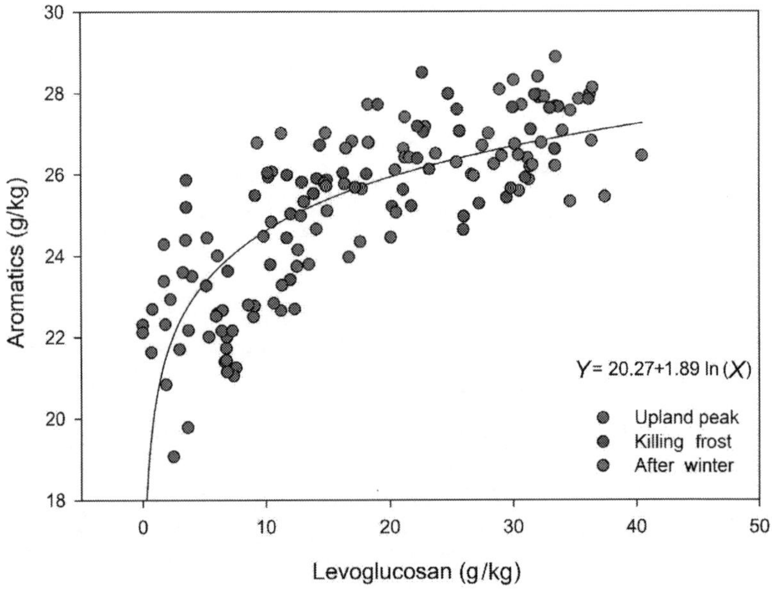

Figure 9.9 Relationship of levoglucosan from noncatalytic pyrolysis to aromatic production from catalytic pyrolysis (Serapiglia et al., 2017a).

increasing selectivites for toluene and xylenes over benzene and naphthalenes. On the other hand, increasing the iron content may influence the conversion of the more productive intermediates over HZSM-5 to the desired aromatic hydrocarbons.

9.4 Implications for optimal harvest time

While delaying switchgrass harvest time can have a negative effect on biomass yield, the ideal time to harvest for optimum biomass quality and conversion potential may be later than peak crop (at anthesis). Harvesting later after a killing frost would also be highly beneficial for longevity and sustainability of the crop, allowing vital nutrients to cycle back below ground (Mitchell and Schmer, 2012). In addition, less ash and metals in the harvested biomass will reduce their effects on both biochemical and thermochemical conversion. Overall increase in the proportion of cell wall polymers in the biomass as a result of the loss of soluble sugars, ash, and extractives as harvest is delayed is highly advantageous. This essentially allows one to produce more usable and convertible material requiring less preprocessing and/or washing to remove undesirable alkaline metals.

Agricultural and pyrolysis studies (Serapiglia et al., 2017a) have demonstrated that delaying harvest time until after a killing frost would improve biomass

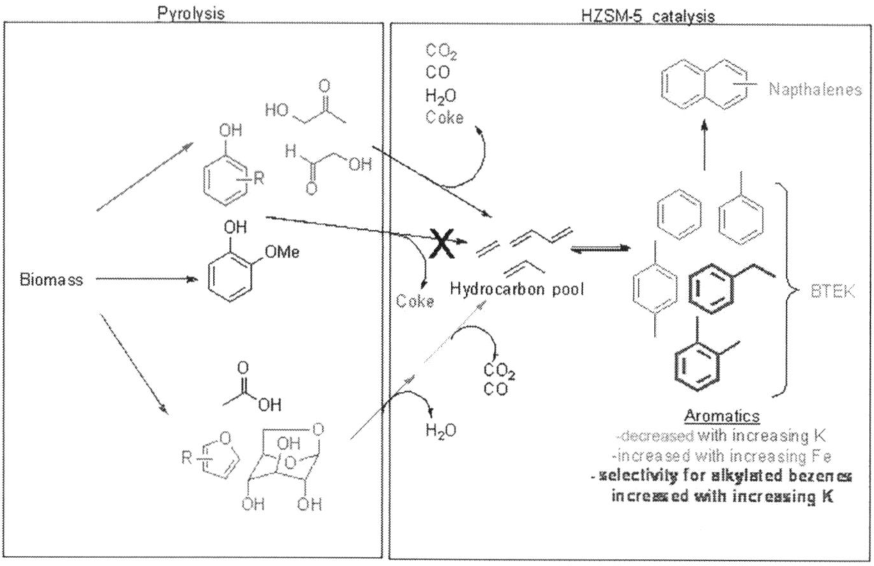

Figure 9.10 Summary of suggested effects of K and Fe on the chemical pathways from biomass to aromatic hydrocarbons by catalytic fast pyrolysis over HZSM-5 (Mullen et al., 2014). Green arrows and products are decreased by increasing K, red arrows and products are increased by increasing K. Green bolded compounds represent those with increased selectivity among BTEK by increasing K. Orange arrows and compound groups indicate those increased by increasing iron content.

composition and potential ethanol yield. When biomass yield potential was taken into consideration and potential ethanol yield was evaluated on a per hectare basis, harvesting after a killing frost may be the best time, even with the loss in biomass yield. This may also be the case to produce bio-oil through fast pyrolysis. The quality of the biomass and the bio-oil is significantly improved by the after-frost harvest with little improvements made by the spring harvest. The K content in the biomass declined over 60% from the upland peak harvest in August to the killing frost harvest in the fall (Serapiglia et al., 2016). This decline in K content had a significant impact on pyrolysis product yield and improving bio-oil quality as harvest time is delayed. Higher levels of ash and K in the biomass have led to reduced bio-oil quality with increased water content and acidity and reduced carbon yield, in addition to increasing gas and char production and reducing overall liquid product yield (Trendewicz et al., 2015).

In conclusion, USDA studies have demonstrated that the time of harvest for the switchgrass bioenergy crop can have significant implications on conversion potential and the quality of the products obtained. Delaying harvest time of the switchgrass crop will lead to the greater production of deoxygenated aromatics, improving the efficiency of the catalytic fast pyrolysis and the bio-oil quality (Fig. 9.11). The data presented here allowed for a further understanding of how changes in biomass

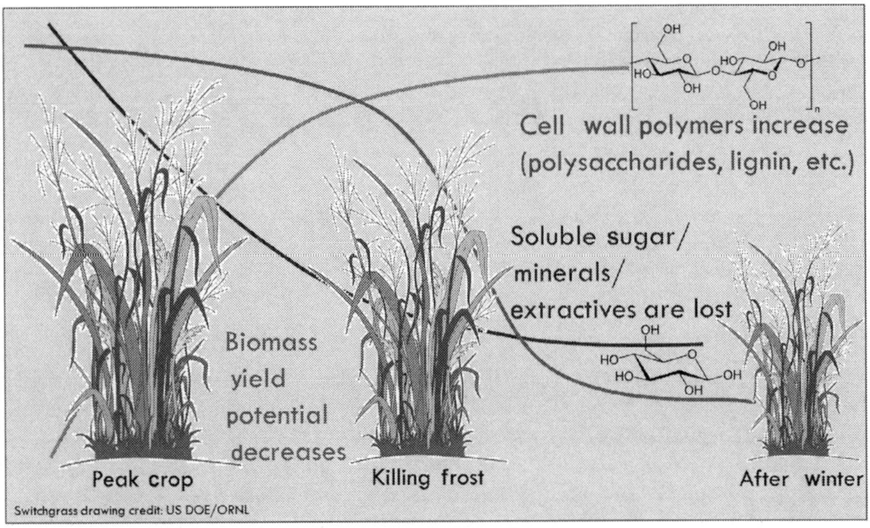

Figure 9.11 Cartoon of mineral and biomass depletion over time leading to harvest.

composition impact pyrolysis and the production of condensable products in the bio-oil (Table 9.2).

9.5 Proteinaceous energy crops

Recently, cyanobacteria and microalgae have been studied as a source for biofuel production as they are among the fastest growing photosynthetic organisms. They can easily adapt to various climate conditions, possess a higher efficiency for CO_2 fixation, and grow at greater productivity rates (ton/ha/year) than terrestrial plants. Algal biomass production does not require arable land and can be cultivated in saline and waste water making algae biomass extremely attractive to the bioeconomy as it poses no competition with food production. Furthermore, microalgae do not follow typical crop schemes and can be produced year-round (Chagas et al., 2016).

High protein biomass produces bio-oil with a greater content of nitrogen and lower content of oxygen than lignocellulosic biomass, following the elemental compositional traits of the feedstock. Compositional characteristics of Spirulina, a cyanobacterium, are presented in Table 9.3. Unlike lignocellulosic biomass, Spirulina is composed primarily of proteins, carbohydrates, and ash with very low content of lipids. In the thermal degradation of algae aromatic hydrocarbons can be produced from the protein fraction, which makes Spirulina, an attractive feedstock to produce bio-oil through the pyrolysis processes. The nitrogen content of Spirulina (10.7%) is considerably high in order of magnitude, compared to lignocellulosic biomass

Table 9.2 Advantages and disadvantages of delayed harvest (U.S. Department of Agriculture).

	Peak crop harvest (summer)	**Killing frost harvest (fall)**	**After winter harvest (spring)**
Pros	High biomass yield potential	High ethanol yield potential Highest energy content in biomass Enrichment of structural carbohydrates Translocation of nutrients to soil	Complete translocation of nutrients Reduced recalcitrance for ethanol production Favorable pyrolysis product yield Provides wintering habitat for wildlife
Cons	Greater levels of nutrient removal requiring fertilizer application in spring Greater levels of impurities for conversion	Lower biomass yield potential Interference with harvesting other commodity crops	Lower biomass yield potential Harvesting problems associated with lodging, wet soils, and spring regrowth@@@

Table 9.3 Characteristics of Spirulina (Chagas et al., 2016).

Ultimate analysis (wt.%) db		**Proximate analysis (wt.%) db**		**Biochemical composition (wt.%) db**	
C	48.05	Moisture	5.60	Crude fat	0.80
H	7.04	Ash	7.94	Crude protein	74.40
N	10.74	Volatile matter	79.38	Carbohydrate	16.86
S	0.72	Fixed carbon	5.28		
O	33.45	HHV (MJ/kg)	22.56		
Inorganic elements (mg/g, db)					
P 10.79 (±0.17)	Ca 0.81 (±0.01)	Mg 2.79 (±0.03)	K 17.10 (±0.28)	Al 0.02 (±0.00)	Na 16.77 (±0.22)
Fe 0.55 (±0.00)	Zn 0.01 (±0.00)	Cu <0.0001			

but the ash is similar to some herbaceous grasses, although it has a low content of K, Na, P, S, and Zn and a high content of Al, Ca, Fe, and Mg. Regardless of the high ash content, it has greater carbon content and calorific value (48.05%, 22.56 MJ/kg) than lignocellulosic biomass.

Pyrolysis of Spirulina yields aromatic hydrocarbons and aliphatic hydrocarbons such as limonene and C17–C20 alkanes, which makes this biomass attractive for production of high-quality bio-oil (Table 9.4). However, oxygenates and nitrogenous compounds are also produced in high quantities with the latter expected of high-protein biomass pyrolysis. When catalytic pyrolysis over zeolite is carried out, aromatic hydrocarbon yields increase, phenol yields decrease, and nitrogenous compound yields remain the same or slightly decrease over that produced by

Table 9.4 List of semiquantified and quantified compounds (Chagas et al., 2016).

Groups	Compounds
Aromatic hydrocarbons	Ethylbenzene[a]
	Styrene[a]
	Toluene[a]
	p-Xylene[a]
	o-Xylene[a]
Nonaromatic hydrocarbons	Heptadecane[a]
	D-Limonene
	3,7,11,15-Tetramethyl-2-hexadecene
	1,2,3,4-Tetrahydro-1,1,6-trimethylnaphthalene
Phenols	Phenol[a]
	o-Cresol[a]
	p-Cresol[a]
	2,4-Dimethylphenol[a]
	4-Ethylphenol[a]
Oxygenates	Furanmethanol[a]
	Acetol[a]
	3-Methylcyclopentanedione
	3-Methylbutanal
	2-Methylbutanal
Nitrogenates	Indole[a]
	3-Methyl-1*H*-indole
	Benzyl nitrile
	Benzenepropanitrile
	Hexadecanamide[a]
	Hexadecanenitrile
	2-Methyl-1*H*-pyrrole
	Hexahydropyrrolo[1,2-α]pyrazine-1,4-dione
	3-Isobutylhexahydropyrrolo[1,2-α]pyrazine-1,4-dione
	Hexahydro-3-phenylmethylpyrrolo[1,2-α]pyrazine

Other quantified aromatic compounds include 1,2,4-trimethylbenzene, benzene, naphthalene, and 2-methylnaphthalene.
[a]Quantified compound.

noncatalytic pyrolysis. Phenol and other oxygenate yields are minimal compared to the yields of hydrocarbons and nitrogenates. At higher catalyst loadings the nitrile or its precursors are converted to aromatics. With CFP over H-ZSM5 the nitrogenous compound yield is most significant and can comprise some 87% of the total fraction represented by hexadecanenitrile.

Overall, CFP of highly proteinaceous energy crops or algal biomass resources gets us closer to petrochemical fuel streams. In a study with Spirulina, a cyanobacteria, very low acidity H-ZSM5 zeolite with a Si/Al ratio of 280 favored the formation of indole and phenols and nonaromatic hydrocarbons, while the highest acidity zeolite [(H-ZSM5(23)] favored the further conversion of these compounds. A likely explanation for this observation is that the lower acidity catalysts are not as active for deoxygenation or denitrogenation but are able to aromatize primary pyrolysis vapors in the presence of the cations. Phenols and indoles can also be possibly formed from intermediate cation trapping with water or a nitrogen species (Mukarakate et al., 2015). It is possible that the lower acid site density in the hihger Si/Al ratio catalysts allows this process to happen over rapid dehydration. Primary phenols and indoles can also be precursors to coke over the more active, lower Si/Al ratio catalysts. The above observations suggest that it is possible to favor one set of chemical species over another for the conversion of algal biomass such as Spirulina by varying catalyst types and loadings in a CFP following suggested chemical pathways depicted in Fig. 9.12.

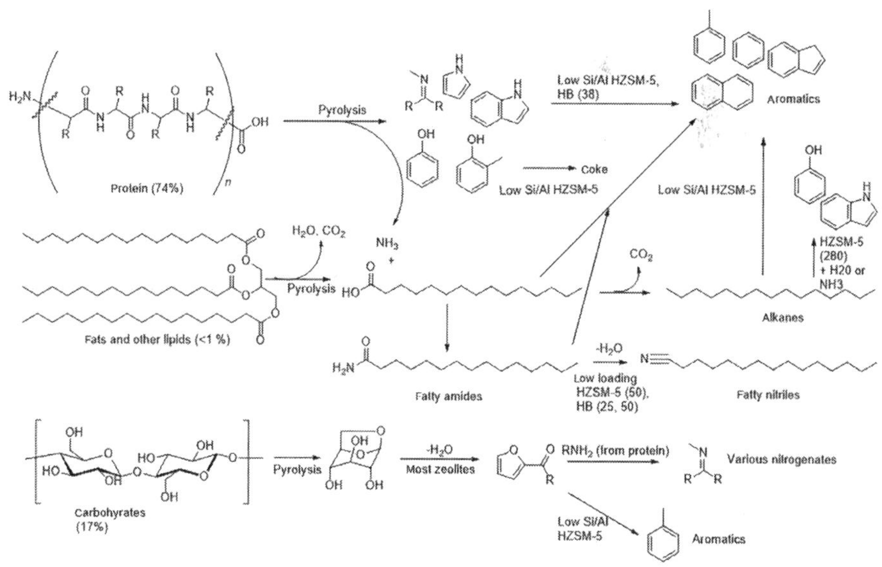

Figure 9.12 Observed reactivity of Spirulina via CFP over various catalysts with proposed chemical pathways. *CFP*, Catalytic fast pyrolysis.

References

Boateng, A.A., Serapiglia, M.J., Mullen, C.A., Dien, B.S., Hashem, F.M., Dadson, R.B., 2015. Bioenergy crops grown for hyperaccumulation of phosphorous in the delmarva peninsula and their biofuels potential. Environ. Manage 150 (1), 39–47.

Casler, M.D., 2012. Switchgrass breeding, genetics, and genomics. In: Monti, A. (Ed.), Switchgrass, Green Energy, and Technology. Springer, London, pp. 29–53. Available from: http://doi.org/10.1007/978-1-4471-2903-5_2.

Casler, M.D., Vogel, K.P., Lee, D.K., Mitchell, R.B., Adler, P.R., Sulc, R.M., et al., 2018. 30 Years of progress toward increased biomass yield of switchgrass and big bluestem. Crop Sci. 58, 1242–1254.

Chagas, B.M.E., Dorado, C., Serapiglia, M.J., Mullen, C.A., Boateng, A.A., Melo, M.A.F., et al., 2016. Catalytic pyrolysis-GC/MS of Spirulina: evaluation of a highly proteinaceous biomass source for production of fuels and chemicals. Fuel 179, 124–143.

Mitchell, R., Schmer, M., 2012. Switchgrass harvest and storage. In: Monti, A. (Ed.), Switchgrass A Valuable Crop for Energy. Springer, London, pp. 113–127. Available from: http://doi.org/10.1007/978-1-4471-2903-5_5.

Mukarakate, C., McBrayer, J.D., Evans, T., Budhi, S., Robichaud, D.J., Iisa, K., et al., 2015. Catalytic fast pyrolysis of biomass: the reactions of water and aromatic intermediates produces phenols. Green Chem. 17, 4217–4227.

Mullen, C.A., Boateng, A.A., Mihalcik, D.J., Goldberg, N.M., 2011. Catalytic fast pyrolysis of white oak wood in a bubbling fluidized bed. Energy Fuels 25 (11), 5444–5451.

Mullen, C.A., Boateng, A.A., Dadson, R.B., Hashem, F.M., 2014. Influence of mineral components of biomass on its conversion to aromatic hydrocarbons by catalytic fast pyrolysis over HZSM-5. Energy Fuels 28 (11), 7014–7024.

Nowakowski, D.J., Jones, J.M., 2008. Uncatalysed and potassium-catalysed pyrolysis of the cell-wall constituents of biomass and their model compounds. J. Anal. Appl. Pyrolysis 83 (1), 12–25.

Serapiglia, M.J., Boateng, A.A., Lee, D.K., Casler, M.D., 2016. Switchgrass harvest time management can impact biomass yield and nutrient content. Crop Sci. 56, 1–11.

Serapiglia, M.J., Mullen, C.A., Boateng, A.A., Dien, B.S., Casler, M.D., 2017a. Impact of harvest time and cultivar on conversion of switchgrass to bio-oils via fast pyrolysis. Bioenerg. Res. 10, 388–399.

Serapiglia, M.J., Boateng, A.A., Lee, D.K., Casler, M.D., 2017a. Impact of harvest time and switchgrass cultivar on sugar release through enzymatic hydrolysis. Bioenerg. Res 10, 377–387.

Trendewicz, A., Evans, R., Dutta, A., Sykes, R., Carpenter, D., Braun, R., 2015. Evaluating the effect of potassium on cellulose pyrolysis reaction kinetics. Biomass Bioenerg 74, 15–25.

Pyrolysis solid coproducts and usage

10

10.1 Introduction

Carbonaceous solid material formed during various thermochemical conversion processes comes in all forms of shapes, porosity, functional groups, and hence applications. When it comes to pyrolysis there are two forms of carbonization processes as we saw earlier, that is, slow pyrolysis (SP) that maximizes carbon, is the traditional route to commercial production of charcoal and fast pyrolysis (FP) that maximizes bio-oil for the purpose of biorefinery but leaves behind a substantial fraction (15–30 wt.%) known as biochar or biocharcoal or biocarbon in recent literature. Aside from solid carbon from the front end of the pyrolysis biorefinery (i.e., FP), there are other solid carbon forms downstream of the pyrolysis step. These include distillation bottoms such as petroleum coke produced in the petrochemical industry. As we saw earlier, useful applications of coproducts can improve the energy/exergy efficiency of the biorefinery, the techno-economics of the value chain, and system's sustainability of the biofuel product life cycle. This chapter looks at solid carbon coproducts associated with FP for fuels and chemicals and their beneficial applications.

FP of biomass can yield as much as 60–70 wt.% bio-oil for use as biocrude for the biorefinery. Accompanying the bio-oil is 20–30 wt.% biochar comprising carbon and ash. The economic viability of the production of pyrolysis oil as a fuel intermediate or as a feed stream for the biorefinery has been tied, in part, to the economic value of the biochar coproduct, and many potential uses have been proposed (Boateng, 2007). The most recent development and, perhaps, most interesting is the charcoal vision proposed by the U.S. Department of Agriculture (USDA), whereby a distributed biorefinery system could potentially produce the biocrude at the farm site and leave the biochar behind for carbon sequestration and storage as well as build soil quality so food and bioenergy crop can be sustainably harvested. Other uses of biochar include process heat and power. In addition, the residual coke bottoms of distillation processes downstream of the pyrolysis step may be used as biorenewable coke for industrial anodes with the potential to replace petroleum coke from fossil resources. Although there is much discussion about the use of biochar for soil amendment, agronomic applications have shown mixed results and the vision to integrate biomass–bioenergy systems that also build soil quality and increase agricultural productivity has not been fully realized. This is because not all biochars are created equal as they vary by their parent biomass and the processes that create them. Our interest here is the utility of biochar or biocarbon associated with pyrolysis biorefinery which means FP. We will examine their characteristics

Pyrolysis of Biomass for Fuels and Chemicals. DOI: https://doi.org/10.1016/B978-0-12-818213-0.00010-2

and some critical applications, their performances, and their potential to improve the techno-economics of the pyrolysis biorefinery.

10.2 Biochar characterization

An elemental mass balance from a FP system that is processing biomass residues collected from corn fields as feedstock (Fig. 10.1) shows that about 30% and 21%

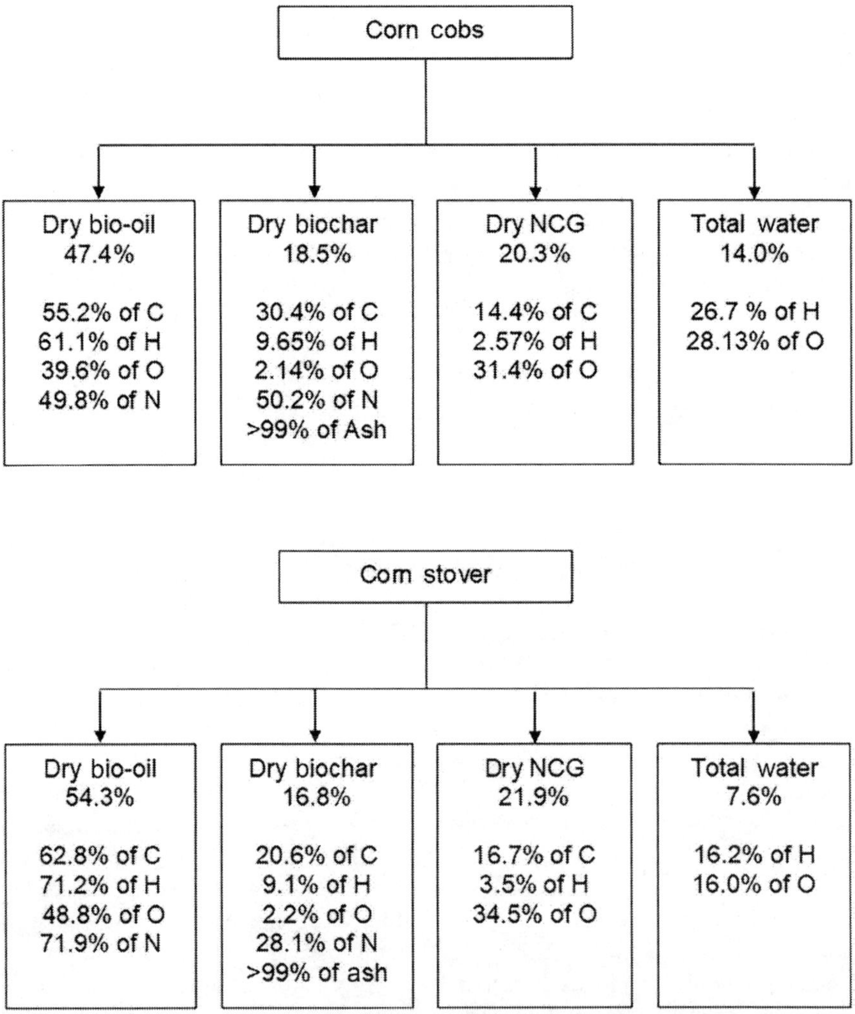

Figure 10.1 Mass and elemental balance (all values for feedstock are 100%) for pyrolysis of corn cobs (top) and corn stover (bottom) (Mullen et al., 2010).

of the carbon in a corn cob and in corn stover, respectively, are converted to biochar (Mullen et al., 2010). Nitrogen, an important nutrient, is split between the bio-oil and biochar amounting to 50:50 in the case of corn cob. Overall, the biochar ash contents show ~13 wt.% for the corn cob and ~32.4 wt.% for the corn stover, a difference reflective of the higher ash content of the parent feedstock. Important plant mineral nutrients such as K, P, Ca, and Mg are sequestered in the ash with representative quantities as shown in Table 10.1.

Biochar pH ranges from neutral to slightly alkaline (Table 10.2). Typical feedstocks used for carbonization including coal, woody biomass, and coconut shells tend to produce more acidic carbons; however, biocarbons made from plant or animal waste streams generally yield higher pH biochars reflective of the ash containing oxides of base metals (Table 10.3).

For FP biochars the BET surface area, as measured by N_2 adsorption, can be nonexistent or very low to the extent of being negligible in values. The values of the BET surface area biochars produced by FP of corn cob and stover only amounted to about 0 and 3.1 m^2/g, respectively (Table 10.2). The mean particle sizes were 513 and 421 μm respectively. Ninety percent of the biochar particles

Table 10.1 Minerals contained in biochar coproduct (mg/g) (Mullen et al., 2010).

	Corn cobs	Corn stover
Si	73.50	196.23
Al	0.21	33.10
Fe	0.65	15.95
Ca	0.97	20.13
Mg	2.15	14.24
Na	0.07	01.07
K	43.35	23.46
Ti	0.01	02.39
Mn	0.05	0.65
P	4.36	12.94
Ba	0.00	0.06
Sr	0.03	0.16
S (inorganic)	0.50	3.24

Table 10.2 Physical and chemical properties: pH, Brunauer, Emmett, and Teller (BET) surface area, and percent moisture content (MC) of corn cob and corn stover biochars (Mullen et al., 2010).

	BET	MC (%)	pH
Corn cob biochar	0	11.1 ± 1.1	7.816
Corn stover biochar	3.10 ± 0.26	8.8 ± 1.4	7.157

Table 10.3 Particle size distribution of biochars (μm) (Mullen et al., 2010).

Biochar source	Mean	Standard	<10%	<25%	<50%	<75%	<90%
Corn cob	708.8	513.3	71.58	265.5	651.3	1043	1479
Corn stover	373.6	420.9	30.35	93.69	212.3	477.8	1008

Figure 10.2 SEM photomicrographs of fast pyrolysis biochar as compared with carbon.

collected "as produced" from FP reactor and collected at the cyclone point were smaller than 1479 and 1008 μm for the corn cob and corn stover feedstock, respectively. Scanning electron microscopy (SEM) photomicrograph of biochar created from FP of switchgrass (SG) depicts an undeveloped surface when compared with carbon produced by SP (Fig. 10.2).

One important characteristic of biochars is their functional groups. Fourier transform infrared (FTIR) spectra of the biochars from SG created by varying time-temperature conditions in a slow-pyrolysis reactor are compared with that created by FP before and after they are subjected to phenol adsorption (Fig. 10.3) (Han et al., 2013). Herein, the C = O stretching is only seen as shoulders on the broad aromatic C–C = C stretching region at 1695–1700 cm^{-1} for FP biochar and that obtained through SP at lower temperature after phenol adsorption. Prior to phenol adsorption, these functional groups are apparently absent in FP biochars as they do for activated charcoal typically considered the standard. The observed pattern supports the hypothesis that levels of acidic functional groups are greatly reduced in biochar samples created by fast or SP. The broad absorption band in the aromatic C–C = C stretching region, 1731–1525 cm^{-1}, is simultaneously diminished when the biochar is created by FP, just as that from high-temperature SP or produced by

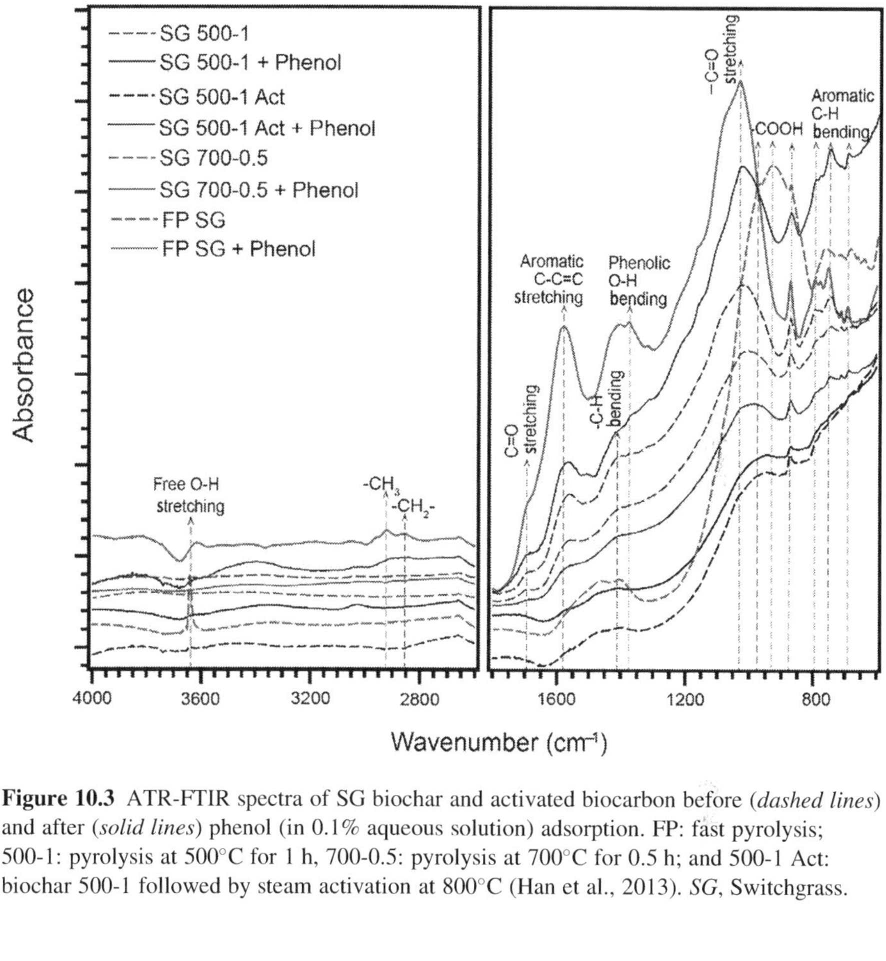

Figure 10.3 ATR-FTIR spectra of SG biochar and activated biocarbon before (*dashed lines*) and after (*solid lines*) phenol (in 0.1% aqueous solution) adsorption. FP: fast pyrolysis; 500-1: pyrolysis at 500°C for 1 h, 700-0.5: pyrolysis at 700°C for 0.5 h; and 500-1 Act: biochar 500-1 followed by steam activation at 800°C (Han et al., 2013). *SG*, Switchgrass.

steam activation. The coexistence of the absorption bands originated from the aromatic C–C = C stretching and phenolic O–H bending upon phenol adsorption is observed for FP biochar as well as SP held at various temperature and time exposures (FP SG, SG 500-1, SG 700-0.5). This is an indication of the binding by phenol molecules. The biochars produced by FP demonstrate most drastic shifts in the typical phenol IR absorption bands.

With regard to their potential as solid fuel, Table 10.4 shows the thermal characteristics of biochars created from SG as compared with Pennsylvania coals. As seen, fixed carbon as well as elemental composition is comparable with coal samples except that biochar has the benefit of not having sulfur. In fact, C on dry-ash-free basis is about twice that of low-rank coal from Pennsylvania.

Table 10.4 Analysis of switchgrass charcoal remaining after pyrolysis experiment compared with some PA coals (Boateng, 2007).

	As received	Dry basis	DAF basis	Schuykill, PA	Indiana, PA
				Raw coal (dry basis)	Raw coal (dry basis)
Proximate analysis (wt.%)					
Moisture	3.78	–	–	–	–
Ash	25.85	26.87	–	59.1	31.7
Volatile matter	28.37	29.48	40.31	8.3	22.3
Fixed C	42.00	43.65	59.69	32.6	46.0
Ultimate analysis (wt.%)					
H	3.99	3.71	5.07	1.2	3.5
C	60.71	63.09	86.27	33.5	57.1
N	0.75	0.78	1.07	0.6	1.7
O	8.70	5.55	7.59	5.1	4.0
S				0.5	1.9
Heating value (kJ/kg)	19,368.60	20,129.20	27,525.20	11,432.50	23,801.90

DAF, Dry ash free; *PA*, Pennsylvania.

10.3 Biochar applications

Overall, the biochar characteristics show some qualities that place them at an advantage toward enhancing soil fertility, absorption of organic pollutants, and its utility as combustion fuel. However, there are also characteristics that place FP biochar at a disadvantage in certain applications; these include high volatile matter, low surface area, and low pore volumes as well as high crystallinity. Some specific applications are presented herein.

10.3.1 Pulverized fuel

An immediate application for FP biochar is its potential use as a renewable solid fuel. The calorific value measured in higher heating value (HHV) for biochar from corn cobs and stover is estimated as 29,968 and 20,969 kJ/kg, respectively; that for SG biochar measures at around 27,525 kJ/kg. These values are comparable to some high-quality coals. Nitrogen levels in the biochar are low (0.5%–1.5%) and sulfur levels considerably lower at 0.02%–0.15% or in some cases nondetect. These are important traits for assuming low NO_x and SO_x emissions from the combustion or gasification of biochars. But how reactive are biochars compared to coal is a question that begs for an answer.

The reactivity of FP biochars toward combustion and/or gasification was assessed by their thermogravimetric data (Boateng, 2007). The degree of biochar conversion may be defined as follows:

$$X = \frac{(m - m_f)}{(m_0 - m_f)} \tag{10.1}$$

where m, m_0, and m_f are the current, initial, and final mass remaining taken as the measured ash content. Conversion plots presented for combustion (air atmosphere) and gasification (CO_2 atmosphere) show moderate reaction rates following the Arrhenius-type reaction kinetics with rates increasing with time and temperature (Fig. 10.4). The extent of reactions in these two atmospheres are about the same; ~50% reacted at 1100°C. Activation energy, E, is estimated at 11.5 and 7.4 kJ/mol respectively. However, their reactivity, defined as

$$R_n = \frac{-(dm/dt)}{(m - m_f)}, \tag{10.2}$$

is about the same as it is directly proportional to the environment temperature and inversely proportional to the exposure time (Fig. 10.4B).

Because of its similarities with pulverized coal, biochar's energy value can only be parred with coal and may be estimated at about \$25/ton based on coal selling

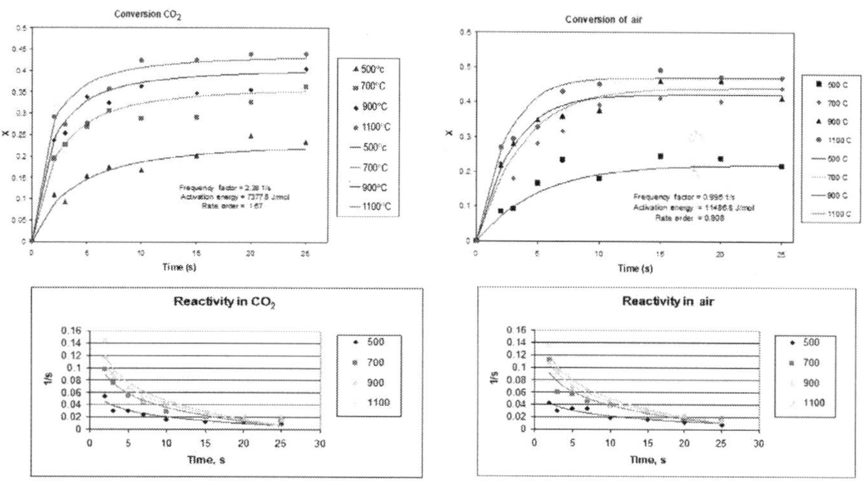

Figure 10.4 (Top): Conversion X of biochar samples in a batch system in CO_2 (gasification) and air (combustion) atmospheres. The points are experimental data and the lines are model, $X = A\exp(E/RT)^n$, where A is the frequency factor, E is the activation energy, and n is the reaction order. (Bottom): Reactivity curves established from conversion of charcoal samples in CO_2 atmosphere and air atmospheres. The temperatures shown in the legend are in degree Celsius (Boateng, 2007).

price of \$1.10/GJ. However, rather than target biochar as fuel commodity, the chemical energy can, rather, be harnessed for the endothermic pyrolysis reaction making the process energy self-sufficient. It has been estimated (Boateng, 2007) that theoretically about 15–20% of the biochar coproduct, based on a typical energy content of 20,000–25,000 kJ/kg range, can fulfill all the energy requirements to produce the 60–70 wt.% bio-oil in a FP system. Further analysis shows an economic advantage to the biorefinery when the biochar is used to provide energy for the drying of the feedstock and for the pyrolysis.

10.3.2 Carbon sequestration

In the light that the world's soils hold more organic carbon than that held by the atmosphere as CO_2 and vegetation, the land is a major factor in carbon sequestration. As the earth is stressed to produce more food, fiber, and energy, more carbon is removed from the ground and emitted to the atmosphere. This impacts the soil's fertility and threatens its long-term effectiveness. It has been argued that biochar is one solution to slowing down or even reversing the process (United Nations Convention to Combat Desertification, UNCCD, Poznan, 2008). Here are the arguments.

The sustainability ramifications of sequestering carbon as biochar may be envisioned by the carbon cycle loops illustrated in Fig. 10.5 using product distribution from FP of soybean straw (Boateng et al., 2010). When biochar is sequestered, minerals elements such as P and K may be returned to the soil, and the transport of these nutrients may be facilitated while carbon is sequestered for hundreds of years. In an on-farm biorefinery using soybean biomass residues (Fig. 10.5), one would expect all the noncondensable gases (NCG) and, perhaps, some bio-oil ($\sim$29 wt.% required, 19 wt.% of overall carbon) will be burned for process heat and the bulk of the bio-oil will be upgraded to transportation fuels. The CO_2 generated by the combustion of NCG, bio-oil, and the transportation fuels will be returned to the replacement soybean plant as CO_2 via photosynthesis. However, to close the carbon cycle loop, the new plant will have to make up for the sequestered carbon by drawing fossil CO_2 from the atmosphere. It is argued that such cycles can potentially place pyrolysis biorefinery in the carbon-negative production of bioenergy while enhancing food production. The cycles discussed show that, all things being equal, there is the possibility that quantitatively 27 wt.% carbon negative operation can result using the pyrolysis data from soybean biomass. The new soybean plant will utilize 73 wt.% of its required carbon from CO_2 produced by the pyrolysis process and the combustion of the resulting fuels, while the 27 wt.% sequestered will be made up from drawing CO_2 from the atmosphere. While this represents a hypothetical and an ideal scenario, under the assumption that there are no net emissions from the cultivation of the biomass, transportation, and bio-refining, it offers a perspective of carbon sequestration potential. It is generally estimated that tilling the land and transporting the soy will release roughly 65 g of CO_2/kg of biomass produced (Sheehan et al., 2009). This amounts to 14 wt.% of the carbon required by the

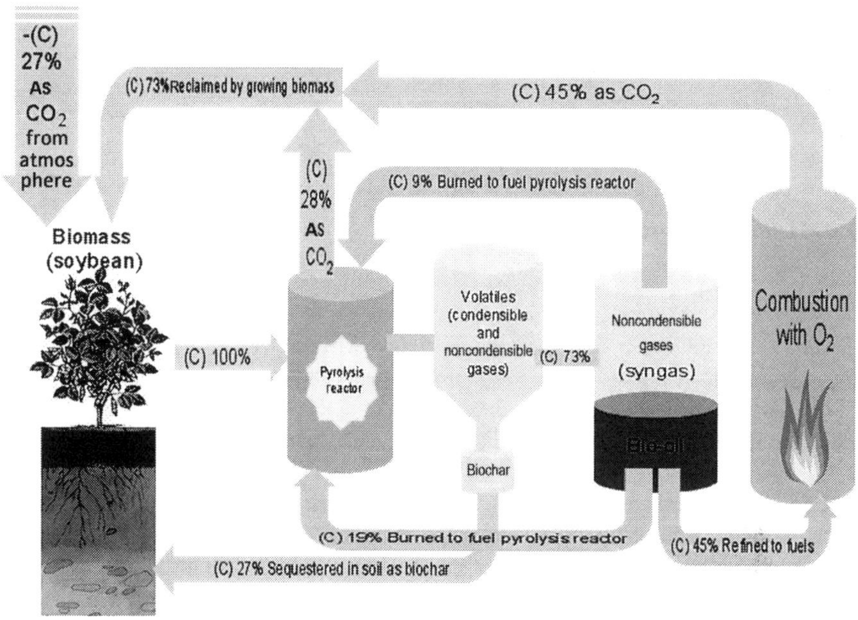

(A)

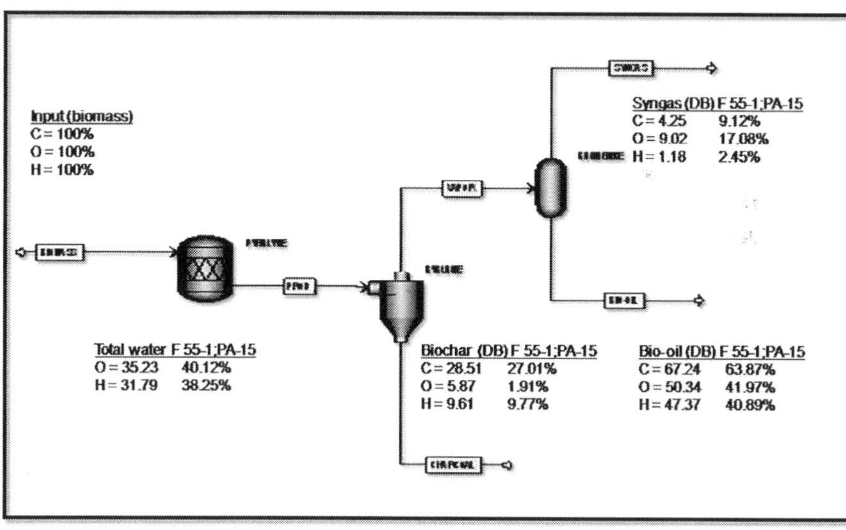

(B)

Figure 10.5 (A) Potential bioenergy production carbon footprint for PA-15 soybean straw pyrolysis. (B) Fast pyrolysis of soybean (USDA F55-1 and PA-15 cultivars) biomass. Carbon and other elemental conversions by mass balance (Boateng et al., 2010).
Source: (A) Adopted from Lehmann, J. 2007.

biomass. When these emissions are factored into the carbon lifecycle of the pyrolysis process, the net atmospheric fossil CO_2 removal will be expected to be in the 13 wt.% range. However, this assessment does not include carbon inputs in the mechanical operation of other unit operations such as tractors, or the energy required to upgrade bio-oil to transportation fuels.

10.3.3 Soil amendment

Biochar additions to soils may mitigate some of the detrimental effects of removing crop residues such as corn cobs and stover from fields and may enhance soil quality. Applying biochar to the soil replaces carbon, nitrogen, and most of the plant nutrients that are removed from the soil with the biomass. Carbon in biochar is highly stable in soil environments and may be sequestered for thousands of years. In soil environments the biochar will initially act as a liming agent raising the soil pH by dissolving the ash and releasing the base cations to the soil solution. Subsequent oxidation of the biochar surfaces will create carboxylate groups such that the biochar becomes a weak acid (Laird, 2008).

In soil environments, biochars with small particle size and high internal porosity are expected to contribute mostly to the enhancement of soil quality. Because of their relatively low surface areas and the coarse particle size distributions, these biochars may not be ideal for soil quality enhancement. On the other hand, the large particle size should enhance stability of the biochar C in soil environments.

10.3.3.1 Fertilizer/plant health

Agronomic benefits arising from biochar additions to degraded soils have been greatly emphasized (Fig. 10.6); so much so that, the European Commission has recently revised the EU legislation on fertilizers, expanding its scope to secondary-raw-material-base to include pyrolysis and gasification materials (Huygens et al., 2019). However, research has shown that negligible and negative agronomic effects have also been exhibited. A review by Spokas et al. (2012) found 50% of agronomic studies on biochar have reported yield increases after biochar additions, with the remaining 50% of the studies reporting alarming decreases to no significant differences. Hardwood biochar (black carbon) produced by traditional SP methods (retort kilns or soil pits) showed the most consistent yield increases when added to soils. But the agronomic impact of FP biochars beyond carbon sequestration is vastly unknown. Here are some interesting developments. The effect of the addition of high and low volatile biochar to soils containing lime and fertilizer components (NPK) on forage growth was tested by the Antals group at the University of Hawaii (Deenik et al., 2011). Visual and statistical analyses (*T*-test) show (Fig. 10.7) that forage growth increases with low-volatile biochars as produced by SP compared to high-volatile biochars that are coproducts of FP.

While application of biochar to soil may achieve several goals with regard to plant health, one of the goals should be unintended effects upon soil biology, including symbioses such as arbuscular mycorrhizas (AM). AM fungi help plants to capture nutrients such as phosphorus, sulfur, nitrogen, and micronutrients from the

Figure 10.6 Biochar deployment. (From USDA).

Figure 10.7 (A) Forage growth in potted plants. (B) Statistical quantification of forage growth. Bars showing same letters are not statistically different. Antal Group (Adapted from the sources: Deenik et al., 2009, 2010).

soil by forming symbiotic relationships with plant roots. USDA experiments examined the interaction of biochar addition and arbuscular mycorrhizal (AM) fungus inoculation upon growth and phosphorus (P) uptake by *Allium porrum* L. (Leek plant) and related these responses to physicochemical properties of the biochars. *A. porrum* seedlings grown with and without *Glomus intraradices*, the largest genus of AM, and either with or without biochar created by pyrolysis of three biomass feedstocks inform the symbiosis. FP biochars greatly reduced colonization of roots by the AM fungus while biochars produced by SP that exhibited higher surface areas were accompanied by higher AM fungus colonization. However, when the FP biochar was pelletized and used in compost, positive results were achieved. Pelletized biochar tested as a carrier for AM fungi in an on-farm system of inoculum production in compost and vermiculite mixtures fared very well (Douds et al., 2014).

The reason on farm production of AM fungi is suitable for vegetable and horticultural crop production is because the inocula may be efficiently mixed into horticultural potting media for plant production in a greenhouse. These inocula are not amenable for use in row crop production because they are not in a form suitable for mechanical application. In experiments conducted in which light expanded clay aggregates (LECA) and pelletized biochar were used in the media for the on-farm production of AM fungus inoculum utilizing compost and vermiculite with *Paspalum notatum* Flugge (Bahia grass) as the nurse host plant, colonization assays failed to detect any infectivity of LECA granules, indicating that the AM fungi did not infest the granules. However, as little as 0.1 g fresh water of biochar was sufficient to produce colonization of test plants. Biochar pellets recovered from the on-farm system used to propagate *Rhizophagus intraradices* exhibited 24 propagules/g fresh water. These results indicate the promise of pelletized biochar as a carrier for AM fungi in inoculum production systems. Scanning electron microscopy showed colonization of biochar pellets by hyphae and spores of putative AM fungi (Fig. 10.8).

10.3.3.2 Ion exchange/contaminants absorption

Another potential use for biochar as soil amendment lies in metal contaminant adsorption. Four metal ions considered environmental pollutants and commonly found in both drinking and wastewaters were chosen for testing Cu^{2+}, Cd^{2+}, Ni^{2+}, and Zn^{2+} adsorption by FP biochars created from corn cob and stover (Table 10.5). Biochars were most effective at removing Cu from solution, followed by Zn, Cd, and Ni. It is possible that the biochars ability to remove metal ions is related to their surface properties with regard to functional groups rather than their surface area (porosity). The larger amounts of inorganic components in biochars such as that created from corn stover could be partially represented as surface charged species possibly leading to some chemisorption phenomena. Lima and Marshall (2005) previously postulated that phosphorous in the form of phosphate groups bridged with the carbon matrix of poultry manure-based biocarbons to aid in the carbon's ability to uptake positively charged metal ions. FP biochars are potentially precursors for remediation of positively charged metal ions, such as copper, cadmium, and zinc just as they do for phenol adsorption. The combination of reduced levels of micropore structure and surface functional groups, observed in the biochars produced by FP, is attributable to the irreversible adsorption/binding mechanism of phenol.

10.3.3.3 Biocidal inactivation

An exploratory study was performed at the USDA to determine the influence of FP and SP biochars on enterohemorrhagic *Escherichia coli* O157:H7 (*E. coli*) in soil

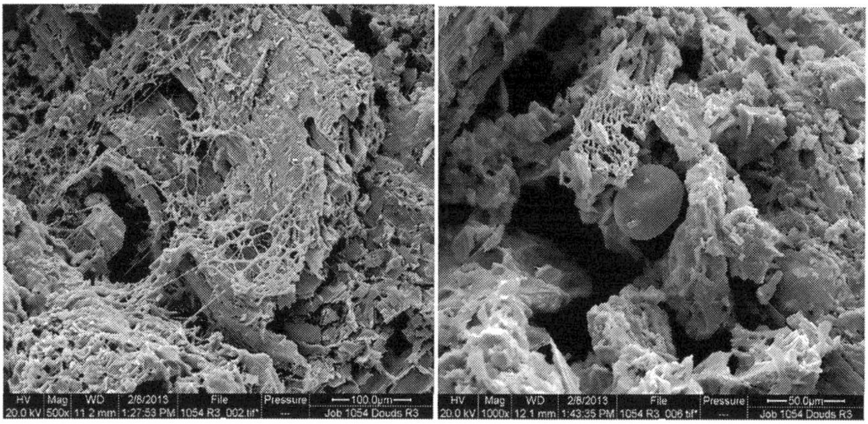

Surface of pellet covered with fungal hyphae   Spore produced within a pore of the biochar

Figure 10.8 Scanning electron micrographs of surface of biochar pellets (2 h at 500°C) incubated in a mesh pouch for one growing season. Mesh bags were surrounded by a 1:4 [v/v] mixture of compost and vermiculite in which grew *Paspalum notatum* plants colonized by the AM fungus *Rhizophagus intraradices*.
Source: USDA.

Table 10.5 Amount of metal ion (mmol/g, dry) adsorbed by biochar from a single metal solution (1 mM).

	Cu^{2+}	Cd^{2+}	Ni^{2+}	Zn^{2+}
Corn cob biochar	0.030 (54.5)	0.021 (22.6)	0.013 (12.3)	0.022 (21.3)
Corn stover biochar	0.045 (80.3)	0.029 (31.1)	0.031 (29.2)	0.038 (37.2)

Percent adsorption in parentheses (Mullen et al., 2010).

(Gurtler et al., 2014). Soil + *E. coli* (inoculated at 7 log CFU/g of soil) + 1 of 12 types of biochar (10% total water: in soil) was stored at 22°C and sampled for 8 weeks. Biochars created by FP of SG and horse litter inactivated 2.8 and 2.1 log CFU/g more *E. coli* than no-biochar soils by day 14. *E. coli* was undetectable by surface plating at weeks 4 and 5 in standard biochars from FP SG, wood oak, and their pelletized forms. Conversely, *E. coli* populations in no-biochar control samples remained as high as 5.8 and 4.0 log CFU/g at weeks 4 and 5, respectively. In addition, three more slow pyrolyses (SP) hardwood pellet biochars (generated at 500°C for 1 hours, or 2 hours, or generated at 700°C for 30 minutes) successfully inactivate greater numbers of *E. coli* than did the no-biochar control samples during weeks 4 and 5. These results suggest that biochar can inactivate *E. coli* O157:H7 in cultivable soil, which might mitigate risks associated with *E. coli* contamination on fresh produce.

Bacterial Composites/Cocktails including five strains of *E. coli* O157:H7 (ATCC 43894, Sakai, 06F00475, 6535, 1484); six strains of *Salmonella enterica* (43894, Sakai,

06F00475, 6535, 1484, 7386); six strains of non-O157 STEC (O26:H11 (00971 Feng), O45:H2 (05−6545), O103:H6 (04162 Feng), O111:H8 (01387 Feng), O121:H19 (O8023 Feng), O145:H18 (07865 Feng); and seven strains of nonpathogenic BSL-1 *E. coli* from ATCC (4157, 117775, 35270, 23848, 25922, 29994, 35218) were tested and sampled weekly up to 12 weeks with results shown in Fig. 10.9.

Results reveal that non-O157 STEC survive poorly in dry control soil (7% moisture), decreasing by 5 logs in 11 days and 7 logs in 47 days. However, it is reduced by 7 logs in only 5 days with addition of 7.5% (w/w) FP biochar from SG. Biochar from FP of SG in soil inactivates 3.5 log more *E. coli* O157:H7 than control soil within 11 days and inactivates 2.0 log more *Salmonella* than control soil within 33 days. Based on these results, the two nonpathogenic *E. coli* O157:H7 strains (ATCC 70028 and 43888) proved to be suitable surrogates for *S. enterica* and *E. coli* O157:H7 in soil/biochar survival studies.

10.4 Bio-green coke

Residues or "bottoms" leftover from bio-oil distillation comprise a significant fraction of the starting bio-oil, in the range of 20−40 wt.%. Hence, conversion into a value-added coproduct is imperative for biorefinery profitability. We discussed the cracking of coke bottom with VGO and explored the synergetic applications earlier,

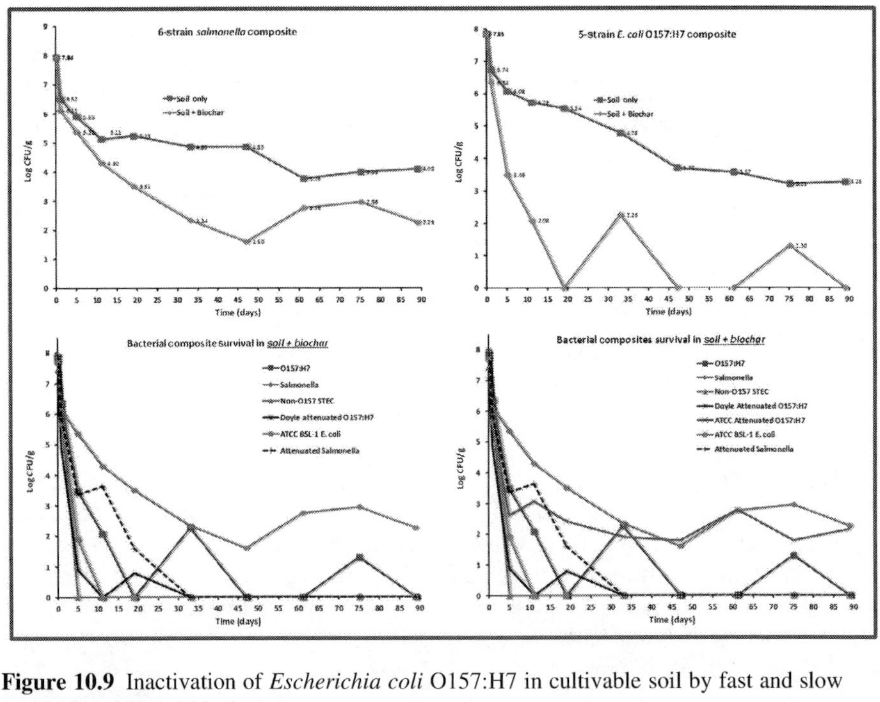

Figure 10.9 Inactivation of *Escherichia coli* O157:H7 in cultivable soil by fast and slow pyrolysis-generated biochars.
Source: USDA.

but the bio-oil distillation bottoms can have a stand-alone application like delayed coke in petroleum refining does. Following the analogous role of petroleum coke, bio-oil distillation bottoms can be further heat treated in an inert atmosphere (> 1000°C, "calcination") to remove all noncarbon atoms and to restructure the polyaromatic carbon into semicrystalline domains. This results in an electrically conductive material (Fig. 10.10) that could be useful as petcoke replacement for aluminum smelting applications. What makes this renewable material especially

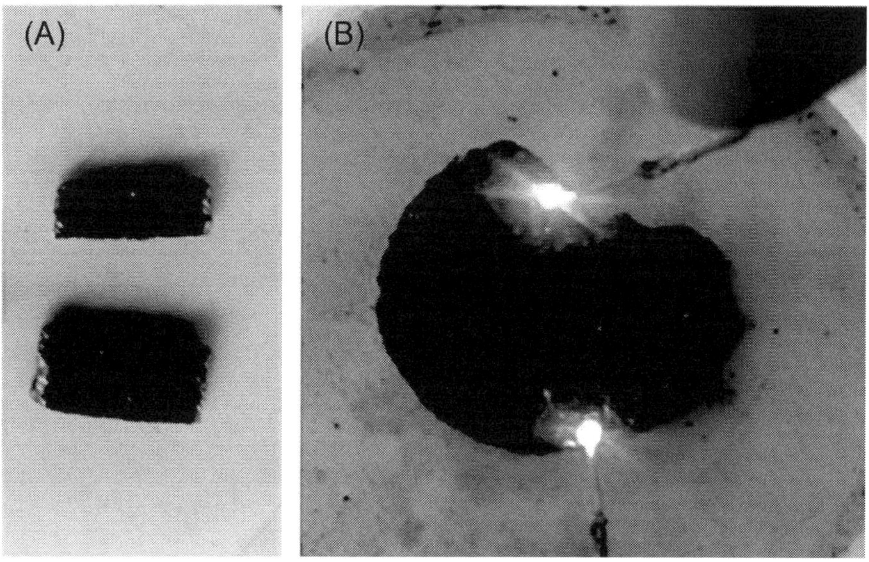

Figure 10.10 Chemical and electrical properties of calcined coke originating from bio-oil distillate bottoms.

Table 10.6 Composition of calcined biocoke, petroleum coke and biochar (Elkasabi et al., 2018).

	Calcined	Petroleum coke		Biochar
	Biocoke	(Raw)	(Calcined)	
wt.% C	97.78	90	>96	63.1
H	0.53	<4	<0.1	3.7
N	0.34	2–3	1–1.5	0.78
O	1.35	1–1.5	0	5.6
S	<0.05	3	3	–
% ash	0.2	<0.4	<0.4	26.9
HHV (MJ/kg)	33.4	31.3	–	20.1
Resistivity (Ω mm)	1.62	2.0E8	1.0	–

HHV, Higher heating value.

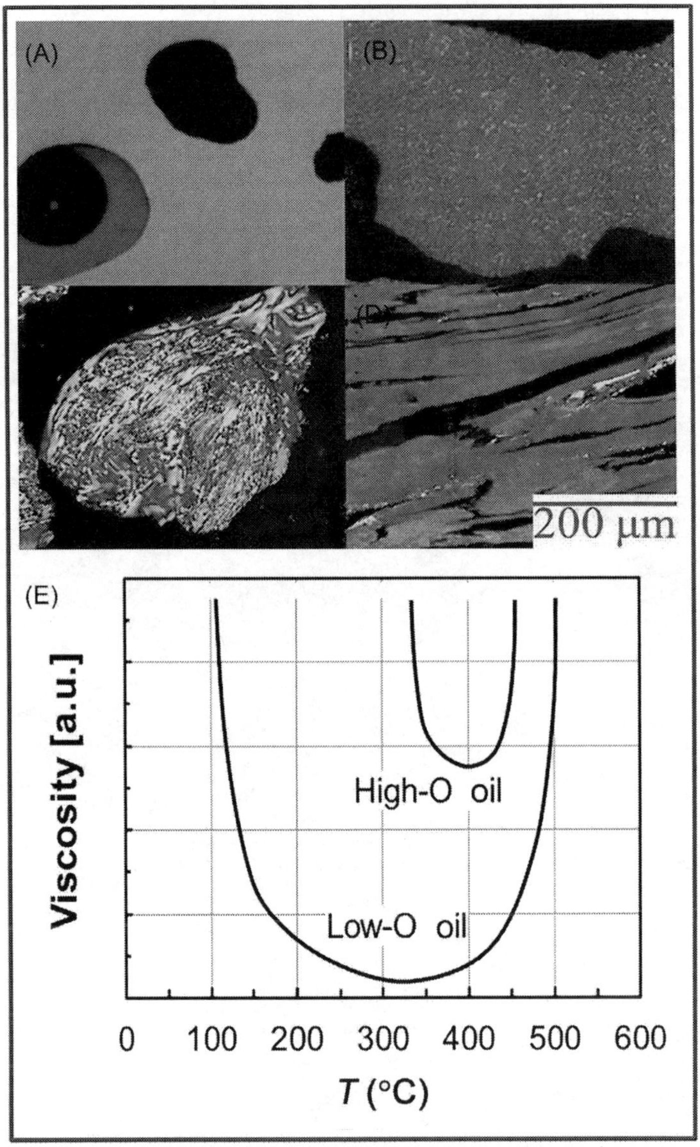

Figure 10.11 Optical microscopy images of coke with: (A) amorphous, (B) isotropic, (C) anisotropic sponge, and (D) anisotropic needle texture. (E) Schematic representation of the relationship between viscosity and temperature of reaction mixture during coking.
Source: Adapted from Monthioux, M., 2002. Structure, texture, and thermal behaviour of polyaromatic solids. In: Carbon Molecules and Materials, Taylor & Francois: London, UK and New York, 2002 (Monthioux, 2002).

attractive is the absence of sulfur and problematic trace metals encountered in some commercial petcoke from petroleum refinery (Table 10.6).

Coke used for making anodes for the electrolytic baths in the aluminum manufacturing industry are selected based on texture. This texture is classified according to the degree of alignment, that is, whether it is amorphous, isotropic, anisotropic sponge, and anisotropic needle (Fig. 10.11). Amorphous coke cannot be used for anodes due to its high coefficient of expansion. Rather, isotropic coke is acceptable in blends with sponge coke for use as in electrodes for electric arc furnaces. If renewable cokes can replace fossil coke, then an industry with a large carbon footprint can turn sustainable. It turns out that the quality of the biocoke made from pyrolysis oil distillation bottoms depends on the quality of the bio-oil precursor. In cokes made from high-O bio-oil, mostly amorphous textures were found. Examples are presented in Fig. 10.12 for coke made from SG-derived oil (23% O in the as-received oil). Elkasabi et al. (2018) showed that biocokes originating from bio-oils with greater than 17 wt.% oxygen content consistently yielded amorphous structures. Very highly stabilized bio-oils derived from reactive pyrolysis of, for example, guayule bagasse yielded coke with isotropic textures with anisotropic

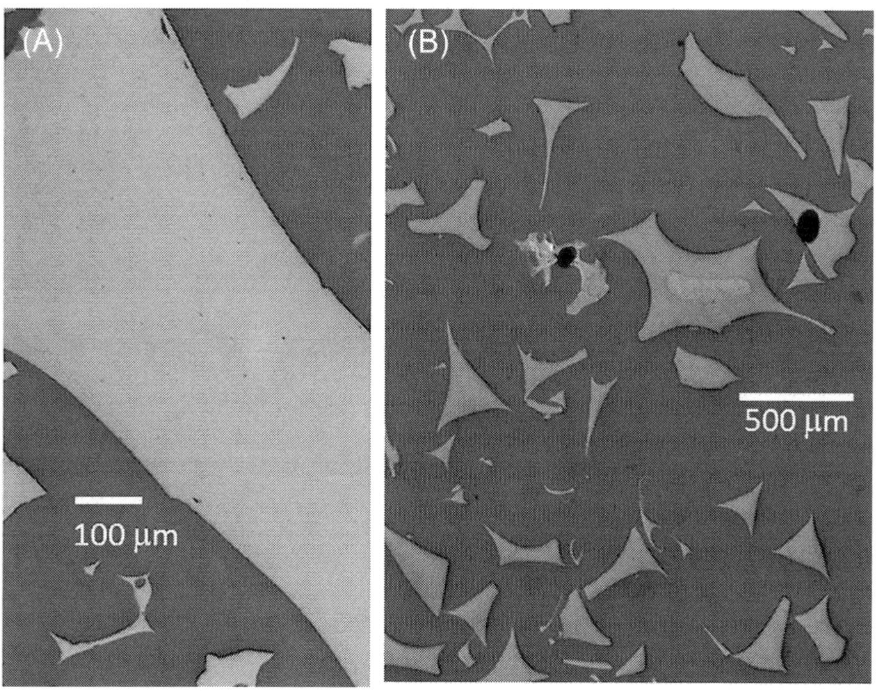

Figure 10.12 Calcined coke produced from higher-oxygen oils (A) higher oxygen oils (horse litter) without staged distillation (B) higher oxygen oils (switchgrass) with staged distillation.

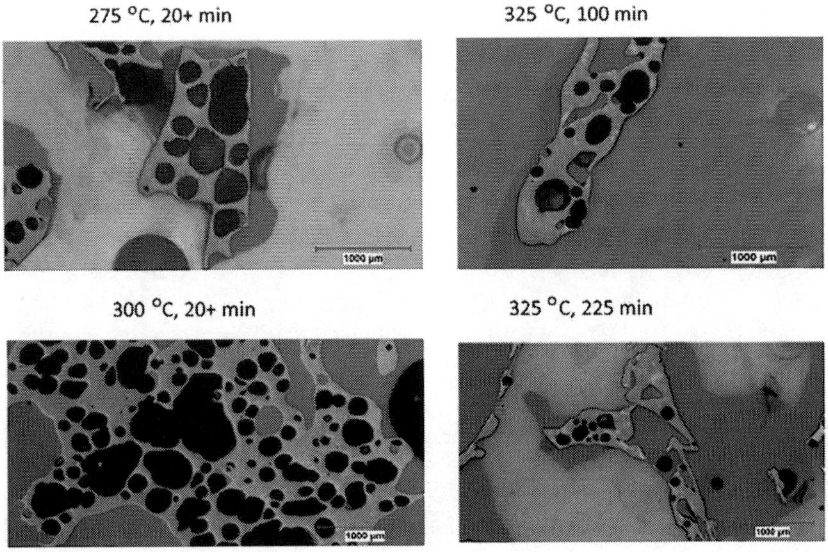

Figure 10.13 Polarized microscopy images of biocoke produced from guayule-based bio-oils showing isotropic/anisotropic texture.

alignment regions after distillation at severe conditions (Fig. 10.13). Guayule-derived coke primarily exhibited more stark differences in color transitions, as shown by the alternating red/blue swirls, indicating a higher degree of anisotropy. The foregoing indicates that the ultimate goal of producing mild oxygen content pyrolysis oils as the biorefinery feeds is beneficial not only to the efficient production of fungible fuels and chemicals as seen earlier but also good for making quality biocoke coproducts.

References

Boateng, A.A., 2007. Characterization and thermal conversion of charcoal derived from fluidized-bed fast pyrolysis oil production of switchgrass. Ind. Eng. Chem. Res. 46 (26), 8857–8862.

Boateng, A.A., Mullen, C.A., Goldberg, N.M., Hicks, K.B., Devine, T.E., Lima, I.M., et al., 2010. Sustainable production of bioenergy and biochar from the straw of high-biomass soybean lines via fast pyrolysis. Environ. Prog. Sustain. Energy 29 (2), 127–130.

Deenik, J.L., Diarra, A., Uehara, G., Campbell, S., Sumiyoshi, Y., Antal Jr., M.J., 2011. Charcoal ash and volatile matter effects on soil Properties and plant growth in an acid ultisol. Soil Sci. 176 (7), 336–345.

Deenik, J.L., McClellan, A.T., Uehara, G., 2009. Biochar volatile matter content effects on plant growth and nitrogen transformations in a tropical soil. Western Nutrient Management Conference, vol. 8. Salt Lake City, UT.

Deenik, J.L., McClellan, T., Uehara, G., Antal, M.J., Campbell, S., 2010. Charcoal volatile matter content influences plant growth and soil nitrogen transformation. Soil Fertility & Plant Nutrition 74, 1259–1270.

Douds, D.D., Lee, J., Uknalis, J., Boateng, A.A., Ziegler-Ulsh, C., 2014. Pelletized biochar as a carrier for AM fungi in the on-farm system of inoculum production. Compost. Sci. Util. 22 (4), 253–262.

Elkasabi, Y., Darmstadt, H., Boateng, A.A., 2018. Renewable biomass-derived coke with texture suitable for aluminum smelting anodes. ACS Sustain. Chem. Eng. 6, 13324–13331.

Gurtler, J.B., Boateng, A.A., Han, Y., Douds, D.D., 2014. Inactivation of *E. coli* O157:H7 in cultivable soil by fast and slow pyrolysis-generated biochar. J. Foodborne Pathog. Dis. 11 (3), 215–223.

Han, Y., Boateng, A.A., Qi, P., Lima, I.M., 2013. Heavy metal and phenol sorption properties of biochars from pyrolyzed switchgrass and woody biomass in correlation of surface structures. Environ. Manage. 118, 196–204.

Huygens, D., Saveyn, H.G.M., Tonini, D., Eder, P., Delgado Sancho, L., 2019. Technical proposals for selected new fertilising materials under the Fertilising Products Regulation (Regulation (EU) 2019/1009). In: EU JRC Science for Policy Report.

Laird, D.A., 2008. The charcoal vision: a win-win scenario for simultaneously producing bioenergy, permanently sequestering carbon, while improving soil and water quality. Agron. J. 100, 178–181.

Lehmann, J., 2007. Optimizing biomass production in a runoff agroforestry system. Front. Ecol. Environ. 7, 381–387.

Lima, I.M., Marshall, W.E., 2005. Adsorption of selected environmentally important metals by poultry manure-based granular activated carbons. Chem. Technol. Biotechnol. 80, 1054–1061.

Monthioux, M., 2002. Structure, texture, and thermal behaviour of polyaromatic solids. Carbon Molecules and Materials. Taylor & Francois, London, UK and New York, p. 2002.

Mullen, C.A., Boateng, A.A., Goldberg, N.M., Lima, I.M., Laird, D.A., Hicks, K.B., 2010. Bio-oil and bio-char production from corn cobs and stover by fast pyrolysis. Biomass Bioenergy 34, 67–74.

Sheehan, J., Camobreco, V., Duffield, J., Graboski, M., Shapouri, H. Life Cycle Inventory of Biodiesel and Petroleum Diesel for Use in an Urban Bus: Final Report. May 1998. P. v. Available <www.nrel.gov/docs/legosti/fy98/24089.pdf>, 19 Aug. 2009.

Spokas, K.A., Cantrell, K.B., Novak, J.M., Archer, D.W., Ippolito, J.A., Collins, H.P., et al., 2012. Biochar: a synthesis of its agronomic impact beyond carbon sequestration. Environ. Qual. 41 (4), 973–989.

United Nations Convention to combat desertification (UNCCD), 2008. 14th session of the Conference of the Parties (COP 14), Poznań, 1–12 December.

Appendix to Chapter 6

(Figs. A6.1–A6.7)

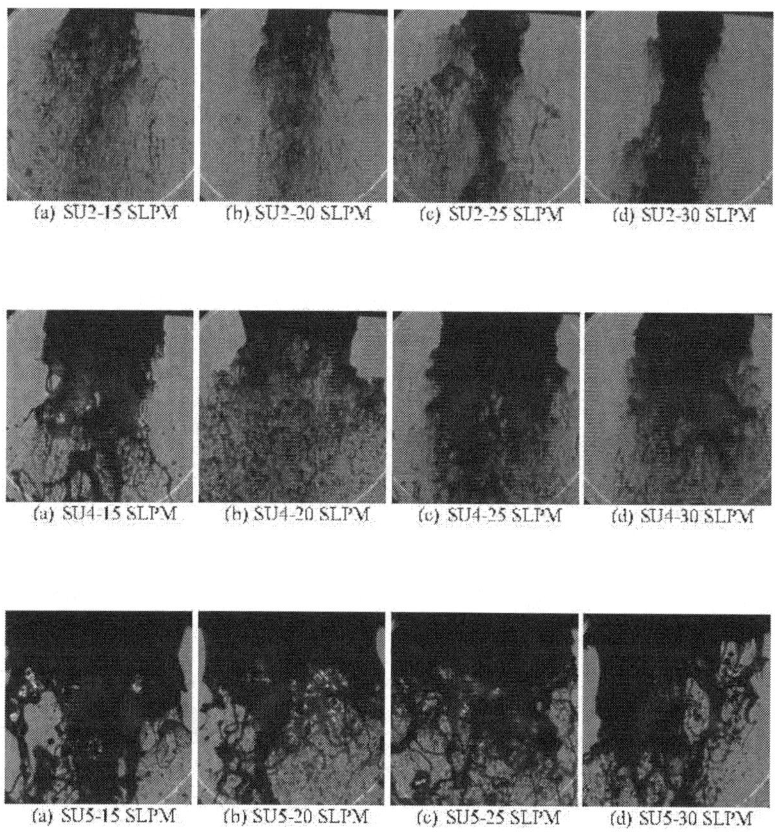

Figure A6.1 30 kW 40:60 bio-oil:ethanol.

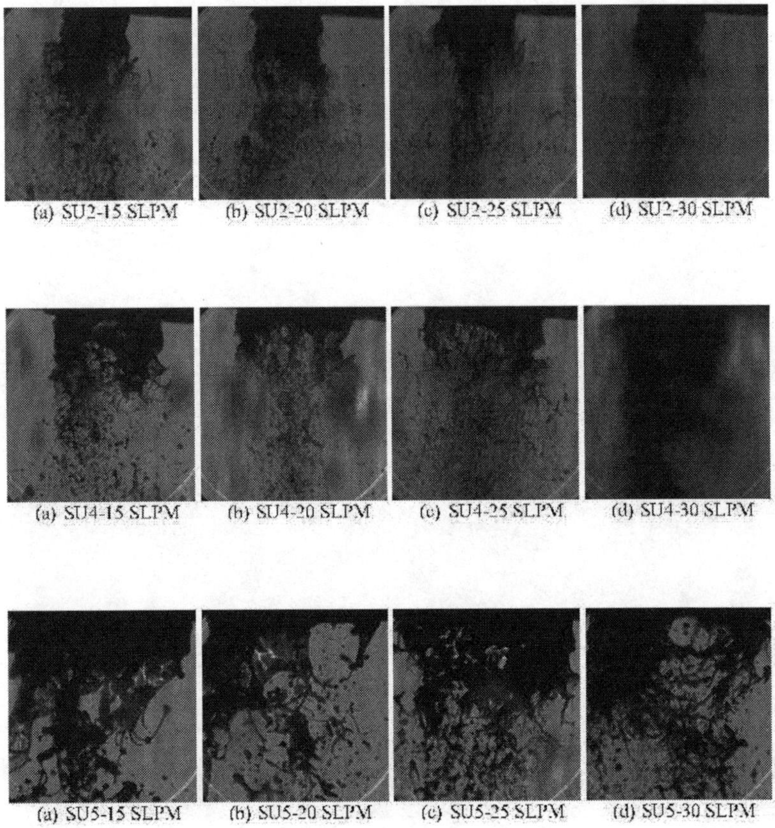

Figure A6.2 30 kW diesel fuel.

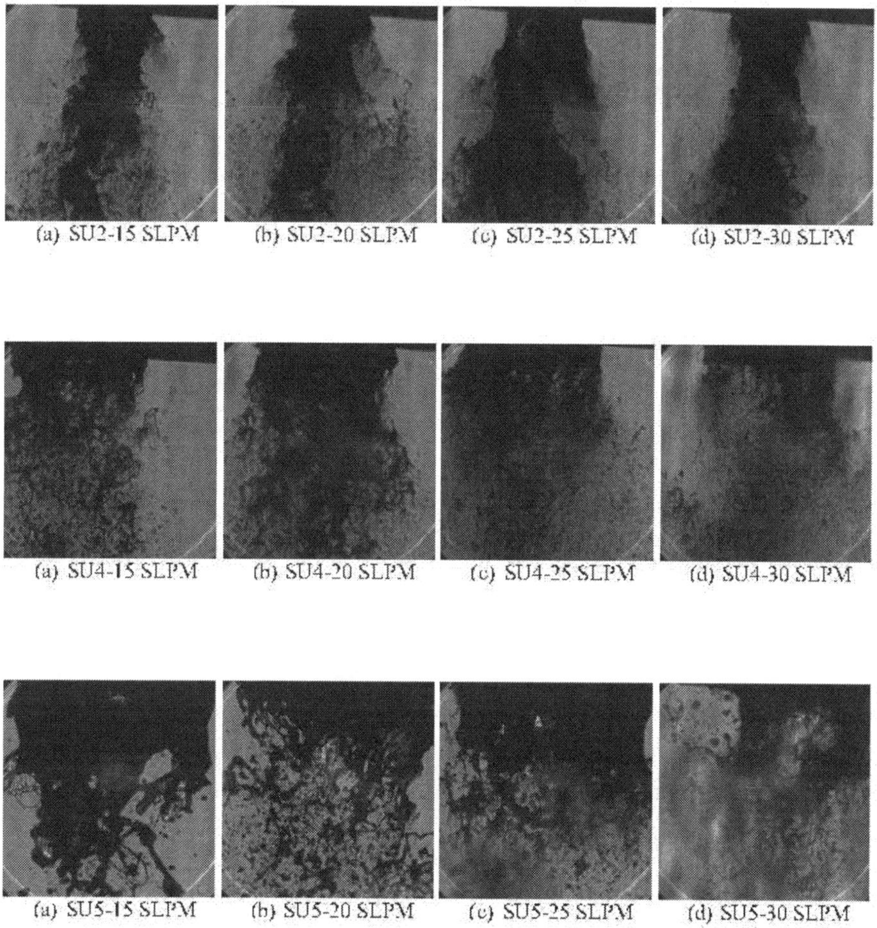

Figure A6.3 30 kW ethanol.

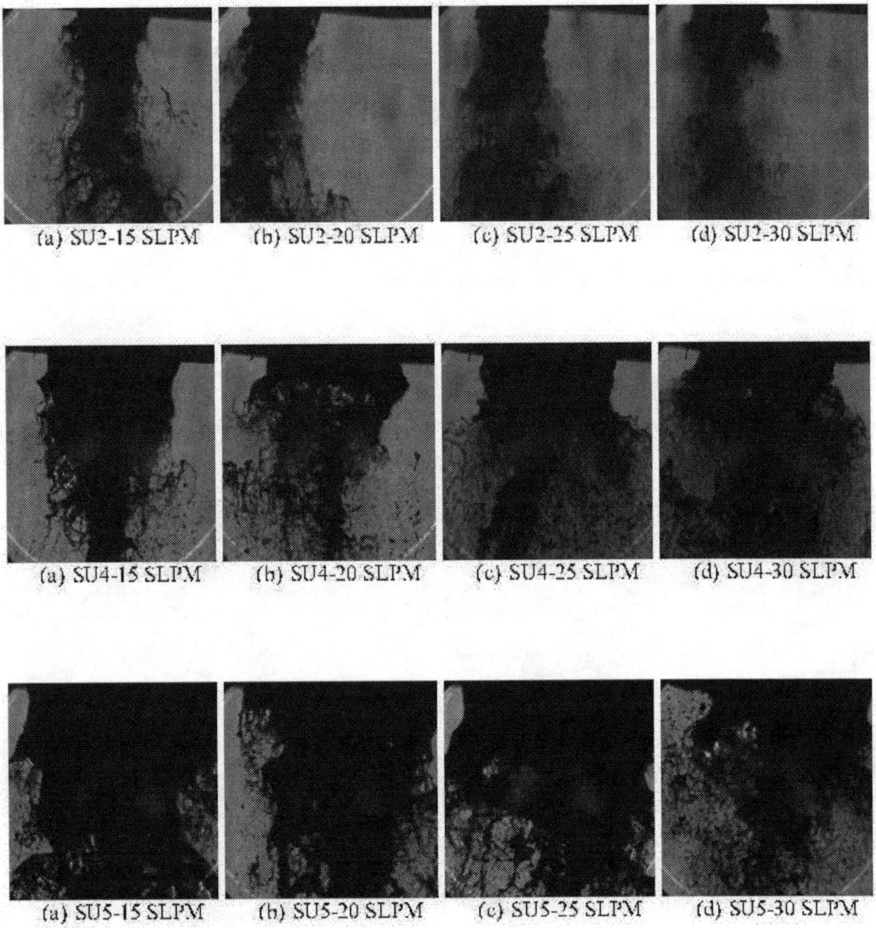

Figure A6.4 50 kW 20:80 bio-oil:EtOH.

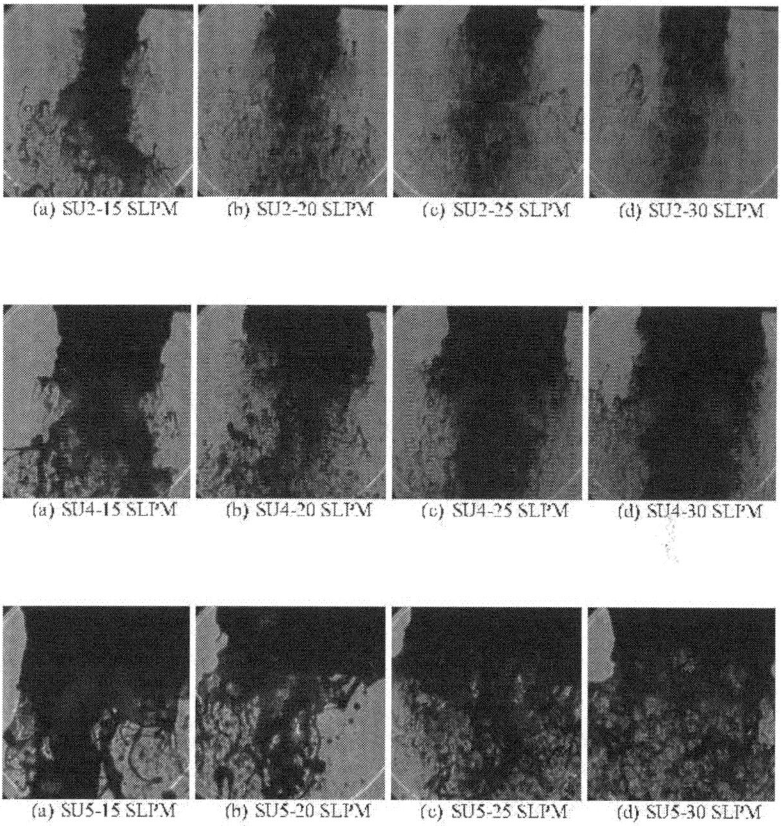

Figure A6.5 50 kW 40:60 bio-oil:EtOH.

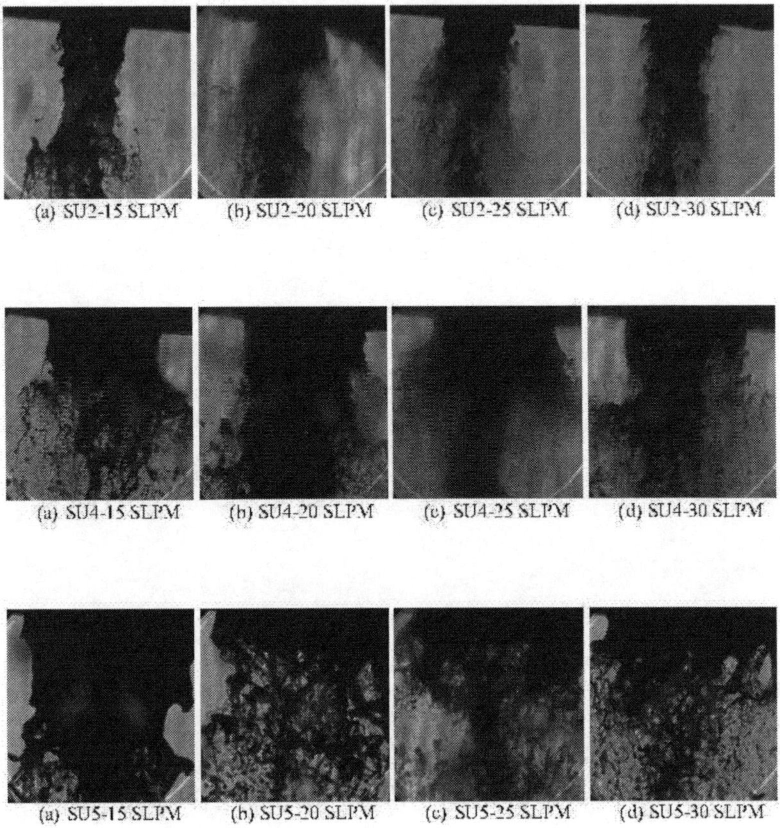

Figure A6.6 50 kW diesel fuel.

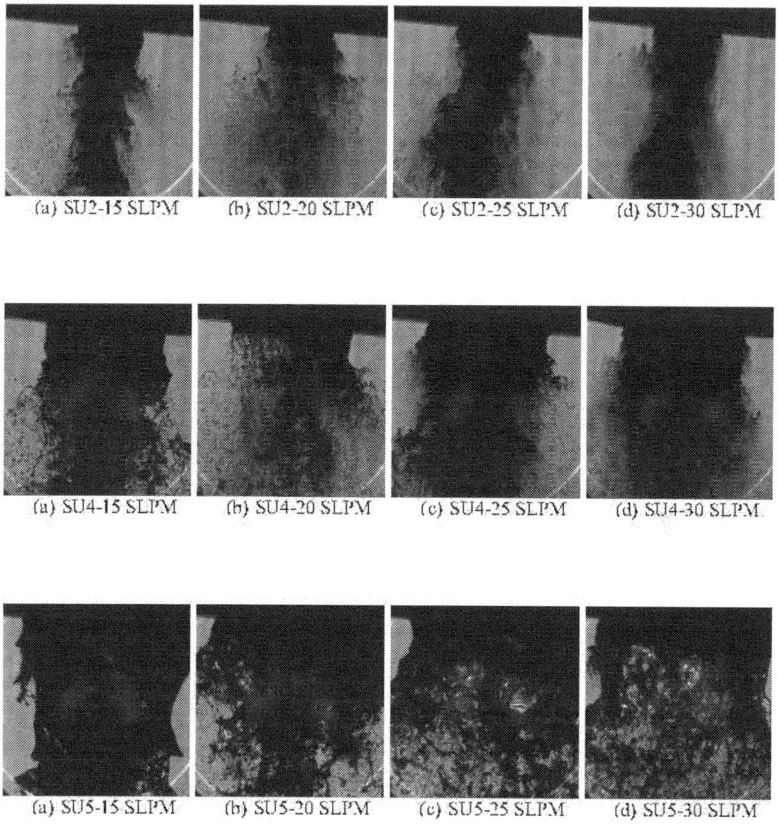

Figure A6.7 50 kW ethanol.

Index

Note: Page numbers followed by "*f*" and "*t*" refer to figures and tables, respectively.

R

Made in the USA
Monee, IL
13 September 2026

63138389R00164